History of Plant Breeding

History of Plant Breeding

Rolf H. J. Schlegel

CRC Press
Taylor & Francis Group
Boca Raton London New York

CRC Press is an imprint of the
Taylor & Francis Group, an **informa** business

CRC Press
Taylor & Francis Group
6000 Broken Sound Parkway NW, Suite 300
Boca Raton, FL 33487-2742

First issued in paperback 2021

© 2018 by Taylor & Francis Group, LLC
CRC Press is an imprint of Taylor & Francis Group, an Informa business

No claim to original U.S. Government works

ISBN 13: 978-1-03-209581-3 (pbk)
ISBN 13: 978-1-138-10676-5 (hbk)

Visit the Taylor & Francis Web site at
http://www.taylorandfrancis.com

and the CRC Press Web site at
http://www.crcpress.com

Publisher's Note
The publisher has gone to great lengths to ensure the quality of this reprint but points out that some imperfections in the original copies may be apparent.

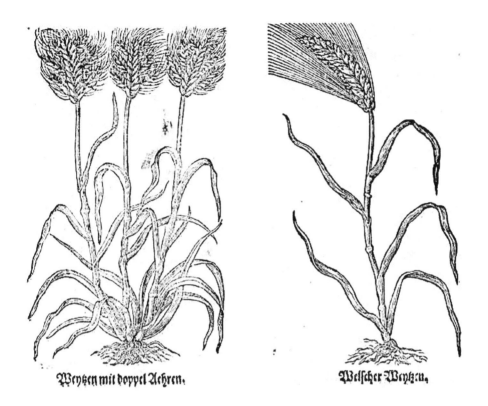

Weytzen mit doppel Aehren.   Welscher Weytzen.

"Who would bring it into being that there are henceforth growing three or four spikes, where previously only one spike stood, he proved its homeland a service, which is to be valued more highly than the deeds of many kings, commanders and poets."

FREDERICK II. The GREAT, King of Prussia (1712–1786)

Contents

Preface

One of the most meaningful moments in the history of mankind was the invention of the agriculture. This drastic change in lifestyle even left traces in our genes when the residents of Europe started to till land and to breed animals during the Neolithic Era. Finds in Cyprus prove that as early as 9000 BC inhabitants from the island were no longer hunter-gatherers, but already ordered fields and kept livestock. But how did this new life come forth here? The people of the so-called preceramic Neolithic era, including seeds and livestock, had apparently reached the mainland at least 65 kilometers away of the island. Accordingly, the south coast of Turkey becomes the starting point in question. There are indications that there were consistent links between Cyprus and the mainland. Ferry traffic could have been developed. The accumulated know-how could have encouraged people to even more extensive tours. This seems to be emerging in the findings. Around 8000 BC, the first domesticated cattle in Cyprus appeared. Circa 7000 BC, they are detectable in Crete, and from 6000 BC, Neolithic emigrants had finally settled on the shores of the Adriatic Sea. The analysis of a 7000-year-old human genome found in Spain demonstrates not only how our forefathers looked at that time but also their way of daily life.

According to UN researchers, demand for agricultural output is projected to grow by at least 70% by 2050. This means that farmers will need to produce 40% higher yields and more food in the next 40 years than they have in the past 10,000 years combined. At the same time, people are increasingly becoming aware of sustainable farming practices as the key to meeting the world's agricultural demands far into the future.

According to the World Bank, since 1960, yield growth has accounted for 92% of the growth of world's cereal production. Genetic improvements have accounted for roughly half the yield growth of major crops. The contribution of genetic improvements to yield growth in developmental countries has been similarly impressive. For example, in India, yield gains in maize and rice increased by 300% since 1940; sorghum, wheat, and soybean yield doubled. This shows the great importance of plant breeding not only for agriculture but also for future societal development on earth.

During the past 50 years, no other branch of biology has developed as fast and comprehensively as genetics, which analyzes the inheritance and development of microorganisms, plants, and animals, including humans. Numerous related disciplines contributed to the knowledge of the basis and structure of heritable factors; their reproduction, modification, and new entities; their intraindividual and interindividual transfer; and their permanent interaction with natural as well as artificial environments. Thus, modern genetics with its many branches became a fundamental discipline of biology. Meanwhile, it also influences different fields of research and stimulates new scientific approaches, including microelectronics.

A key figure in the agricultural and horticultural history of the United States was ▶ Liberty H. BAILEY (1858–1924), a student of W. J. BEAL at Michigan State University, East Lansing, Michigan. Plant breeding was at the core of his work. It can be said that he invented the term "plant breeding" and created the discipline and identity of plant breeding as a profession (PERKINS 1997). In 1891, he published *Philosophy of Crossing Plants* and three years later *Plant Breeding; Being Five Lectures upon the Amelioration of Domestic Plants*. His maize activities at the Michigan Agricultural College laid the foundations for hybrid breeding in maize. A publication in 1876 describes the growing of 2 maize varieties together for cross. He removed the pollen-bearing male flowers at the top of the plant so that only the female could be fertilized by the pollen of the other plant. This seed is the first-generation hybrid (cf. Section 3.4).

Plant breeding is one of those fields that has been strongly driven forward by genetics, particularly during the past century. It seems to be just the beginning of a tremendous progress of targeted modification, reconstruction, and design of plants besides the other challenges of breeding.

Nevertheless, plant breeding has also happened independently of the development of scientific biology, philosophy, and politics. The introduction and improvement of crop plants was a simple way to (better) feed men and animals; that is, a continuous process of optimization of plants for specific environments and utilization. Not only does the nineteenth century clearly show that breeding activities and progress were performed without the knowledge of Mendelian laws, but also does not ignore that breeding is a social process. To take the most obvious sense in which this is true, there would be no professional breeders without education and the societal support for research stations, universities, and private businesses within which breeding takes place.

This book is a modest approach to gather the many data, information, persons, methods, and historical developments presently spread in a wide range of references into one volume. There is no similar book available on the market in English, German, French, or Spanish. The author felt that it is time and that a certain need exists for this in the literature. This topic is usually mentioned in the introductory chapters of textbooks, papers, and/or lessons. In addition, this is done mostly in consequence of progress in genetics, not from an original point of view.

At present, it is difficult to provide students with a comprehensive historical summary without pointing at the numerous literature references. In addition, it always gets more difficult in spite of the rapid development of present plant breeding and biotechnology.

Since plant breeding and adequate research become more and more a global task of private enterprises and several national and international organizations, the exchange of information and communication predominantly takes place in the English language in this sphere. This book can therefore be of advantage for a worldwide readership. Students of plant breeding, genetics, biotechnology, or biology might not be the only beneficiaries, but also breeders, teachers, or other interested persons.

The abundance of results, methods, and crops make it more difficult for the author to consider the complete area of breeding history and to convey all the knowledge based on his own experiences. However, it does offer the advantage of a more systematic representation as compared to many single contributions by different authors and views. The first approach was successful, and the author hopes the second edition will find its readers as well.

Acknowledgment

I express my thanks to Mariana Atanasova, MSc, of Doubroudja Agricultural Institute, General Toshevo/Varna (Bulgaria); Dr. H. Knüpfer and W. Mühlenberg, Institute of Plant Genetics and Crop Plant Research, Gatersleben (Germany); Prof. T. Lelley, Institute of Agrobiotechnology, Tulln (Austria); Dr. B. Leithold, Institute of Plant Breeding, Martin Luther University (Halle); G. Lautenbach, Julius Kühn Institute, Quedlinburg (Germany); T. Heidrich, E. Benary Samenzucht GmbH (Germany); Dr. Rolf Bielau and Dr. Martin Stein, Quedlinburg (Germany), E. Maul, Julius Kühn Institut, Siebeldingen (Germany); Monique Muller-Abrahams, Stellenbosch University (South Africa); Prof. R. McIntosh, University of Sydney (Australia); Dr. Elena Klestkina, Institute of Cytology and Genetics, Novosibirsk (Russia); John Stolarczyk, World Carrot Museum (UK); Monika Jurišič Hlevnjak, Maribor (Slovenia); Dr. Karen Forster, Yale University (USA); Dr. Avinash Kumar Srivastava, Indian Institute of Pulses Research, Kanpur (India); Dr. Garry T. Horner, Iowa State University (USA); Dr. Stacy A. Bonos, Rutgers University (USA); Dr. H. J. Braun and Clyde R. Beaver III, CIMMYT (Mexico); and Dr. Theodore Hymowitz, University of Illinois (USA), for substantial contributions to the manuscript and proofreading, as well as for providing several photographs.

Author

Rolf H. J. Schlegel, PhD, DSc, is a professor of cytogenetics and applied genetics with over 50 years of experience in research and teaching of advanced genetics and plant breeding in Germany and Bulgaria. Prof. Schlegel is the author of more than 150 research papers and other scientific contributions, co-coordinator of international research projects, and has been a scientific consultant at the Bulgarian Academy of Agricultural Sciences for several years. He received his master's degree in agriculture and plant breeding and PhD and DSc in genetics and cytogenetics from the Martin Luther University of Halle–Wittenberg, Germany. Later, he became head of the Laboratory of Chromosome Manipulation and the Department of Applied Genetics and Genetic Resources at the Institute of Plant Genetics and Crop Plant Research, Gatersleben, Germany, and the head of the Genebank at the Institute of Wheat and Sunflower Research, General Toshevo/Varna, as well as at the Institute of Plant Biotechnology and Genetic Engineering, Sofia, Bulgaria. He worked as R&D director in a private company in Germany, and is now retired.

His latest book contributions are *Encyclopedic Dictionary of Plant Breeding and Related Subjects* (Harworth Press Inc., Philadelphia, Pennsylvania, 2003) and "Rye (*Secale cereale* L.)—A Younger Crop Plant with Bright Future" Eds. R. J. Singh and P. P. Jauhar, in *Genetic Resources, Chromosome Engineering, and Crop Improvement: Cereals*, Volume 2 (CRC Press, Boca Raton, Florida, 2005), *Dictionary of Plant Breeding* (2nd Ed., CRC Press, Boca Raton, Florida, 2009), and *Rye: – Genetics, Breeding & Cultivation* (CRC Press, Boca Raton, Florida, 2013) received significant international attention.

User's Guide

This book provides a representative selection of information from the huge amount of data on the history of plant breeding, genetics, and methods as well as institutions and persons associated with the development of breeding and breeding research. A chronological representation was used in principle. However, the variety of parallel developments in terms of countries and cultures, the many temporal cross-references according to methods and persons, sometimes breaks the chronological order.

To limit the contents of the book strictly on the historical aspects, the explanation of technical terms or methods was avoided when possible, although they are often needed for understanding. These terms are included in the "Dictionary of Plant Breeding" (Schlegel 2014). All scientific names are given in Italics. Special designation, titles, citations, and variety names are set with quotation marks.

When Greek letters were necessary in association with some words, they were translated in English. When possible, city names, institutes, and organization were spelled in the national language. Cross-referenced terms and names are indicated by the symbol ▶.

Names of scientists and/or family names are in all capital letters. However, when used adjectivally (e.g., Mendelian) or as a part of variety names (e.g., Tschermak's Weisshafer), only the first letter is capitalized.

Explanations to a given person within the "Gallery of Breeders" may be more or less extensive depending on their importance. A semicolon has simply separated several information.

Cross-references have been provided wherever necessary for economizing space, demonstrating interrelationships, and organizing the material in a clear manner.

1 Introduction

In 1994, German archeologists made a sensational discovery in Anatolia. The place is called Göbekli Tepe. It was a Stone Age settlement with monumental buildings, which are three times as old as the first Egyptian pyramids. Parts of the settlement are 12,000 years old and show 50 tons of heavy megaliths. And this happened during the Middle Stone Age, where only hunters and gatherers were suspected. So, who lived here? How the settlement was built. This leaves only one conclusion. There lived settled inhabitants. And they were able to sufficiently feed themselves on that place.

It is not only a coincidence that one can find in the same region the origin of the einkorn wheat, which is a wild form including a fragile spindle of the spike. Geneticists have now shown that in this group of grasses there was a change in a single (Q) gene, which is located on chromosome 5A, in a single plant, which (among other characteristics) caused a firm attachment of the spikelets to the spindle. This sort of wheat could easily be harvested and was suitable for threshing. It was the prerequisite for preparing bread for the people. This spontaneous mutation in einkorn (*Triticum boeoticum*) can be traced back to the Göbekli Tepe region. The mutation was not beneficial for the evolution of einkorn wheat. It would have a negative selection value. It benefits people, however. As a result, these early settlers must have selected and multiplied the einkorn wheat with tough spindle, that is, *Triticum monococcum*. They were probably the first plant breeders on earth.

Thus, more than 12,000 years ago, the first husbandmen started to process grains laboriously from wild grasses—a fateful invention. About 6000 years later, this culture arrived in Europe. It was the beginning of global warming after ice times. In favorable regions, for example, the Fertile Crescent, plant growth "exploded," including cereals, legumes, and others (Table 1.1).

The use of cereal grains made our forefathers less dependent on hunting and collecting. It offered food to more and more people. This way of life was without alternative, considering bigger population of people in villages and an increase in births. At some point in time then it came to cereal cropping with various types of specialization.

The origin of new information on agriculture, horticulture, and plant breeding derives from two traditions: empirical and experimental. The roots of empiricism derive from the efforts of Neolithic farmers, Hellenic root diggers, medieval peasants, farmers, and gardeners everywhere to obtain practical solutions to problems of crop and livestock production. By the way, the Neolithic populations, which colonized Europe approximately 9000 years ago, presumably migrated from Near East to Anatolia and from there to Central Europe through Thrace and the Balkans. An alternative route would have been island hopping across the Southern European coast. Recent DNA studies on humans point to a striking structure correlating genes with geography around the Mediterranean Sea with characteristic east to west clines of gene flow. The gene flow from Anatolia to Europe was through Dodecanese, Crete, and the Southern European coast, compatible with the hypothesis that a maritime coastal route was mainly used for the migration of Neolithic farmers to Europe (PASCHOU et al. 2014). Before 4500 years they approached northern Europe already as farmers (Figure 1.1). Latest findings, based on diagnostic biomarker lipids and $\delta^{13}C$ values of preserved fatty acids, revealed a transition at nearly 2500 BC from the exploitation of aquatic organisms to the processing of ruminant products, specifically milk, confirming that farming was practiced at high latitudes (CRAMP et al. 2014).

The accumulated successes and improvements passed orally from parent to child, from artisan to apprentice, have become embedded in human consciousness via legend, craft secrets, and folk wisdom. This information is now stored in tales, almanacs, herbs, and histories and has become part of our common culture.

TABLE 1.1

Major Steps in Plant Breeding over 12,000 Years of Development and Continuous Influence by Scientific Achievements

	Time Scale from Neolithic to Twenty-First Century	
	10,000 BC **...1900 AD ...** **... 2000 AD...**	

Basic stages strongly directing breeding advances

Crop plant design and genome editing

Gene engineering

Biotechnology

Computer-aided breeding

Using tissue and cell techniques

Induced mutations for variability

Selection by quantitative approaches

Breeding by genetic laws

Hybridization of genera, species, and varieties

Selection by experience

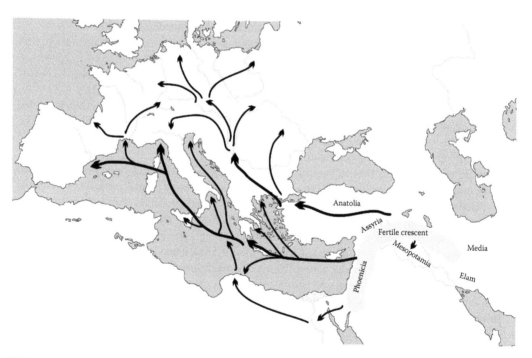

FIGURE 1.1 Distribution of agriculture from Near East to Europe until 5000 BC. (Courtesy of R. SCHLEGEL.)

More that practices and skills were involved as improved germplasm was selected and preserved via seed and graft from harvest to harvest and generation to generation. The sum total of these technologies makes up the traditional lore of agriculture, horticulture, and breeding. It represents a monumental achievement of our forebears.

Without the knowledge about the development of a scientific discipline, nobody is able to judge the recent achievements and to weigh the future chances. Otherwise, one could overestimate the presence too much. The plant breeding does not form any exception at this. Who traces its way back recognizes that the performances and breeding of crop plants are based on centuries- and even millennium-old experiences, such as, for barley (Table 1.1).

Crop plants are a wide and somewhat ambiguous term for many plants grown for food and other purposes regardless of their status as domesticate. The total number of plant species, which are cultivated as agricultural, forest, or horticultural crops, can be estimated to be close to 7000 botanical species. Nevertheless, it is estimated that 30 species only "feed the world" because the major crops comprise a very limited number of species. The basic types of recent crops have arisen this way. Of course, for other crops—such as sugarbeet or triticale—this step cannot be dated back so long. On the other hand, one can follow how some plants will just in the real meaning become cultivated plants (some fodder, ornamental, or industrial crops).

Most of the crop plants are thus the result of a long development process. They are derived from wild types, which are known quite well. Others are the result of spontaneous crosses unifying the genes of two independent species within a hybrid. Under specific circumstances, the linkage of genes remains stable without a following segregation or dissociation. Bread wheat, domesticated plum, and rapeseed have evolved in this way.

The wild types differ from the cultivated plants not only in terms of yield but also in terms of many characteristics important for their existence under growing conditions untouched by man. So, wild cereals show a brittle rachis of the spike and special shaped awns in order to dig in the soil segments of the spike or panicle. Seeds and fruits of wild species are commonly small. The ripen fruits shatter the seeds, such as legumes, linseed, or poppy.

To prevent post-ripening sprouting in dry areas or frost-endangered climates, some wild species form hard-shelled seeds or show dormancy, a resting condition with reduced metabolic rate found in nongerminating seeds and nongrowing buds. Some seeds contain bitter substances against damage caused by game (lupines). Typical are also small and badly shaped tubers or beets of root crops. The wild forms are substantially more unpretentious than the cultivated plants in their requirements of climate and soil. They grow slowly and ripen more unevenly. Wild species are pubescent more frequently. This awards them the characteristics of a very rough, resistant, and solid plant already externally. However, they are not in principle more resistant than crop plants. There are also susceptible wild plants and resistant crops.

It can be summarized that wild plants are adapted to produce sufficient offspring, that is, to maintain the species. A luxuriant shape usually is not required for this—often it is even adverse.

In general, the crop plants are differentiated from wild species by missing of typical wild characteristics, such as seed shattering, brittle spikes, bitter fruits, or branched roots, in addition to traits useful in agriculture, horticulture, or forestry.

The resulted cultivated plant is no longer able to exist under natural environments because of such changes. It is now more or less dependent on man to look after its propagation by sowing, harvesting, and threshing. In some crops, the grade of dependence has progressed particularly far. For example, in maize, the seeds sit so tightly at the cob that self-seeding is impossible, whereas other cereals still shatter the grains when overripe or late harvested. On the other hand, this shows that crop plants still can have wild features, for example, burst of pods in legumes or rapeseed, seed shattering in cereals, dropping of fruits in fruit trees, long germination period in parsley, bitterness in cucumbers, deep rooting, or multigerms in sugarbeet.

The oldest plant breeder is already mentioned in the Bible. "Noah, a man of the soil, proceeded to plant a vineyard..." after surviving the Flood. (Genesis 9:20–27)

To be classified as a cultivar, it is important that the plant shows either the majority or the essential traits different from the wild characteristics and/or modified for human utilization. What it is still missing can or has to be improved by subsequent breeding.

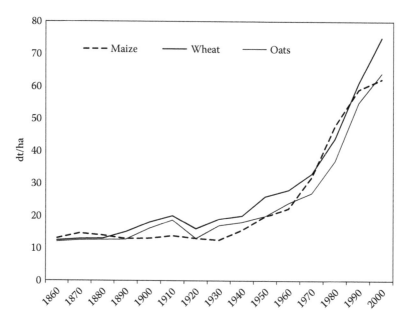

FIGURE 1.2   The development of average yields in grain maize (USA), wheat, and oats (Germany) over a 100-year period (compiled from data of Schmalz 1969 and Allard 1999a): during 1860–1900, almost no progress by utilization of allogamous populations in maize and landraces in wheat and oats; during 1900–1930, slow progress by utilization of cross and combination breeding; and during 1940–2000, continuous yield progress by using combination and mutation breeding of wheat and oats; during 1940–1960, significant progress in maize using double crosses; during 1970–2000, tremendous progress in maize by the application of single crosses. (Courtesy of R. SCHLEGEL. With Permission.)

Plant breeding, however, does not confine itself to the supplement of useful traits but also on the improvement of already available characteristics. Genetic engineering makes feasible even an interspecific and an intergeneric transfer of alien characteristics, as well as the creation of novel traits (Chapter 5).

In some cases, intensive breeding has led to the situation in which cultivars lost too much of the original wild characteristics. Breeding toward re-introgression of wild characteristics is one way out of the dilemma. The enhancement of the dormancy in order to reduce preharvest sprouting became such a task in rye or oat improvement.

The origin of cultivated plants is basically a process of displacement of wild characteristics and an enrichment of suitable traits—a process that began before thousands of years, for example, for wheat, barley, or millet. Nevertheless, plant breeding as an artificial version of natural evolution, involving artificial selection of desired plant characteristics and artificial generation of genetic variation. It complements other farming innovations (such as introduction of new crops, grafting, changed crop rotations and tillage practices, irrigation, and integrated pest management) for improving crop productivity and land stewardship. During the period from 1930 onward, crop breeding, in concert with these other innovations, has led to spectacular increases in crop yields especially of cereal grains (Figure 1.2). Plant breeding is now practiced worldwide by both government institutions and commercial enterprises. International development agencies believe that breeding new crops is important for ensuring food security and developing practices of sustainable agriculture through the development of crops suitable for minimizing agriculture's impact on the environment.

2 Plant Breeding Since 10,000 Years BC

Recent archaeological evidence has begun to undermine this model pushing back the date of the first appearance of plant agriculture. The best example of this being the archaeological site Ohalo II in Syria, where more than 90,000 plant fragments from 23,000 years ago show that wild cereals were being gathered over 10,000 years earlier than previously thought, and before the last glacial maximum (18,000–15,000 years ago).

Agriculture had begun in only a very few places around the world: at least four places in the Old World and three in the New World have been identified. The migration of the early people around the world has, of course, also spread their efforts for food as well as agriculture (Figure 2.1). The oldest region is in the Old World at the eastern end of the Mediterranean Sea extending through Syria, Turkey, Iraq, and Iran. Two separate places are located in China, one in the south along the Yangtze River and another in the north in the valley of the Yellow River.

The fourth region is in a band in Africa extending south of the Sahara. The transition was gradual in each separate case, and the plants that were domesticated were distinctive of the particular region. These places were called Centers of Origin, but it is now clear that there are at least 10 such centers instead of 6 claimed by N. I. VAVILOV (Chapter 9). Wheat was developed in the Middle East, apparently in the northern portion of the Fertile Crescent: the lands immediately east of the Mediterranean Sea. It was apparently brought to the western hemisphere in historic times by Europeans, post-Columbus. Current dating puts its origin at 10,000–10,500 years ago. The same area seems to have given rise to rye maybe even as early as 13,000 (at Abu Hureyra, Euphrates), barley at 10,000, and figs at 11,400 years ago.

RIEHL et al. (2013) concluded that the first peasants settled in the Fertile Crescent between Euphrates, Tigris, and the Mediterranean over 11,000 years ago. There they began to grow grain for the first time. From there, agriculture spread over large parts of Europe and Asia.

It was unclear, however, where our ancestors first began to make wild plants into crops by breeding. The German–Iranian researcher team has discovered decisive indications in the foothills of the Zagros Mountains in Iran: in a Stone Age settlement, they found 11,700-year-old relics of wild and semidomesticated grain precursors, including wheat, barley, and large-seeded legumes. This shows that the cultivation of grain did not begin in a single center, but almost simultaneously at several points in the Fertile Crescent.

About 9000 years ago, people in the Middle East knew how to cultivate grapes and produce an alcoholic drink from them. It began in Turkey's Taurus Mountains and the Caucasus Mountains. Later, the wine culture spread to the west—but it is unclear how and when exactly. Cruising sailors from Canaan, as well as the Phoenicians and Greeks, who are also sailors, are regarded as essential for the journey around the world. It is estimated that around 800 BC, the Etruscans from Central Italy came into contact with the Phoenicians, which is reflected in their culture as an increase in "orientalization." At that time, the wine culture began to flourish in Italy (Figure 2.2).

Two types of millet are traced to interior China (8000); and rice and foxnut (*Euryale ferox*) are traced to coastal China at that same time. India is the home of mungbeans, horse gram, and other millets at 4500. New Guinea folks contributed yams, bananas, and taro at roughly 7000.

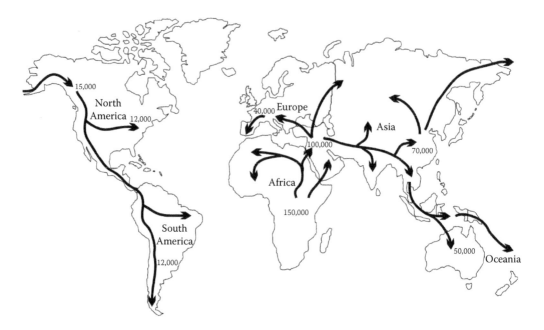

FIGURE 2.1 The migration of early humans around the world during the last Ice Age. (Courtesy of R. SCHLEGEL. With Permission.)

FIGURE 2.2 Etruscan limestone pressing platform unearthed at the ancient coastal port site of Lattara in southern France provide the earliest known archaeological evidence of grape wine from this country and points to the origins of French wine-making around 500–400 BC. (Courtesy of MICHEL PY, © l'Unité de Fouilles et de Recherches Archéologiques de Lattes. With Permission.)

Sub-Saharan Africa contributed sorghum at about 4000, pearl millet at 3000 and African rice at 2000. Five centers are now recognized in the New World. Peanuts at 8500, manioc at 8000, and chili peppers at 6100, from the southwestern Amazon basin, form one group. From the Andes Mountains (basically the Peru/Bolivia border) come potatoes at 7000 and quinoa at 5000. From northwestern South America come moschata squash at 10,000, arrowroot at 9,000, yam (separate species from New Guinea) at 6,000, cotton at 6,000, sweet potatoes and lima beans at 6,500, and leren (*Calathea allouia*) at 10,000. Central America (roughly Costa Rica) seems to be the source of pepo squash at 10,000, maize at roughly 8,500, and common beans at 4,000. The area south of the Great Lakes of northern America, centering roughly on Louisville, Kentucky (USA), seems to have contributed sunflower and a separate domestication of pepo squash at 5000 and chenopods and marsh elder at 4000.

Humans have been hunters and gatherers for 99% of the 2 million years our species has roamed the earth. Only in the last 12,000 years have people become agriculturists. The dates of domesticated plants and animals vary with the regions, but most predate the sixth millennium BC and the earliest may date from 10,000 BC. According to carbon dating, wheat and barley were domesticated in the Middle East in the eighth millennium BC, millet and rice in China and southeastern Asia by 5500 BC, and squash in Mexico at about 8000 BC. Legumes found in Thessaly and Macedonia are dated as early as 6000 BC. Flax was grown and apparently woven into textiles early in the Neolithic period (4500–1800 BC). However, latest research demonstrates earliest plant breeding in *Moraceae* (KISLEV et al. 2006). It was generally accepted that the fig tree was domesticated in the Near East some 6500 years ago. The authors report the discovery of nine carbonized fig fruits and hundreds of drupelets stored in Gilgal I, an early Neolithic village, located in the Lower Jordan Valley, which dates to 11,400–11,200 years ago. These edible fruits were gathered from parthenocarpic trees grown from intentionally planted branches. Hence, fig trees could have been the first domesticated plant of the Neolithic Revolution, which preceded cereal domestication by about a thousand years!

The most sweeping technology change for humans occurred in prehistory: the use of tools, the discovery of fire, and the invention of agriculture. In addition, climatic changes during the late Pleistocene, with the consequent shift of vegetation as a whole, and encouragement of certain plant species in particular to proper and spread have been suggested as another reason for the emergence of agriculture.

Since humans became settled, plant breeding developed as a part of agriculture. Cultivation, the first step toward domestication, began, at least for the seed crops, with planting of harvested seeds by man to provide a new crop. The harvested seeds represent a selected sample of the total variability, biased toward those characters of the population particularly attractive to man and/or having the least-efficient mechanisms for seed dispersal. For example, there is a QTL locus QGD-$4BL$ on chromosome 4B that likely fixed during the process of wheat domestication, favors spikelets with seeds of uniform size and synchronous germination. Seed planted by man is to some extent protected from the pressure of natural selection and this, together with changes in the population size, will lead to changes in variability of the crop through time.

At least six regions of domestication have been identified, including Central America, the southern Andes, the Near East, Africa (Sahel and Ethiopia), Southeast Asia, and China (cf. N. I. VAVILOV). Agriculture is thus one of the few inventions that can be traced back to several locations. From these foci, agriculture was progressively disseminated to other regions, including, for example, Europe and North America.

There are two theories on the origin of agriculture: a single origin with diffusion versus multiple independent origins. The single origin plus the diffusion concept is articulated in a thoughtful essay by CARTER (1977). However, the presently accepted dogma is that agriculture arose as independent invention in various parts of the world at different times. There is evidence that agriculture originated in the Mideast over 10,000 years ago.

At the old cultural centers of the world, such as China, Egypt, Asia Minor, around the Mediterranean Sea or Central and South America, since thousands of year's agriculture, often with irrigation, was practiced. The importance that one attached to wheat, rice, soya, millet, or sorghum can be followed from a regulation of the Chinese emperor CHEN-NUNG around 2700 BC. It describes the custom of annual spring sowing ceremonially carried out by the emperor himself. The oldest excavation of cereals with characters of a cultivar was made in Jarmo (Kurdistan/Iraq) (HELBAEK 1959). It was dated to seventh millennium BC. The wild species *Triticum dicoccoides* and the cultivated type *T. dicoccum* were identified. As compared with the natural conditions, already the earliest agricultural activities brought modified environments for the plants cultivated. Plant species first recognized by man as worth growing—and probably used already at the time of collectors—are classified as primary crop plants, such as wheat, barley, millet, rice, soybean, cotton, potato, or maize.

But the changed conditions did not directly modify the wild types toward a cultivar. It behaved rather so that those genotypes became predominant among the diverse mixture of phenotypes, which showed a better performance, better adaptation, and better propagation under the culture conditions of man. This is called a choice or selection process. It happens also in nature, but the selection contributes exclusively to the maintenance of the species not to the improvement of particular traits for human utilization. In spite of the geographically diverse distribution of the domestication centers, a remarkably similar set of traits can be identified that have been selected in widely different crops. These traits jointly make up the so-called *domestication syndrome* (HAMMER 1984). They result from the selection of spontaneous mutations that occurred in wild populations and were selected at various stages of growth of these wild plants, for example, spikelet nonshattering in *Avena abyssinica* grown in barley fields or pod indehiscence and flat seed in weed *Vicia sativa* in lentil fields. Another example is the initially rare semi-tough-rachized (domestic) phenotypes of wild einkorn (*Triticum monococcum*) that could have achieved fixation within 20–30 generations of selection.

The tough rachis mutant is caused by a single recessive allele, and this mutant is easily identifiable in the archaeological specimens as a jagged scar on the chaff of the plant noting an abscission (shedding of a body part) as opposed to the smooth abscission scar associated with the wild-type brittle rachis. Simply counting the proportion of chaff types in a sample gives a direct measure of frequency of the two different gene types in the plant. Studies have shown that the tough rachis mutant appeared some 9250 years ago and had not reached fixation over 3000 years later even when the spread of agriculture into Europe was well underway. Studies like these have shown that the rise in the domestication syndrome was a slow process and that plant traits appeared in slow sequence not together over a short period.

The inheritance of traits has been investigated numerous times. Initially, they were analyzed as Mendelian traits because many of them display qualitative variation and discrete phenotypic segregation classes. More recently, for a limited number of crops (maize, bean, tomato, rice, and millet), these same traits have been analyzed by quantitative trait locus (QTL) approaches, which are more powerful because they allow a genome-wide analysis of the influence on several traits at the same time. It was shown that their state is often controlled by recessive alleles at one or, at most, two or three loci. Furthermore, the joint involvement of these genes accounts for most of the phenotypic variation, suggesting a high heritability. Finally, many of the genes are located in a limited number of linkage groups (chromosomes), and, on these linkage groups, are sometimes closely, although not tightly, linked.

When man began to mechanize the harvest by using first primitive tools, bigger spikes were preferentially selected than smaller ones. From the Romans, a harvest wagon is known that was pushed into a stand of cereals by the help of a donkey and two farmers (Figure 2.3). By finger-like tools on the front side of the wagon, the spikes were pulled off and fell in a basket. This led to the selection of larger spikes and the reduction of genotypes showing smaller growth habit. The consequence was a stepwise improvement in the cultivars.

In addition, one can assume that some growers tried to get an improvement of traits and species also by conscious choice. Another observation probably led to an increase in the cultivated species. Some species became more enriched and showed better growth in soils closer to early settlements than

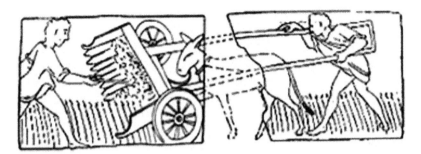

FIGURE 2.3   Reconstruction of a Gallo-Roman mow wagon (the Trevires' harvester, an ingenuous piece of agricultural machinery described notably by C. PLINIUS, the Elder (23–79 AD): "In the vast fields of Gaul, large harvesters, with blades fixed at one end, are pushed on their two wheels across the crop by an ox yoked in the opposite direction. Uprooted in this way, the ears of corn fall into the harvester;" found in 1958 on two limestone reliefs in Buzenol and Arlon (Belgium) dated to the first half of the third century AD. (From The Musee Luxembourgeois, Arlon mow wagon already mentioned in 170 BC. With Permission.)

those planted in distant and poorer soils. By time, the area of settlement was enriched with nutrients from animal and human excrements and waste. Those plants species not recognized as worthy to be collected or grown on poor places were now detected as suitable crop plants. Hemp, poppy, castor, cabbage, beets, mangold, carrots, parsley, and some pharmaceutical plants became cultivated species.

They particularly like nutrients and are called *anthropochores*. One of the latest plant species that seems so to become a crop plant is the stinging nettle (*Urtica dioica*, Urticaceae). There are breeding approaches for different end uses.

In this context, a third type of origin of cultivated plants has to be mentioned. Within the fields of primary crop plants grew many weeds, for example, brittle wild rye occurred within the stands of emmer wheat. As long these weeds were poor wild types, they were eliminated from the cropped plants because of the brittleness, dispersal fruits, small grain size, and so on during the harvest. However, as soon the accompanying weeds became adapted to the ripening time and cropping conditions of the main crop, or showed other inheritable features useful for man, the time was ready even to use them as crop plants, often when the main crop suffered from drought or winterkilling. Important crop plants, such as rye, oats, buckwheat, pea, lentil or white mustard, gold-of-pleasure (*Camelina sativa*, Brassicaceae), rocket salad (*Eruca sativa* syn. *E. vesicara*, Brassicaceae), and spergularia (*Spergularia maxima*, Caryophyllaceae) originated within emmer wheat and barley, emmer wheat and linseed, respectively.

Particularly, the development of rye (*Secale cereale*) can be traced back over 5000 years. With the distribution of wheat toward northern regions, the rye was superior to wheat in the sub-Alpine and steppe climates. Because of its cold and nutritional tolerance, it became the predominant cereal. It reached northern Europe about third to second millennium BC. The deterioration of climate about 3500–3000 years ago promoted its distribution in central Europe. Despite poor quality of flour, wheat became the most important cereal crop of Celts, Teutons, and Slaves. A similar way of agricultural introduction is true for the cultivated oat (*Avena sativa*). Prehistoric excavations seem to confirm that it came as a weed with barley to Western Europe about 2500 years ago.

The spatial scale of development of agricultural societies widely correlates with the distribution of domesticated crops. Agriculture appears to have spread at speeds averaging 1–5 km per year or 20 km per generation, having started around seven to six millennium BC in southeastern Europe and being completed around third millennium in northwestern Europe.

As weed in maize and bean fields, tomato (*Lycopersicon esculentum*) was distributed by humans from central Peru to Mexico. Only in Mexico, it was cultivated; although it remained for long as a garden weed. The anthropochoric hemp (*Cannabis sativa*) took the way as a crop plant from a weed of the Altai region of Central Asia to an accompanying plant in nomadic settlements.

Crops that arose from primary crops are designated as *secondary crop plants*. Again, new cultivated plants can of course develop within secondary crops. Common corncockle (*Agrostemma gith-
ago*, Caryophyllaceae) and rye brome (*Bromus secalinus*, Gramineae) are two such crops that show
evolved characters under the influence of recent cultivation within rye concerning tough rachis or
grain size. Within common oat (*Avena sativa*), grown on the poor soils of northwest Europe, some
wild oats (*A. fatua, A. strigosa*, or *A. elatior*) achieved cultivated characters as well.

Other events occurred next to these elapsed one during long periods and are hardly accessible for a
direct observation. Surely, the man consciously affected the genesis of cultivated plants already early
by preference of most useful phenotypes, best seeds, and vegetative propagules for re-seeding. When
we know also today that the plants looking best do not ever yield the best descendants, it is not controversial that this practice led to improved cultivars and races, although it was useless in the singles case.

2.1   THE OLD WORLD

Agriculture, horticulture, stock farming, fishery, and medicine of the old Mesopotamia show extensive and versatile biological knowledge. Mesopotamian civilizations are largely based on Semitic
populations that existed between the Tigris and Euphrates Rivers, now present-day Iraq. The climate is winter wet and summer dry, particularly suitable for livestock rearing and large-scale cereal
cultivation. It is the source of wild wheat and barley as well as sheep and goats. The area includes
the Fertile Crescent—present-day Israel, Jordan, Lebanon, Syria, Iraq, and Iran as well as all of
Asia southwest of Russia and the Black Sea and west of India and Afghanistan. The spelt wheat
(*Triticum spelta*)—the precursor of common hexaploid wheat (*T. aestivum*)—seemed to be originated from different places of Iran and northern Caucasian regions (KUCKUCK 1959).

The agricultural history of Mesopotamia can be inferred from many sources including cuneiform
tablets and inscriptions, as well as archaeological remains. Agriculture and horticulture of this region
is richly annotated in Biblical sources. Sumerians and Akkadians recorded long lists of plant names.
In the so-called Garden Book of the Babylonian emperor MARDUK-APAL-IDELINA (eighth century
BC), cultivated plants are named: barley, emmer and common wheat, durra millet (*Sorghum vulgare*
or *S. bicolor*), beans, peas, lentil, onions, garlic, leak, beets, radish, cucumbers, melons, sesame,
linseed, the carob (*Ceratonia siliqua*), olive, almond, and pomegranate; from the south (Egypt) came
the date; from the southeast (India) came the fig, pomegranate, and citron; from the north to northeast
came the roses, lily, grapevine, apple, pear, peach, pistachio, plum, mulberry, quince, and walnut.

2.1.1   SUMERIA

Sumeria was one of the advanced cultures born in the fourth millennium BC, probably from non-
Semitic populations of the East. The "Garden of Eden" or often "Paradise" is the biblical "garden of
God" that is described most notably in the Book of Genesis Chapters 2 and 3. Much like the records
of the great flood, the creation story, and the confusion of languages, the story of Eden echoes the
Mesopotamian myth of a king, as a primordial man, who is placed in a "divine garden" to guard
the tree of life.

In "Gilgamesh," similar events are portrayed as later in the Bible. "Gilgamesh," originally
"Bilgamesh" is the main character in the "Epic of Gilgamesh," an Akkadian poem that is considered
the first great work of literature, and in earlier Sumerian poems. Recent research has located this place
at the ancient Sumer, namely in the northwestern Persian Gulf or Gulf of Basra (ZARINS 1992).

> And a river went out of Eden to water the garden; and from thence it was parted, and became into four
> heads.

> The name of the first is Pison: that is it which compasseth the whole land of Havilah, where there
> is gold;

And the gold of that land is good: there is bdellium and the onyx stone.

And the name of the second river is Gihon: the same is it that compasseth the whole land of Ethiopia.

And the name of the third river is Hiddekel: that is it which goeth toward the east of Assyria. And the fourth river is Euphrates. (Genesis 2:10–14)

After the Ice Time, this region was not flooded yet and obviously a vigorous and prosperian place of early settlers. This is most likely accompanied by the earliest agricultural and selective activities of people.

Until recently, Sumer was a lost culture unknown to Herodotus (484–424 BC). He describes the exuberant of the full growth of barley and wheat, with 200- to 300-fold harvests thriving (!). He centered Sumeria in the Euphrates Valley in the Chaldean plains. It contained the ancient city of Ur, three times its capital. The Sumerians were the first to develop writing (3000 BC) in the form of cuneiform script etched on soft clay tablets, which were allowed to harden into a permanent record. Probably due to frequent drought periods after the Ice Age, the Sumerians introduced canals and were among the first systematic agriculturists. Primitive sickles characterize the early farmers. By 3000 BC, there were extensive irrigation systems branching out from the Euphrates River that were controlled by a network of dams and canals. The main canals were lined with burned brick and the joints sealed with asphalt. At its peak, 10,000 square miles were irrigated. The legendary Sargon I., known as Sargon the great (2334–2279 BC), founded the Akkadian–Sumerian Empire. In a tale similar to that of *Moses* a thousand years later, he is discovered in a reed bask.

The river bore me away and bore me to Akki the irrigator who received me in the goodness of his heart and reared me in boyhood. Akki the irrigator made me a gardener. My service as a gardener was pleasing to ISTAR and I became King.

2.1.2 MESOPOTAMIA AND BABYLONIA

There are rich literary sources for Mesopotamian agriculture. A cuneiform text from Nippur called "The Dialogue between the Hoe and the Plow" is a source of agricultural information dating between 1900 and 1600 BC, but may well have older origins, perhaps belonging to the Ur III period (~2100–2000 BC). In the second millennium BC, the great civilization along the Euphrates known as Babylonia formed from the union of Akkadian and Sumerians, with Babylon as its capital (Table 2.1). The historic figures include HAMMURABI (~1750 BC) and NEBUCHADNEZZAR, who was the King of Babylon (605–562 BC). The Code of HAMMURABI contains many laws concerning agricultural crop practices, such as irrigation and fermentation. They produced beer from barley and wheat as well as wine from grapes and dates. Vinegar was a byproduct.

The Hanging Gardens of Babylon, long considered one of the seven wonders of the ancient world, was supposedly built for NEBUCHADNEZZAR's home-sick bride (Figure 2.3). Spiral pumps irrigated gardens 20 m high, with the royal chambers located under the terraces. The latest archaeological research revealed Nineveh (today Kurdish Iraq) as the true site of the Hanging Gardens. After the conquest and destruction of Babylon by SENNACHERIB in 698 BC, its capital Nineveh was regarded as the "New Babylon." Until the modern era there was persistent confusion regarding the location of the legendary Hanging Gardens of Babylon. Even earlier discoveries in Babylon itself had disclosed that SENNACHERIB after the conquest of Babylon renamed the city gates in Nineveh after the name of the Babylonian city gates.

In this relief, trees are shown, which can be seen on roof gardens above a colonnade—just as the classic descriptions describe the "Babylonian Gardens" (Figure 2.4). Nineveh was also an ancient Assyrian city on the eastern bank of the Tigris River, and capital of the Neo-Assyrian Empire.

TABLE 2.1

Time Scale of the Old Kingdoms and Dynasties of Orient Associated with Progress in Agriculture and Technology

BC	Egypt	Syria-Palestine	Anatolia	Mesopotamia	Iran
6000		Jericho	Catal Hüyük	Jarmo	
5500			Hacilar		
5000					Balkun I
					Sialk I
4500	Faijum "A" and Taska culture			Hassuna	Susa I
4000				Samarra	
				Half	
3500	Badari culture and Amra culture and Girza culture	Ghassul		Obeid; first hieroglyphics of a foot, hand, and flail carved in limestone of Kish (near Babylon)	
3000	Thinit Kingdom, until 2600 BC	Khuera	Troy	Predynastic time and intermediate period, until 2700 BC	Susa II
			Beycessultan		Giyan VI
2500	Old Kindom		Alaca	First Early dynasty	Elamite Kingdom
				Second Early dynasty, until 2350 BC	
				Akkadian Kingdom, until 2150 BC	
				Gutaea I	
2000	1st intermediate period, until 1991 BC	Mardikh	Küitepe	Ur III	Elamite Kingdom
				Isin-Larsa	
				Babylonian dynasty	
1500	Middle kingdom and 2nd intermediate period, until 1570 BC and new kingdom, until 1085 BC; 20th dynasty, until 1169 BC (of Ramses III)	Amorit Kingdoms Mari Alalach Ugarit Hasor	Old Hittite Kingdom New Hittite Kingdom	Kassite Dynasty Assyrian Kingdoms (Figure 2.1)	Elamite Kingdom
1000	Late period	Syro-Hittites Phoenician Israelites	Phrygian Urartu	New Babylonian Period	Iranian invasions Elamite Kingdom, until 612 BC
500		Assyrian and Babylonian Kingdoms			

Source: Modified after GARBINI, G., *Alte Kulturen des vorderen Orients*, Verl. Bertelsmann, München, Germany, 1974.

FIGURE 2.4 Artistic interpretation of the Hanging Gardens of Babylon against the backdrop of the Tower of Babylon, probably from the 19th century; it is a copy of a bas relief from the North Palace of Ashurbanipal (669–631 BC) at Nineveh shows a luxurious garden watered by an aqueduct. (Modified from LAYARD, A. E., Discoveries among the ruins of Nineveh and Babylon. G. P. PUTNAM & Co., New York, 1853.)

It was even the largest city in the world for some 50 years until, after a bitter period of civil war in Assyria itself, it was sacked by an unusual coalition of former subject peoples, the Babylonians, Medes, Persians, Chaldeans, Scythians, and Cimmerians in 612 BC. Babylonian agriculture images include a plow containing a seed drill, beer drinking, viaducts, channels, and water lifting with a shadul (some 400 years before ARCHIMEDES).

A cuneiform tablet considered the restoration of a document from 1500 BC from the ancient Sumerian site of Nippur might be the first farmer's almanac. It consists of a series of instructions addressed by a farmer to his son guiding him throughout the year's agricultural activities. A document tablet from the same period described a myth (INANNA AND SHUKALLITUDA: The Gardener's Mortal Sin) and reveals the agricultural and horticultural techniques of windbreaks, planting shade trees in a garden or grove to protect plants from wind and sun.

An Assyrian herbal in the seventh century BC named 900–1000 plants. Examination of clay tablets in the library of King ASSURBANIPAL of Assyria (668–626 BC) identified 250 vegetable drugs including asafetida, calamus, cannabis, castor, crocus, galbanum, glycyrhiza, hellebore, mandragon, mentha, myrrh, opium, pine turpentine, styrax, and thymus. Their special growing and selection can be assumed.

A cuneiform tablet from about 1300 BC shows a map of fields and irrigation canals (Table 2.1). Sumerians and Babylonians knew the sexual dimorphism and dioecious habit of the date palms. They artificially pollinated the female flowers either by hand (Figure 2.5) or by hanging the male flower bunch in the crowns of the fertile trees. The number of male trees could be kept small and the fruit setting of the females could be increased. Recent planting of palm trees distributes about one male plant among 100 females (SWINGLE 1913).[1]

---

[1] ▶ W. T. Swingle (1871–1952). Through his interest in citrus plants, the American botanist came to deal intensively with Chinese botany. He procured over 100,000 Chinese books for the Library of Congress and established a personal library on Chinese agriculture. His materials are now housed in the Walter Tennyson Swingle Collection at the University of Miami.

FIGURE 2.5   Relief of eagle-headed winged figure standing between two sacred trees, ca. 883–859 BC, Alabaster, 84 13/16 × 83 1/8 in. (215.5 × 211.2 cm). Brooklyn Museum, Purchased with funds given by HAGOP KEVORKIAN and the Kevorkian Foundation, 55.156. (Courtesy of Brooklyn Museum, New York. With Permission.)

In the history of breeding, this is the first documented evidence for guided *allogamy* by man opening the chance for breeding new varieties and broadening the genotypic variability for selection. ROBERTS (1929) reported that in four Sahara oases, more than 400 varieties of date palms could be observed distinguished by the size, shape, and taste of fruits. In Mesopotamia and Egypt, no other dioecious species was grown at that time, which was better known concerning their breeding characters and handling.

Recently, a 2000 years old date seed has been successfully germinated. In 1973, an ancient date seed (*Phoenix dactylifera*) was excavated from a historical mountainside fortress, on the shores of the Dead Sea, Masada, and the radiocarbon dating showed the germination to the first century Common Era. The growth and development of the seedling over 26 months was compatible with the normal date seedlings propagated from the modern seeds (SALLON et al. 2008). The oldest documented seed to be grown previously was a 1300-year-old lotus.

Possibly, maize is another example for early breeding-like approaches. Although confessed, no record of this is available. However, primitive varieties of maize were found in the Bat cave of South America dated back to 4000 BC (see Section 2.2).

2.1.3   JUDEA

Most is known about the culture of Judea (1200–587 BC) because of the impact of many books of the Bible, which have come to us almost intact. In the Biblical literature, common agricultural and horticultural practices are discussed, but the interpretations are usually religious or moral. However, the writings can also be read from a reverse point of view. A reading of the scriptures tells us much about the horticulture and agriculture of that period. The basic agricultural roots of these "desert" people are well represented, although in the early times, the annual rainfall was higher and the soils more fertile than nowadays. More than hundred plants, plant products, and agricultural technology are referred to in many of the verses. Genetics can even be found in the Old Testament of the Bible. In *Leviticus* 19:19, the Israelites are warned not to let their cattle breed randomly and not to sow their fields with mingled (contaminated) seed. About breeding activities, it is written further:

> Yet I had planted thee a noble vine, wholly a right seed: how then art thou turned into the degenerate plant of a strange vine unto me? (Jeremiah 2:21)

> If a man shall cause a field or vineyard to be eaten, and shall put in his beast, and shall feed in another man's field; of the best of his own field, and of the best of his own vineyard, shall he make restitution. (Exodus 22:5)

> I am the true vine, and my father is the husbandman. Every branch in me that beareth not fruit he taketh away: and every branch that beareth fruit, he purgeth it, that it may bring forth more fruit... As the branch cannot bear fruit in it, except it abide in the vine; no more can ye, except ye abide in me. I am the vine, ye are the branches: He that abideth in me and I in him, the same bringeth forth much fruit: for without me ye can do nothing. If a man abide not in me, he is cast forth as a branch, and is withered; and men gather them and cast them into the fire, and they are burned (JOHN 15:1–6). In *Amos 7:14* it is mentioned:

> I was no prophet, neither was I a prophet's son, but I was a herdsman and a gatherer of sycamore fruit.

In a related context, it is thought that the fruits of wild figs were artificially ripened. The fruit depends on pollination by a wasp (*Ceratosolin arabicus*). When sycamores were introduced to Egypt, apparently the wasp was not introduced and the seeds were not produced. To ripen the fruit without pollination, the ancient system was to "scrape" the fruit (THEOPHRASTUS):

> It cannot ripen unless it is scraped but they scrape it with iron claws, the fruit thus scraped ripens in four days.

The practice is still carried out in Egypt and Cyprus. The wounding acts to increase ethylene, which induces ripening. Ethylene is the most recent addition to the list of plant hormones. 2-(Chloroethyl) phosphoric acid—an ethylene-generating substance—is now commercially used to induce ripening, as along with other uses, for example, for latex flow in rubber.

Sometimes, the Bible can be true! Excavations (in an old house of Gilgal I) along the Jordan valley in 2006 demonstrated a 11,400-year-old Neolithic agriculture. Fig trees and not cereals seem to be the oldest crops. Together with wild oats, wild barley, and acorns, figs were kept in the Neolithic storages. At the beginning, people gathered wild plants and practiced first planting of the crops. This type of living was developed 12,000 years ago and is kept until present day in some Near East regions.

Moreover, it was found that the first figs cropped set fruits without pollination and did not shatter but remained on the tree until full ripeness. However, these trees did not produce seeds. Therefore, the ancient farmers had to propagate the trees by clonal sprouts.

2.1.4 EGYPT

Until now, the conventional wisdom was that the first groups of modern humans left Africa roughly 70,000 years ago, stopping in the Middle East *en route* to Europe, Asia, and beyond.

Then, about 3000 years ago, a group of farmers from the Middle East and the present-day Turkey came back to the Horn of Africa. It probably brought back crops like wheat, barley, and lentils to Africa.

> Carrots were recognized as one of the plants in the garden of the Babylonian king Merodach-Baladan II († 694 BC). Monumental inscriptions claim that he introduced alfalfa, garlic, leeks, cress, lettuce, and many spices into the region.

Egyptian civilization dates back to the dawn of civilization and the remnants exist in a continuous 6000-year-old record (Table 2.1). The artistic genius engendered by the Egyptian civilization, the superb condition of many burial chambers, and the dry climate have made it possible to reconstruct a history of agricultural technology. Ancient Egypt is shown to be the source of much of the agricultural technology of the occident. From 4000 to 3000 BC, these mingled peoples of the Nile valley formed a government, constructed the first pyramids, and established a highly advanced agricultural technology. The ancient names for Egypt underscore the relation between the land, the people, and its agriculture. These include "Ta-meri," the beloved land cultivated by the hoe, "Ta Akht," the land of flood and fertile soil, "Kmt," the black soil, "Tamhi," the land of the flax plant, "Nht," the land of the sycamore fig tree, and "Misr," the safe and civilized country. The name, Egypt, was derived from the name of the Earth God, "Ge," or from "Agpt," referring to the land covered with floodwaters.

The ancient Egyptian god of vegetation OSIRIS is credited with introducing the skills of agriculture to the Egyptians. He became the god of the dead and the underworld, following his slaying by his brother SET and the restoration to life by his wife and sister ISIS. OSIRIS is sometimes depicted with a green skin in paintings or in statues made from green stone, reflecting his aspects of agriculture, vegetation, fertility, and resurrection. He was recorded as the first to make mankind give up cannibalism and credited with introducing culture of the vine and fermentation of its fruit to produce wine. This may be the oldest account of a biochemical process. The legend of OSIRIS and ISIS dates back to at least 2400 BC, as recounted by the Greek historian PLUTARCH (46–120 AD). The historian Diodorus SICULUS (~First century BC) included OSIRIS among those who had been men, immortalized by virtue of their sagacity and good works. The story may be derived from the Sumerian goddess of fertility, ISHTAR, who could grant crops and children to her devotees. ISHTAR'S son TAMMUZ rose from the dead each year, after the hot summer months, to make the land green again.

Knowledge of crops of the ancient Egypt can be deduced from the artistic record, but definite proof comes from the desiccated remains of plants themselves. The chief ancient grain crops, used for bread and beer, were barley and various wheats including the diploid einkorn, the tetraploid emmer and durum wheats, and the hexaploid spelt and bread wheats. One of the ancient cereals of Egypt—not clearly classified as *Triticum turgidum*—now marketed as "kamut"[2] has recently been introduced to the USA and Europe (cf. Section 7.3.1.1).

The vegetable crops of ancient Egypt included a number of root crops, leafy salad crops, legumes, and various cucurbits (Figure 2.6). The ancient root crops, such as the pungent alliums (garlic, *Allium sativum* and onion, *A. cepa*) as well as radish (*Raphanus sativum*) continue to be

---

[2] "Kamut" derives from the ancient Egyptian word for wheat. It is marketed as a new cereal; however, it is an ancient relative of modern durum wheat (*Triticum durum*). It is thought to have evolved contemporarily with the free-threshing tetraploid wheats. It is also claimed that it is related to *T. turgidum*, which also includes the closely related durum wheat. It was originally identified as *T. polonicum*. Some other taxonomists believe it is *T. turanicum*, commonly called Khorasan wheat. Although its true history and taxonomy are still in dispute, the great taste, texture, and nutritional qualities as well as its hypoallergenic properties are unequivocal. It is two to three times the size of common wheat with 20–40% more protein and is higher in lipids, amino acids, vitamins, and minerals.

FIGURE 2.6 A farmer harvests onions in a painting in the ancient Egyptian "Bird Tomb" of Neferherenptah, at Saqqara (5th dynasty, 2494–2345 BC, ca. 2310 BC). (Courtesy of Tour Egypt, Nasr City, Cairo, Egypt.)

very popular in modern Egypt. Among the leafy salad crops were lettuce (*Lactuca sativa*) and parsley (*Petroselinum crispum*). The carrot dates back about 5000 years ago, when the root was found to be growing in the area now known as Afghanistan. Temple drawings from Egypt in 2000 BC show a purple plant. Some Egyptologists believe it to be a purple carrot. Egyptian papyruses containing information about treatment with carrot and its seeds were found in pharaoh crypts. Throughout the centuries, Arab merchants traveled the trade routes of Arabia, Asia, and Africa, bringing home to their villages the seeds of the purple carrot. During these years, the vegetable appeared in a variety of hues ranging from purple to white, pale yellow, red, green/yellow, and black (but never orange!— that came about in the fifteenth century).

There were a number of pulses, such as cowpea (*Vigna unguiculata*), broad bean (*Vicia faba*), chickpea (*Cicer arietinum*), and lentils (*Lens culinaris*). The cucurbits included cucumber (*Cucurbita sativa*), melons (*Cucumis melo*), gourds (*Lagenaria* spp.), and later watermelon (*Citrullus lanatus*). The date and doum palm (*Hyphaene thebaica*), also known as the gingerbread palm, as well as the sycamore fig (*Ficus sycomorus*) are considered pre-dynastic Egyptian fruits, although the sycamore is not indigenous. The sycamore fig or the fig-mulberry (because the leaves resemble those of the mulberry), sycamore, or sycomore, is a fig species that has been cultivated since ancient times. The ancient Egyptians seized upon a sycamore fig, whose pollinator was absent. Regularly, this species should not have yielded a single ripe fig. But farmers worked out that they could trick the tree into ripening its figs by gashing them with a blade. Before long, the figs were a mainstay of Egyptian agriculture. Farmers even trained monkeys to climb trees and harvest them. To the north and east, the Egyptian fig's sweeter cousin, *F. carica*, became an important food to several other ancient civilizations. The Sumerian King URUKAGINA wrote about them nearly 5000 years ago. King NEBUCHADNEZZAR II had them planted in the hanging gardens of Babylon. King SOLOMON of Israel praised them in a song. The ancient Greeks and Romans said figs were heaven-sent.

The jujube and grape were known since the Old Kingdom; the carob and pomegranate were introduced in the Middle Kingdom the olive and apple appear in the new Kingdom; and the peach and pear date to the Graeco-Roman period.

About 2000 species of flowering and aromatic plants have been found in tombs. An exquisite bas-relief, depicting a visual representation of the fragrance from essential oils being extracted from an herb, is found on the walls of the PHILAE Temple. Herb and spice plants, important for culinary, cosmetic, medical, and religious uses, were continually introduced. Pharaohs were horticulturally sophisticated and collectors. From foreign campaigns, they brought back exotic trees and plants to be grown in their palace or temple gardens.

Black peppercorns were found stuffed in the nostrils of RAMESSES II, placed there as part of the mummification rituals shortly after his death in 1213 BC. Little else is known about the use of pepper in ancient Egypt and how it reached the Nile from India. Queen HATSHEPSUT (ruled between 1479 and 1458 BC) organized a plant expedition delivering living myrrh trees from Punt (northeastern Africa) for the terraced gardens of her Temple at Deir el-Bahri in nearly 1500 BC, and she kept records about the plants discovered (BREASTED 1951). It could be the first expedition in order to collect plant genetic diversity and to keep it like a genebank material (Figure 2.7). Herbs, spices, aromatics, and medical plants include "ami" or Ethiopian cumin (*Carum copticum*), anise (*Pimpinella asisum*), caper (*Capparis spinosa*), coriander (*Coriandrum sativum*), cumin (*Cuminum cyminum*), dill (*Anethum graveolens*), fennel (*Foeniculum vulgare*), fenugreek (*Trigonella foenum graecum*), marjoram (*Origanum majorana*), mint (*Mentha spicata, M. sativa*), mustard (*Sinapis alba*), rosemary (*Rosmarinus officinalis*), safflower (*Carthamus tinctorius*), and thyme (*Thymus acinos*). At the Hatshepsut mortuary temple, Deir el-Bahri, Egypt, there are inscriptions explaining botanical characters of the myrrh trees (Figure 2.8).

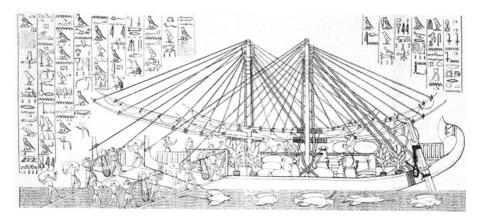

FIGURE 2.7 Loading Egyptian boats. Relief showing plant expedition of Queen HATSHEPSUT (ruled between 1479–1458 BC) to Punt (Northeastern Africa) at the Hatshepsut mortuary temple, Deir el-Bahri, Egypt. (Modified from NAVILLE, E., *The temple of Deir el Bahari: End of northern half and southern half of the middle platform*. Egypt Exploration Fund, London, 1898.)

FIGURE 2.8 Sycamore trees transported in baskets from the shore to the ship at Punt for planting in Egypt; Drawing from relief at the Hatshepsut mortuary temple, Deir el-Bahri, Egypt, after the expedition to Punt by Hatschepsut. (Modified from Edwards, A., *Pharaohs, Fellahs and Explorers*, Cambridge Library Collection, Harper & Brothers, New York, 1891. With Permission.)

Her nephew THOTHMES III (1450 BC) had the curious plants brought from Syria carved upon the walls of the Temple of Amen in Karnak from which iris can be recognized (Table 2.1).

Industrial and fiber crops were important in ancient Egypt. Oil of the castor bean (*Ricinus communis*) was used for illumination and also as a medical plant, flax (*Linum usitatissimum*) for linen and oil, henna (*Lawsonia inermis*) for dye, papyrus (*Cyperus papyrus*) for paper, aquatic lotus (*Nymphaea lotus* and *N. caerulea*) for seed and decoration, and acacias (*Acacia* spp.) for gums and oils.

The ancient Egyptians loved flowers, as evidenced by murals portraying court ladies wearing Egyptian lotus blooms, container-grown plants, and funeral garlands. RAMSES III (1198–1166 BC) founded pleasure grounds with

> …wide places for walking with all kinds of sweet fruit trees, laden with fruit, a sacred way, beautiful with flowers of all lands.

Small shrubs were grown in large earthenware pots, the forerunner of the present potted plant industry. Plant dyes were an important part of the cosmetic arts. Aromatic ingredients from flowers were incorporated into oils and fats for use in perfumes, unguents, and ointments (Table 2.1).

The basic tools of agriculture, the axe, the hoe, the plow, are independent Egyptian inventions. The prototype hoe can be seen as a modification of a forked branch, whereas the more developed form has a hafted wooden blade. The plow was at first a modification of the hoe, originally drawn through the ground, perhaps first by a man with a rope, but drawn by a pair of oxen by the Old Kingdom. Later, metal plowshares were added. In the New Kingdom, handles were lashed by ladder-like cross pieces and the shaft was bound to a double yoke over the oxen's horns. Sowing followed plowing. Often, the sowers scattered seed in front of the plow, so that the oxen treaded it in while fine seed as flax was shaken directly into the furrows. If the seed was sown after flooding, sheep, goats, or swine were driven to tread in the seed. HERODOTUS described the seeding technology as follows:

> …for they have not the toil of breaking up the furrow with the plow, nor of hoeing, nor of any other work which all other men must labor at to obtain a crop of corn; but when the river has come of its own accord and irrigated their fields, and having irrigated them has subsided, then each man sows his land and turns his swine into it; and when the seed has been trodden into by the swine he waits for harvest time: then … he gathers it in (DURANT 1954).

Surveyors measured fields for purposes of tax collection. Large stones were used to establish property boundaries. Moral teachings included the maxim:

> Remove not the landmark on the boundaries of the sown, nor shift the position of the measuring-cord.

Harvest and postharvest handling of grain were favorite themes in the Egyptian art. Early sickles, used to cut wheat, had flint teeth set in a wooden or bone haft followed by curved sickles with a short handgrip. Metal sickles were common in the New Kingdom. Wheat was bound into sheaves and loaded onto donkeys for storage or later carried in net baskets. Fruit was collected and packed in shallow baskets, artfully arranged. Evidence of grain storage dates to the Neolithic time in buried baskets or earthenware jars. Later, the storage of grain and other provisions became a state function and communal silos and granaries were constructed. In the temple of Abu Simbel built by King RAMSES II (XVIII. dynasty), the following words are caved:

> I (the God PTAH) give to thee (RAMSES II.) constant harvests, to feed the two lands at all times; the sheaves thereof are like the sand of the shore, their granaries approach heaven, and their grain-heaps are like mountains (DARBY et al. 1976).

The Roman world considered Egypt the "breadbasket of the world." Grapes were much appreciated in ancient Egypt. There are abundant depictions of grapes, grape culture, and wine making (Table 2.1). C. PLINIUS (23–79 AD) reports that vines were grown directly on the field surfaces, but there are many representations of trellises and arbors. The round arbor was a common feature between the New Kingdom and the Graeco-Roman period. Greek authors confirmed that various cultivars of *Vitis vinifera* were developed in Egypt, for example, the variety "Gutedel" (in German; "Chasselas" in French) is still well distributed in Europe. The latter is characterized by red-veined and five-lopped leaves found on 4000–year-old reliefs and walls at Luxor. It was brought by the Phoenicians from the region of Lebanon to the Rhone valley of France and later further north.

The beginnings of biotechnology are directly traced to the manufacture of bread and wine. The harvest, threshing, grinding of grain to flour, and subsequent sieving are abundantly illustrated. The grinding of grain was handled with a hand mill called a saddle quern by the housewife, but eventually grinding was carried out on a large scale by millers. Fermentation by the use of leaven, a mass of yeast, was a development that changed the making of bread. By 1200 BC, over 30 different forms of bread and cakes were mentioned!

Alcoholic fermentation was carried out in pots with bread or flour to make beer, or with sugary fruit juices, particularly grape but also dates and pomegranates, to make wine, or with honey to make mead. Wine is specified as early as the first dynasty and was associated with HORUS, the falcon-headed god, son of ISIS, the Great Mother, loyal sister and wife of OSIRIS, god of the beneficent Nile. About 1200–1500 BC, the complete wine-making process is illustrated, beginning with grape harvest from arbors, treading by workers maintaining their balance by holding hanging cords attached to a frame, and the squeezing of the sediment collected in cloth bags, with fermentation carried out in amphorae. After fermentation, wine vessels were sealed with plugs of straw and clay designed to prevent bursting from gas accumulation and impressed with official stamps containing the year of the king's rule, the district, the town, and the name of the wine. By the Graeco-Roman periods, there were literally hundreds of wine types from grapes, indicating intense genetic selection.

2.2 PLANT CULTIVATION IN ASIA SINCE NEOLITHIC TIMES

Asia is one of the continents in which civilization first developed and where humans learned to live by cultivating plants. In fact, more than half of the world's edible food crops originated in Asia. The prehistoric discovery that certain plants are edible or have curative powers and others are inedible or cause harm is the origin of the healing professions and its practitioners and the plant sciences—botany, agriculture, and horticulture. Despite the fact that Asian agricultural technology had and continues to have an enormous impact on the world, its agricultural history has seemed mysterious to the West and ignored. For example, Sogdian tribes in Central Asia—the predecessors of Tadzhiks—already 3000 years ago bred apricot trees whose fruits showed a sugar content of 70%. The fruits dried on the tree without dropping down.

The huge continent is often lumped with Europe and the Mid-East as Eurasia, but there is no precise demarcation to this large landmass. However, eastern Asia, which is known to the West as the Orient, is ecologically separated from the western part of the continent by formidable boundaries: deserts and massive mountain ranges. Although contact between east and west is very ancient, it was largely indirect through trade by sea via India and Arabia and overland through the Silk Road (Table 2.1). Recent evidence confirms even a pre-Silk-Road trade: before 3000 BC, societies of western Asia were cultivating wheat and societies of China were cultivating broomcorn millet; these are early nodes of the world's agriculture. By searching for early cereals in the vast lands of Asia, a breakthrough at Begash in southeast Kazakhstan was considered. Here, high-precision recovery and dating have revealed the presence of both wheat and millet in the later third millennium BC. Moreover, the context, a cremation burial, raises the suggestion that these grains might signal a ritual rather than a subsistence commodity (FRACHETTI et al. 2010).

The world's traditional agronomic and horticultural crops must be considered as an introduction from Asia. In contrast, as a result of glaciation during the Ice Age, the northern areas of the world, such as Siberia, Northern Europe, Canada, and the Continental United States contain few native species and have contributed relatively few world crops, pasture grasses, sunflower,[3] and some small fruits, such as strawberries, blueberry, cranberry, and lingonberry, are examples. Many of the berries grown today commercially were recently hybridized from wild berry plants and bushes that grew as native plants on many continents since ancient historical times, such as the strawberry plants, blueberry plants, and raspberry plants, leading to the development of hybrid berries grown today such as the boysenberry plant, loganberry plant, and youngberry plants that are crosses between, blackberry, *Rubus* sp., and the red raspberry, *Rubus idaeus*, the latter hybrid berry plants have only been in existence for a short time. Most modern blueberry hybrid bushes have only been available as USDA (USA) releases for about 50 years, and the enormous berry plant production has created huge agricultural fortunes for modern growers of raspberry plants, blueberry bushes, blackberry bushes and vines, and endless fields of strawberry plants. It is well known that the raspberry plant was used as food in ancient cultures, and parts of raspberry bushes were used to make a medicinal tea.

As a consequence, Asian germplasm resources must be particularly considered for future crop improvement. SAUER (1952) has long speculated that the origin of plant domestication occurred in southeastern Asia, but archeological evidence for agriculture appears later than that of the Mideast. The major reason for this is that the oldest civilizations about which we have much information emerged in the Fertile Crescent. C. GORMAN (REED 1977), based on the evidence of crop remains in the Spirit Cave in Thailand, has pushed the evidence for agricultural beginnings as early as 10,000–14,000 years ago. There is now almost consensus that plant and animal domestication emerged in China at least 7000–9000 years ago. Of course, even with an independent origin, the diffusion of information and germplasm strongly influenced the agricultural development.

The earliest evidence of agriculture is found in northern China from 5000 to 7000 BC, although an excavation in South Korea in 2001 dates back 59 cultivated grains of rice more 14,000 years. The P'EI-LO-KANG and related cultures occupied the loess lands and domesticated foxtail millet (*Setaria italica*) and panic millet (*Panicum miliaceum*). Large Neolithic villages (YANG-SHAO culture)

---

[3] In Europe, the sunflower plant was initially used as an ornamental, but the literature mentions sunflower cultivated for oil production by 1769. Most of the credit is given to Peter the Great (1672–1725), Czar of Russia. By 1830, the manufacture of sunflower oil was done on a commercial scale. The Russian Orthodox Church increased its popularity by forbidding most oil foods from being consumed during Lent; because sunflower was not on the prohibited list, it gained immediate popularity as a food. By the early nineteenth century, Russian farmers were growing over 800,000 ha of sunflower. During that time, two specific types had been identified: oil-type for oil production and a large variety for direct human consumption. Russian government research programs were implemented. ► V. S. PUSTOVOIT developed a successful breeding program at Krasnodar (Russia): the oil content and yield were increased significantly. Today, the world's most prestigious scientific award for research on sunflower is known as the Pustovoit Award. By the late nineteenth century, Russian sunflower seed found its way into the USA. By 1880, seed companies were advertising the "Mammoth Russian" sunflower seed in catalogs. In 1970, this particular seed name was still being offered in the USA, nearly 100 years later. A likely source of this seed movement to North America may have been Russian immigrants. The first commercial use of the sunflower crop in the USA was as silage feed for poultry. In 1926, the Missouri Sunflower Growers' Association participated in what was probably the first processing of sunflower seed into oil. Canada started the first official government sunflower-breeding program in 1930. The basic plant-breeding material that was utilized came from Mennonite immigrants from Russia. Acreage spread because of increasing oil demand. By 1946, Canadian farmers built a small crushing plant. Acreage spread into Minnesota and North Dakota. In 1964, the Government of Canada licensed the Russian cultivar "Peredovik." This variety produced high yields and high oil content. Acreage in the USA escalated in the late 1970s to over 2 million ha because of high European demand for sunflower oil. This European demand had been stimulated by Russian exports of sunflower oil in the previous decades, during which time concerns about cholesterol had reduced demand for animal fats. However, the Russians (and, later, Bulgarians) could no longer supply the growing demand for sunflower, and European companies looked elsewhere. Europeans imported sunflower seed that was then crushed in European mills. Western Europe continues to be a large consumer of sunflower oil today, but it depends on its own production and breeding.

have been uncovered, yielding pottery, hoes, polished axes, and weights for digging sticks. Neolithic cultures flourished from the Korean Peninsula, Manchuria to Vietnam, and large settlements began. Artifacts include beautiful painted pottery including storage jars with seed. Crops include bamboo for shoots, persimmon, grass seed, walnut, pine nut, chestnut, and mulberry.

By 4000 BC, there were large farming villages. Evidence of rice cultivation appeared in the lower Yangtze Valley. Chinese civilization was to be based on rice and millet. *Brassica* seeds were found in pots and hemp (*Cannabis sativa*) was grown as an edible seed and as a fiber plant for clothing. The anesthetic properties were known. Mulberry and silkworm culture had begun along with first known domestication of insects, predating honeybee culture by thousands of years.

Agriculture spread to Manchuria by 3500 BC. Millets were predominant. Rice was farmed from Taiwan to Central India before 2500 BC. Melons, sesame, and broad bean were cultivated. Wheat and barley were introduced from Afghanistan.

The founder of Chinese agriculture and medical botany is the mythical emperor SHEN-NUNG (2737–2697 BC). In the first millennium, he was known as the "Divine Cultivator" of the five grains (rice, soya, wheat, and two millets), inventor of the plow, soil tests for suitable crops, and the originator of ceremonials associated with sowing vegetables and grains.

CONFUCIUS lived from 551 to 470 BC, considered as the author of "Book of Songs." It includes 300 traditional songs of the ZHOU Dynasty (1066–221 BC), with hundreds of references to food, and providing an agricultural picture of the age. Mention is made of clearing artemisia, thistles, and weeds. Fiber crops were silk, hemp, and kudzu. The staple food was millet. "Book of Songs" mentions 44 food plants (as compared with the Bible, which names 29). These include

Grains: Millets (*Panicum* and *Setaria*), barley, rice.

Vegetables: Kudzu (*Pueraria lobata*), hemp (*Cannabis sativa*), Chinese cabbage (*Brassica* species), Chinese chives (*Allium tuberosum*), daylily (*Hemerocallis flava*), bottle gourd (*Lagenaria*), melon (*Cucumis melo*), soybean (*Glycine max*), lotus (*Nelumbo* species), yarro (*Achillea siberica*), mugwort (*Artemisia*), motherwort (*Leonurus* species), mallow (*Malva verticillata*), huanlan (*Metaplexis stauntoni*), plantain (*Plantago major*), poke (*Phytolacca acinosa*), huo (*Rhynchosia* species), cocklebur (*Xanthium strumarium*), sweet flag (*Acorus gramineus*), water plantain (*Alisma plantageo*), water fern (*Marsilea quandrifolia*), elm (*Ulmus*), and bamboo (*Bambusa* ssp.).

Fruits and nuts: Peach (*Prunus persica*), pear (*Pyrus* species), plum (*P. salicina*), Japanese apricot (*P. mume*), Chinese jujube (*Ziziphus jujuba* and *Z. spinosa*), raisin tree (*Hovenia dulcis*), Chinese chestnut (*Castanea mollissima*), white mulberry (*Morus alba*), oak (*Quercus spp.*), pine (*Pinus ssp.*), brown pepper (*Zanthoxylum piperitum*), and Japanese quince (*Chaenomeles japonica*).

From third to second century BC, there was increase in the complexity in culture. A machine to winnow grain was invented. A machine was used to control the separation of chaff and grain, where air movement was generated by a fan. Various types of wheelbarrows, unknown in the West until the eleventh century AD, appear in China in the first century BC. At this time, seed was sown by hand along the ridges with extensive hoeing to destroy weeds and create a soil mulch to conserve moisture. A multitube seed drill dates to this period.

2.2.1 THE OLD CHINA

Since 7700 years rice has been grown by Chinese, as the fossils show. Already during the early stone age, forests and bushland were burned down in order to gain land for cropping. Fields were even irrigated. Not only wild rice was discovered but also advanced rice. Rice was brought to Europe by the nomads from Central Asia. Steppe nomads also ensured the exchange of grain between China and Europe (Section 2.4). That the Chinese had wheat by 1500 BC and that oranges (*Citrus sinensis*) and peaches (*Prunus persica*) had arrived in Europe from China by the Roman times attests to the power of the lively trade route.

In the first century, FAN SHENG-CHIH'S "Agricultural Manual" describes multiple cropping, that is, winter wheat or barley followed by millet. Pretreatment of seed, for example, steeped in fertilizer made from cooked bones, manure, or silk worm debris, to which aconite or other plant poisons added is known. Irrigation of rice, water trapping for dry land fields in the north was practiced. Cultivation in pots and pot irrigation can be found. Ridge cultivation, scheduling of fertilization, watering and planting, organic matter recycling, and soil adaptation to crops were described and/or can be reconstructed. Around 1100 BC, during the SHANG dynasty (1600–1046 BC), TSUNG emperors commanded the breeding of early varieties of rice based on the accessions of Burma (100 instead of 180 days for vegetation period).

Crops and foods mentioned from the HAN dynasty (206 BC–220 AD) include wheat, barley, glutinous and spiked millets, soybeans, rice, hemp, and *Vigna*, as well as gourds, taro, mulberries, Artemisia, melons, scallions, perilla, sesame, elm (leaves and seeds eaten), mustard greens (Chinese cabbage), mallow (Malva species), leeks, onions, water peppers (an aquatic green similar to watercress), and unidentified herbs. Other foods include lotus, longan, litchi, cinnamon, fagara or Chinese pepper (*Zanthoxylum*), magnolia buds, peonies, rush shoots, galangal, daylilies, true oranges, grape, chestnuts, water caltrop (*Trapa bicornis*), bamboo shoots, sugarcane, honey, assorted wild herbs, and wild ginger. Small beans appear to be adzuki bean or red bean (*Vigna angularis*).

Peasant agriculture was based on small, walled communities with contiguous fields. Plows were made of wood, stone, and then iron, but the basic tool was the hoe. Soil was amended with natural fertilizer.

Cereals were chiefly rice and millet, with some wheat and barley. The major beverage was tea, first used as a medicine, but made a ubiquitous beverage by TANG dynasty (618–907 AD). Vegetables were widely consumed, including soybean and its sprouts, garlic, onion, many crucifers, and berries and fruits including peach and Asian pears. There is even a report about the cultivation of "mikan." It produces sweet, seedless, and easy-peeling fruits, which much resemble mandarins but belong to a different citrus species (*Citrus unshiu*). It was listed as a tribute item for Imperial consumption in the TANG Dynasty. The best record of the cultivation of this variety in ancient China is from Jijia Julu, written by HAN YAN, the governor of the region, and published in 1178.

The oldest extant Chinese book on agriculture is "Chimin yaoshu," written by JIA SIXIA in 535 AD. The contents of JIA'S sixth-century book include sections on land preparation, seeding, cultivation, orchard management, forestry, and animal husbandry. The book also includes peripherally related content covering the trade and culinary uses for crops.

2.2.1.1 Tibet

The human history in the Tibetan highlands is already 20,000 years old. Data from this period suggest that people hunted here occasionally—but probably not permanently. CHEN et al. (2015) from the Chinese University of Lanzhou are now investigating the artefacts, animal bones, and plant remedies from 53 sites in the northeast of the Tibetan plateau.

According to their evaluations, the human history in the region is divided into three phases. During the early period, hunters and collectors occasionally hunted in the highlands, reaching altitudes of about 4300 m above sea level. During the second phase, between 5200 and 3600 years ago, millet farmers had established themselves in the area of the middle course of the Yellow River and spread more and more upstream. They also reached areas in the northeast of the Tibetan plateau. But they did not reach an altitude of 2500 m. The reason: millet is a comparatively frost-susceptible plant species—the cultivation in the mountain regions therefore came up at height limits. Nevertheless, the selection of more robust millets led to this adjustment.

But an important change had already taken place toward the end of the second phase. According to dates: about 4000 years ago, barley and wheat entered agriculture at the upper reaches of the Yellow River. These two crops originally came from the Fertile Crescent. In addition to the cultivation of millet, they now also enriched food production in the lower reaches of the northeast of the Tibetan plateau. Especially, barley should become the key factor of phase 3 in the settlement history. The recent landraces are the result of about 4000 year of selection. They opened up areas up to 4700 m above sea level. This led to the development of Tibetan culture. Even the famous capital of Tibet, Lhasa, is situated at 3650 m above sea level. And still, the Tibetan cuisine is characterized by a basic food: the stirrup corn barley.

2.2.2 MEDIEVAL TO EARLY MODERN CHINA

The agricultural development in China after the HAN period was influenced by the influx of new crops from Central Asia and India, such as tea. By the TANG dynasty (618–907 AD), the "Golden Age," crops such as spinach, sugar beet, lettuce, almond, and fig entered China from Central Asia and the Mideast, whereas palm sugar (jaggery), date, yam (*Dioscorea* ssp.), new types of rice, taro, myrobalan plum, citrus, cassia, banana, *Canarium*, and litchi had entered from the south. Distillation, a Chinese invention, appeared and was spread to Europe by Arab cultures. In SONG dynasty (960–1269 AD), food production became rational and scientific. New crop cultivars were introduced, including short season rice ("Champa") and green lentils (mungbean) from India. Watermelons and sorghum became mainstays of north China's dry landscape. Cotton and sugarcane from Central Asia was widely grown and used. "Champa" was the neighboring Indochinese state (modern-day Vietnam) from where the early-maturing rice was derived and got its name. Those advanced "Champa" rice significantly improved the yields in central China. It was a government-sponsored early "green revolution" feeding a lot of people, especially in the vulnerable border areas.

At the end of MING dynasty (1368–1644 AD), New World crops became well known throughout China. Maize, sweet potato, potato, tomato, and peanut profoundly affected Chinese agriculture. This food became the habit of the Chinese by the end of MING dynasty. Rice made up approximately 70% of grain cultivated, wheat and millet the remainder.

By QING period (1644–1912 AD), intensive agriculture produced food in surplus. Root crops such as sweet potato and potato became increasingly abundant. Maize and tomato cultivation increased. Indigenous Chinese species, particularly ornamentals, such as the tea rose, but also food species such as Chinese gooseberry, kiwi, and Asian plums, entered the West.

Particularly, horticulture became embedded in the culture of China and Japan through the establishment of rural retreats and urban gardens. Cultivating flowers was considered one of the seven arts and assumed mystic importance. The peach blossom became the emblem of spring, the lotus of summer, chrysanthemum of autumn, and the narcissus of winter. Plum blossoms symbolized beauty and bamboo stood for long life. Lotus and peony were especially prized. By the eleventh century AD, there were 39 cultivars of tree peony and 35 cultivars of chrysanthemum.

2.2.2.1 Japan

One of Japan's eternal problems is that there is very little arable land. As consequence, high yield has been recognized important for perhaps longer than anywhere outside Europe. Production had to be intensified. During period 1603–1867, under the rule of TOKUGAWA shogunate, Japan experienced some of the features of the European agricultural revolution, with widespread publication of books on farming. With MEIJI restoration in 1868, the farmers were freed from the restrictive feudalism. There is an important person from the Nara perfection, NAOZO NAKAMURA (1819–1882), who bred several rice varieties. He collected landraces from all over Japan, assessed each population, and then distributed the best. Peasants too were active in selection, particularly in early-maturing genotypes, that allowed double-cropping or harvest before typhoon season. Experimental stations were also established during MEIJI era (Table 3.1).

2.2.3 India

The origin of the subcontinent is based on an ancient collision of a landmass with Asia to produce the Himalayas, a natural boundary between China and India, even for the distribution of plant species. The climate varies from the wet monsoons to deserts, from the snows of the Himalayas to the hot jungles of the south. India never was a single nation. Some of the original invaders were called Aryans or tillers, who derive from peoples from the shores of the Caspian Sea, whose western branch populated Europe. It can be traced back by etymological comparisons.

The Indus valley is considered one of the areas in which agriculture was discovered. Beginnings appear in the sixth millennium BC. The remains of ancient cities rivaling those of Sumer, Babylonia, and Egypt are found in the two sites, Mohaenjo-dao and Harappa, which date about 3000 BC (Table 2.1). They include pottery, coins, seals that bear undecipherable inscriptions, weapons, and a huge granary on a brick base with an air-drying system. The original peoples of southern India were agriculturists who tilled the soil, raised crops of barley, wheat (rice was a later introduction), pea, and cotton, and occasionally, sesame and mustard, as well as neem.[4]

From these beginnings, cultivation in India extended to rice, pulses, millet, vegetables, and fruits. Dams for irrigation were constructed as early the first century AD. An Indian cuisine developed that is rich in spices, such as curry, ginger, cloves, and cinnamon. One of the early crops was cotton, known to the ancients as the tree-grown wool. Evidence of herbal use in India predates the first millennium BC, with listings of plants for therapeutic value dating to 500 BC.

Trade between Asia and the Near East is very ancient. Evidence of silk strands in an Egyptian mummy in the tenth century BC suggests very early exchanges of goods between China and Egypt, probably from the overland routes that included Persia. It is known from the ancient spice trade between the Far East and Egypt from the Biblical story of Joseph in "Genesis" and repeated allusion to spices, such as cinnamon and cassia. These valuable spices, which originated in the Maluccas or Spice Islands, reached the Near East overland or by sea via India.

The Silk Road was undoubtedly the route of Asian crops, such as citrus and peach, to the West. In fact, peach was first domesticated in ancient China around 4000 BC. It tasted earthy and slightly salty, like a lentil (FAUST and TIMON 1995). Peaches used to be small, cherry-like fruits with little flesh. But after thousands of years of selective breeding, peaches are now 64 times larger, 27% juicier, and 4% sweeter: peach so named because it sojourned in Persia on the way and mistakenly thought to have originated there. It was imported to Rome during the second century BC.

Monks in the sixth century AD have been reported to have brought back mulberry leaves and caterpillars to introduce the silk industry to the West. Sugarcane drifted to the West via the Arabs, an exchange that was to have an enormous effect on the history of the New World by the development of the plantation system based on African slaves.

The movement of crops was not one way. During the Age of Exploration in the sixteenth century AD, colonial excursions of the West on the East was pursued by Portugal, which established trading Centers in India as a result of the voyage of Vasco da GAMA (1429–1624 AD). Spanish and Portuguese introduced the New World crops, which spread rapidly, mostly from Manila. Sweet potato introduced in the latter half of the sixteenth century AD was the most important introduction

---

[4] Neem (*syn* niem *syn* neem *syn* margosa *syn* nimtree *syn* Indian lilac) is the Indian (Hindi) name of the tree. Its leaves and seeds are known for insecticidal properties and seeds for retarding nitrification due to the action of triterpenes present in them, which have nitrification inhibiting properties. The first indication that neem was being used as a medical treatment was about 4,500 years ago. It was the high point of the Indian Harappa culture, one of the great civilizations of the ancient world; in these ancient texts, neem is mentioned in almost 100 entries for treating a wide range of diseases and symptoms, most of which continue to vex humanity. Neem is a tropical evergreen native to India and Burma and grown in southeast Asia and western Africa; it can live up to 200 years, *Azadirachta indica syn Melia indica* (Meliaceae).

and was well known by 1594 AD as "camotl," the Aztec name, or by Chinese names such as "calle chin-shu" (golden tuber), "pai-shu" (white tuber), "fan-shu" (Barbarian tuber), or "kan-shu" (sweet tuber). Peanut is first mentioned in 1538 AD and maize in 1555 AD. Other New World crops included tobacco and various crops including tomato, guava, papaya, and yam bean. Throughout the centuries, there has been a continuous exchange of plant material, germplasm, and agricultural knowledge.

2.3 CROPPING PLANTS IN ANCIENT AMERICA

Three great civilizations were found, known today as Aztec, Maya, and Inca. These were great monumental cultures—similar in many respects to the Egyptian civilization of 2000 BC—with enormous temples in the form of pyramids, pictorial writing, a system of cities and government, a developed agriculture, and a magnificent art. Archaeological evidence indicates that the New and Old Worlds are connected—through the Bering Straits—and the first peoples of the New World migrated about more than 50,000 years ago. Modern molecular studies of human DNA variability revealed that last South America was settled. Sporadic settlements by the people of Oceania before the Spanish or the Chinese and Vikings in North America did not influence agriculture.

Aztec is derived from "Aztlan" ("white land"), probably northwest Mexico, where by tradition the tribe originated. It is also known as "Tenocha." It gave the name to "Tenochtitlan," a city founded by the Aztecs on an island in Lake Texcoco in the Valley of Mexico (now Mexico City). The Aztecs arrived late in Mexico, about 1168 AD. They were the heirs of previous cultures. Aztec herbals described hallucinogenic plants, such as peyote. The life revolved around the cultivation of maize (called "milpa"). Milpa culture remained unchanged for 3000 years. The maize was planted in March in holes 10–12 cm deep. No fertilizer was used, except human feces. The Incas had bird manure (guano) and limited irrigation. Recent studies dated the first domesticated maize, from a cave in Oaxaca (Mexico), to about 6300 years ago. One of the earliest experiments in "genetic engineering" took place about 7500 years ago and resulted in the first corn on the cob. Thus, the period between 8990 and 6980 BC can be taken for the onset of plant domestication (in southern part). Neolithic farmers selected specific strains of the wild grass, which eventually resulted in a plant that produced a tightly knotted clump of nutritious seeds on a cob.

However, agricultural activities could be traced back to more than 10,000 years. Peanut (7840 years), manioc, pumpkin (9240), and cotton (5490 years) seeds were found along the Ñanchoc valley, north of Peru, together with irrigating systems, fields, and storages.

The study found the farmers were unwittingly modifying a genetic-control region in the grass, which caused long tassels of its seeds to shorten into edible ears that could be harvested more easily.

Sunflower was the other common crop among American Indian tribes throughout North America. Evidence suggests that the plant was cultivated by Indians in the present-day Arizona and New Mexico about 3000 BC. The domestication likely happened independently in North America for food by selecting big seeds and in Mexico as ornamental plant. Some archaeologists suggest that sunflower may have been domesticated before maize. Sunflower was used in many ways throughout the various Indian tribes. Seed was ground or pounded into flour for cakes, mush or bread. Some tribes mixed the meal with other vegetables such as beans, squash, and maize. The seed was also cracked and eaten as a snack. There are references of squeezing the oil from the seed and using the oil to make bread. Nonfood uses include purple dye for textiles, body painting, and other decorations. Parts of the plant were used as medicines ranging from snakebite to other body ointments. The oil of the seed was used on the skin and hair. The dried stalk was used as a building material. The plant and the seeds were widely used in ceremonies. This exotic North American plant was taken to Europe by Spanish explorers sometime around 1500 AD. The plant became widespread throughout the present-day Western Europe, mainly as an ornamental, but some medicinal uses were also developed. Sunflower became very popular as a cultivated plant in the eighteenth century.

In temperate zones, beans and squash were put in the same hole—the maize acted as a support for the climbing beans. Maize and beans complement to form a complete staple food. Maize alone causes deficiencies in the essential amino acid lysine and beans alone are deficient in the sulfur-containing amino acids, such as cysteine and methionine. The mixture of beans and tortillas make a complete protein food. Beans and maize also complement agriculturally, with maize providing a support for beans and beans, an N-fixing plant, providing the added nitrogen to the soil. The Aztecs had little irrigation technology, and thus depended on rain. Other plants, such as sweet potato (warmer valleys), tomato, chili pepper, amaranth (pigweed), pineapple, avocado (ahuacatl), chicle-zapote, or chocolate (cacao), were grown in the gardens and small fields.

The "chinampas system" predated the Aztecs and became the basis of their agriculture. Maize produced several crops a year. Before planting, the people scooped rich mud from the bottom of the lakes, loaded it in a canoe, and so fertilized the "floating" fields. Seed nurseries were near the end of the canals. Today, chinampas[5] get seven crops per plot per year. Two are maize. Others may be beans, chili, tomato, and amaranth. There were elaborate gardens and ceremonies. The Spanish conqueror H. CORTEZ (1485–1547) was welcomed by the chieftains bearing bouquets of various kinds of floral ornament baskets of flowers.

Recent reports show also a high-resolution 2150-year paleo-ecological record from a French Guianan coastal savanna where the pre-Columbian savanna peoples practiced raised-field agriculture. Raised-field farmers limited burning to improve agricultural production (IRIARTE et al. 2012).

Maize was also the basis of Mayan civilization. Each person was allotted plots (land was communal property), which was hand weeded. Stalks were bent at harvest to deter birds. Grain was preserved in storage bins and underground granaries. Water was provided from reservoirs and wells. The yield was deduced from the present-day statistics for subsistence agriculture, that is, 12 acres/farmer = 168 bushels (=14 bushels/acre). Other Mayan crops were Tepary beans (*Phaseolus acutifolius*), squash (*Cucurbita moschata, C. mixta*), pumpkin, chili pepper, sweet potato, sweet cassava, chicham (turnip-like), papaya, avocado, achiote (source of food color), gourds for bowls, balche (strong alkaloid for mead), hemp for fiber, sapodilla (chewing gum tree, also used for adhesives), copal, Brazil wood, palms, cacao. The best varieties of cotton in the recent days of the USA and Russia derived from Mexican farmers and their Mayan ancestors. Many of the local crops can be identified in pottery. Staples include potato and maize. Peru is the center of origin for potato. Tubers were preserved to a freeze-dried product ("chuñu") by continued freezing combined with squeezing the tubers by walking on them.

The basis of Inca society in Peru was the farmer–soldier (as in antique Rome). First, fruits were given as a religious offering to the local shrine ("huaca"). The seasonal working of Inca farmlands is illustrated in a series of monthly drawings by a Peruvian of Indian–Spanish descent in 1580.

---

[5] Elevated rectangular fields are called "chinampas." Scholars of raised-field agriculture generally concur that the Mayan people of the lowlands of southern Yucatan in Mexico and parts of Belize and Guatemala developed the system, whose adoption by other agriculture-based societies then proceeded northward. To construct the chinampas system, the Aztecs removed and piled up aquatic vegetation and muck to create horticultural platforms flanked by waterways and drainage canals ("zanjas") by dredging one vertical meter of canal debris every 1–4 years. Trees such as willow and alder, which often grow in symbiosis with a nitrogen-fixing actinomycete (Frankia spp.), are planted around the islands' perimeters. The trees provide shade, increase diversity, and anchor the soil in the wet environment, thus reducing erosion. By the First century AD, raised-field agriculture was the modus operandi of food production in Teotihuacan, near present-day Mexico City. By the sixteenth century, the chinampas supplied grains, vegetables, and fruits to a quarter million Aztecs in Teotihuacan.

The chinampas system was still used in Teotihuacan in the late nineteenth century. Functional examples of the system persist today in Xochimilco in Mexico City and southwest Tlaxcala State, Mexico. Similar systems flourished in present-day Peru, Bolivia, and Ecuador well before Columbus' arrival in the New World. Popular use of the system began to decline at the time of the Spanish Conquest. Factors that influenced this decline include salinization, population pressures, and inequitable access to technologies that affect labor use, such as plows. Recent studies are reviving interest in the chinampas system as a way to sustain food production in specific ecological conditions with minimal imported inputs.

The Incas called maize "sara." The many types included sweet maize ("choclo") or parching maize ("kollo sara"). There is still maize grown in the highlands of Peru showing very big kernels (3–4 times bigger than common once). It belongs to the group of "cuzco" maize. Crop remains of excavations documented that the Inca farmers were already growing this variety more than 1500 years ago. Maize cultivation swept through the Americas in the millennia following its domestication from Teosinte,[6] a wild ancestor from Mexico's tropical Balsas river valley, some 9000 years ago. At the Waynuna site (north of Arequipa on the western slope of the Andes), maize starch grains were the most common plant remains on the grinding stones. Phytoliths derived from the leaves of maize provided evidence that maize was grown at the site. The shape and grinding damage of maize particles suggests that two races of maize—one used as flour and another popcorn or dent corn variety—were probably grown and processed at the site. The Waynuna house is older than any of the other sites in Peru where maize has been found and sets back the date of maize cultivation and processing in the region by nearly 1000 years. From Teosinte with less than 20 mm long spike, about 5 to 10 hard kernels, with taste like potato, from a limited area of Central America, and divided into eight varieties, domestication led to a judge maize plant with an ear of about 200 mm length, about hundreds of sweet, refreshing, juicy kernels, diversified in more than 200 varieties, with white, yellow, red, purple, and blue-black seeds, as well as grown now in more than 200 countries around the world.

Quinoa, a chenopod, is a spinach-like plant whose dry seed is consumed as a grain. As amaranth, it is still consumed in South America and is now being reconsidered in the USA and Europe and may have industrial prospects because of its unique starch properties. Many other fruits and vegetables were cropped on terraces, such as chili pepper, tomato, beans, squash, pumpkins, wild gherkins, papaya, avocado, cherimoya, guava, granadilla, peanut, manioc, pineapple, soursop, and sweet potato.

For example, a wide variety of chili peppers were cultivated and used in cooking. Recently, a starch from chili peppers was identified on ancient pottery and stone tools that is diagnostic of groups of chili species. The starches were found at various archaeological sites, including from about 6500 years ago in Ecuador, that is, these starch microfossils have been found at seven sites dating from 6000 years before present to European contact and ranging from the Bahamas to southern Peru. The starch grain assemblages demonstrate that maize and chilies occurred together as an ancient and widespread Neotropical plant food complex that predates pottery in some regions (PERRY et al. 2007).

Prehistoric findings prove that cassava (*Manihot esculenta*) has also been cultivated in tropical America for several thousand years, as can be seen from the clay pots from the trenches of Paracas (on the arid coast of Peru), where the roots are represented stylized, or the remains of manihot were found. Similar findings were made at Nazca in the south of Peru. A ball-shaped vessel of the culture of Moche (fourth to eighth century AD) shows two armed warriors streaming through the fields planted with maize trees and cassava (Figure 2.9). As a result of the very simple possibility of spreading many species through stalk pieces, even the simplest farmers succeeded in a rapid increase. The Spanish chroniclers COBO and CIEZA described the cultivation of bitter and sweet varieties in the sixteenth century, and MAGALLANES DE GANDAVO described the use of the "tipiti squeezer"[7] to obtain flour without the poisonous sap. Bitter *cassava* contains poisons compounds that can be fatal.

---

[6] From the time of its discovery until 1896, teosinte was known principally to a few botanists who had preserved some dried specimens in European herbaria and bestowed upon it the Latin name, *Euchlaena mexicana*. Teosinte was placed in the genus *Euchlaena* rather than in *Zea* with maize because the structure of its ear is so profoundly different from that of maize that nineteenth-century botanists did not appreciate the close relationship between these plants. When the first maize × teosinte hybrids were discovered in the late 1800s, they were not recognized as hybrids but were considered a new and distinct species, that is, *Zea canina*. A Mexican agronomist J. Segura made the first experimental maize × teosinte crosses, demonstrating that *Zea canina* was a maize × teosinte hybrid and thereby implying that maize and teosinte were much more closely related than previously thought.

[7] A 1.5–2 m long, tubular and conical container made of braided palm fibers. Indigenous people had developed a process to remove the toxins via a process of soaking, boiling, squeezing, and cooking the cassava.

FIGURE 2.9 Spherical vessel from the culture of Moche (Peru of eighth century) with the depiction of a cassava tuber (shaped structures of the root in the upper part of the vessel). (Courtesy of Museo Nacional de Antropología y Arqueología, Lima, Peru, 2016. With Permission.)

2.4 THE GREEK AND ROMAN WORLD

Around 5000 years ago the regional separation of major crop plants was quite clear: In China, there were millet and rice, while the origin of wheat was unclear. In the western hemisphere, it was the reverse. Wheat was supposed to be a native crop, while the origin of millet was unclear. Now reveal archaeological finds in Kazakhstan the crucial helpers in exchange of cereal crops: the steppe nomads of Central Asia (Spengler et al. 2014). They grew as early as 2800 BC both cereals, wheat and millet. It shows that the nomads already operated in the Bronze Age agriculture—even if only seasonally. On the other hand, they brought in their movements the types of cereals in the regions where they were not known, so shaping the agricultural changes throughout Asia.

The Eurasian steppe is traditionally the realm of nomadic horsemen. The itinerant ranchers dominated large parts of Central Asia for thousands of years, and it developed a high culture. They were among the first who established two-wheeled chariots; their metal processing was artful in the Bronze Age and surpassed even that of China; they imported and imitated. From the third millennium BC, these mobile ranchers played an important role in the intra-Asian transport of raw materials and commodities such as copper, ceramic, and bronze objects. Contrary to the popular belief, these nomads were also not always on the go. Some steppe cultures even built large, fortified cities, others changed between solid summer and winter quarters.

The earliest evidence of agriculture in the steppes and mountain regions of Central Asia is only from about 800 BC. Now, it is known that the steppe nomads already cultivated grains in

the highlands of Kazakhstan some 2000 years earlier. In the ruins of two only seasonally inhab-
ited settlements from the early Bronze Age, the researchers came across 4800-year-old remains of
barley, wheat, and millet grains. The high density of these grains indicates that they were cultivated
locally. These is the first evidence that the ranchers in Central Asia have been cultivating cereal
since 2800 BC.

One of the two settlements, Tasbas, was in a fertile valley, approximately 1500 m above sea
level. Here, the nomads probably held over the summer, and built during this time of grain. In the
fall, after harvest, the Bronze Age ranchers then went down into the plane and survived the winter
nomadic. With 550 m above sea level, the second camp was excavatcd, Begash. It served more as a
winter camp where the nomads emaciated from their inventories. Because the ground was dry here,
the vegetation was more typical steppe barren.

The findings reveal even more. For they also show that the steppe nomads played a crucial role in
the spread of the various domesticated cereals and their use. Although wheat was cultivated in the
Middle East about 6000 years ago already, this cereal did not reach China until around 1500 years
later (Section 2.2.1). For the Chinese domesticated millet already about 8000 years ago—a crop that
was known Europe and western Asia only about 4000 years later. The Bronze Age inhabitants of
the two Central Asian settlements knew and used both cereals by 2800 BC. They played a key role
in the dissemination of agricultural practices in Asia.

There is evidence of civilization appearing in the Bronze Age (3200–2000 BC) in mainland
Greece and neighboring islands. A recent report suggests that Bronze Age Greeks could have begun
baking high-quality bread as early as 3300 BC. That is centuries before experts have believed the
types of wheat needed to produce bread similar to modern varieties existed. Other findings also
suggest wheat has been domesticated twice and in two different places, rather than only once as was
previously believed, the two experts said. Wheat was thought to have first been domesticated about
8000 BC in the Fertile Crescent, an area that crosses parts of modern day Iraq, Iran, and Jordan. But
a parallel development appears to have been going on in northern Greece. Another discovery in the
ancient wheat were DNA markers, suggesting a husked variety known as spelt, which experts previously did not think had become known in the Mediterranean area until the first century AD. Husked
wheat, which produces bread similar to modern "stone-ground" varieties, contains a hard covering
that is commonly broken off with a stone grinder.

The period between 2000 and 1600 BC marks the arrival of the first Greek-speaking Indo-
European populations. By this time, a complex urban civilization developed in Crete. By 1600 BC,
these is evidence of royal tombs at Mycenae. The Mycenaean kingdoms developed an agriculture
including irrigation and the draining of Lake Coapis. The culture reached its high point in the fifth
century BC. The time frame of 750–450 BC is known as the era of Hellenism. This was the great
period of Greek colonization probably instigated by the shortage of arable land on the Greek mainland. Colonialization extended in the Mediterranean region as far west as Spain and as far east as
northern boundary of the Black Sea. This period coincides with innovations in all fields of thought
and technology. Hellenism is associated with the flourishing of the arts and sciences.

The basic techniques of agriculture and horticulture were well established in the ancient cultures,
that is, Antiquity (from 3000 BC to 500 AD). The progress of Antiquity was the creation a written
record of agricultural achievements. It also contained the seeds for the beginnings of what one can
call scientific studies. The agricultural and horticultural achievements include:

- Planting and cultivation technology involving plowing seed bed preparation and planting
- Irrigation technology including water storage in dams and ponds; channeling of water
  above and below ground; water-lifting technology including "shaduf," ARCHIMEDES
  screw, or "sakieh"
- Basic technology of storing agricultural products in granaries, underground storage, cave
  storages
- Fertilization and crop rotations

- Basic propagation technology, such as seed handling, grafting, layerage, and cuttage
- Improvement of crops by selection and clonal propagation
- Basic development of food technology, such as fermentation (bread and wine), drying, or pickling
- Development of gardens and parks

Greek culture, based on the domination of ideas rather than technology *per se*, spread throughout the entire Mediterranean basis and had a powerful influence on Roman culture. Today Greek and Latin are the basis of scientific English in botany, biology, and agriculture.

The culture of the developing West is based on a fusion of Greek culture, Babylonian and Egyptian science, and Semitic religion. The art is typified by idyllic realism, includes depictions of gods, animals, particularly horses, and plants, and includes agricultural practices. DEMOCRITUS of Abdera (460–370 BC) was the founder of the atomic theory. He also had theories on the nature of plants. For example, he thought plant diversity was due to differences in the atoms of which they were composed.

HIPPOCRATES of Kos (460–377 BC), considered the originator of a Greek school of healing, was the first to clearly expound the concept that diseases had natural causes. He particularly noted the influence of food and diet on health, recommending moderation. The use of drugs was not ignored. Between 200 and 400 herbs were mentioned by him (Table 2.1).

THEOPHRASTOS of Eresus, city of Lesbos (371–287 BC), is claimed as founder of the botanical sciences and thus known as the Father of Botany. He was a student of ARISTOTELES (384–322 BC). From THEOPHRASTOS several books remained, for example, "Historica plantarum" and "De causis plantarum." In continuation of the biological work of his teacher ARISTOTELES, he described and classified different plant species. Beside the medical doctor DIEUCHES, he for the first time described oats. He introduced numerous terms, for example, "carpos" (the fruit) or "pericarpion" (the mature ovary wall). They are kept until recent time. He already taught that higher plants are sexually propagating. However, the knowledge got lost during late Antiquity, until CAMERARIUS again demonstrated different sexes in plants (see Section 2.7).

ARISTOTELES' description about the diversity of crop plants is particularly remarkable. In "Historica plantarum" can be read:

> By the way, of both (wheat and barley) there are two different types concerning grains, spikes, shape and even the effects. In barley there is a 2-rowed type but also 3-, 4-, 5- and 6-rowed types. The Indian barley produces tillers. Some types show big and limp spikes, others are smaller and dense. Even the seeds of barley are either more round and smaller or longer and bigger. Some are white, others are reddish.

In "De causis plantarum," THEOPHRASTOS discusses the problem of heritable modifications as follows:

> Sometimes changes happen on fruits sporadically, rarely on entire trees, and this the prophets call miracles, and it is seen as malformation against the nature, when it is happen in few cases but not if it occurs more frequent. Some herbs return to the wild shape too when they are without care; tiphia and zea (einkorn and spelt wheat) however turn into wheat within three years if they are sown without glumes, and this is associated with changes of growth conditions on site; because the annuals also change by time…

> Possibly the local conditions produce the plant species, either all or some of them. Because they unify the species or differentiate them with respect to usefulness as in the Thrakian wheat the polyglumeness and late germination, both are caused by winter frosts. And therefore in the other (regions) the early sown Thrakian wheat comes up late and grows late and in turn of other (regions) sown germinates late. Then the habit has become like nature.

Herewith, the problem of "Transmutatio frumentorum" was named for the first time and with great obviousness—a problem that caused many discussions throughout the medieval time till present (DITTRICH 1959).

The term "inheritance" was already in use, however, with complete different meaning. In Antiquity, it meant the production of descendants of the same type and with similar features. In this context, there were already "eugenic" measures for improvement of governmental staff.

In "Historica plantarum," there is a detailed description of plant flowers. It was recognized that some flowers are sterile. In cucumber and lemon trees, those flowers are sterile, exhibiting an outgrowth as a distaff. Those missing the outgrowth are fertile. That seems to be the oldest note on a carpel (JESSEN 1864).

CAPELLE (1949) investigates the botanical considerations of THEOPHRASTOS concerning the origin of plant malformations. THEOPHRASTOS pointed to three types of malformations:

- Malformations caused by human treatment, that is, the agricultural and horticultural activities and/or missing care.
- Malformation caused by environmental effects, for example, temperature, wind, drought, and so on.
- Malformations caused by the plant itself.

According to his philosophy and the philosophy of Antiquity—nothing happens without reason in nature—the latter type of malformations was classified as "incident against the nature."

The herbal "De Materia Medica" by P. DIOSCORIDES of Anazarba, a Roman army physician, wrote in the year 65 AD, one of the most famous books ever written, was slavishly referred to, copied, and commented on for 1500 years. DIOSCORIDES did make an effort to systematize knowledge with plants originally grouped by form and origin, a practice continued until ▶ LINNAEUS, although, in 1700, J. P. de TOURNEFORT (Aix-en-Provence, 1656–Paris, 1708) already developed the first generic classification of plants by separating clearly between genus and species, published in his "Institutiones Rei Herbariae."

Although the Romans historically did not promote very much the understanding of biological inheritance, their performances on applied inheritance, selection, breeding, and horticulture were outstanding. C. PLINIUS, the Elder (23–79 AD), the prolific author of "Historia naturalis," compiled a monumental encyclopedic treatment of science and ignorance. His coverage of the natural world is the best-known and most widely referred source book on classical natural history. He records that KRATEUAS, a Greek herbalist and physician to MITHRIDATES VI, King of Pontus from 120–63 BC described the nature of herbs and painted them in color, creating the first illustrated herbal.

On the other hand, the Romans demonstrated excellent knowledge of crop varieties. The Italian and native of Spain, L. J. M. COLUMELLA, described during the first century AD in his book "De re rustica" several wheat varieties, among them also "siligo" wheat (*Triticum monococcum*). Known were four varieties of spelt wheat (*T. spelta*) differentiated by color, quality, and grain weight. He also recommended the single-spike selection for cereals in his book. Although the Egyptians grew already asparagus 3000 ago, as early as 200 BC, the Romans had how-to-grow directions for asparagus. They enjoyed it in season and were the first to preserve it by freezing. In the first century, fast chariots and runners took asparagus from the Tiber River area to the snowline of the Alps where it was kept for 6 months until the Feast of Epicurus. Roman emperors maintained special asparagus fleets to gather and carry the choicest spears to the empire. The characteristics of asparagus were so well-known to the ancients that Emperor CAESAR AUGUSTUS described "haste" to his underlings as being "velocius quam asparagi conquantur" (quicker than you can cook asparagus).

C. PLINIUS characterized 15 beech, 15 olive, 4 pine, 4 quince, 7 peach, 12 plum, 30 apple, 41 pear, 29 fig, 18 chestnut, 11 hazelnut, and 91 grape varieties. He also mentioned a fig cultivar "dottato" that is still cultured in some regions of Italy. It is an example of an ancient clone demonstrating the durability and efficiency of horizontal resistance. COLUMELLA tells that there are so many grape varieties as grains of sand in the Sahara desert. He knew that grapes grown on other places than the native lose their "character." M. T. VARRO (116–27 BC), P. V. M. VERGIL (70–19 BC), and COLUMELLA published instruction for best seed preparation. VARRO claimed:

In order to have the best seeds for sowing, the best spikes have to be separately threshed.

("Quae seges grandissima atque optima fuerit seorsum in aream recerni opotet spicas, ut semen optimum habeas" De re rustica Lib. I, 52.)

This could be brought in context with a single spike selection and positive mass selection of recent breeding methods. The rules given by VARRO are supplemented by COLUMELLA:

I will add the instruction that immediately after harvest one has to care about good seed on the threshing floor. Namely, one must, as CELSUS correctly mentioned, when cereals harvested of reasonable quality, the good spikes collect individually and thus look after the future. When better seeds were harvested, so grains have to winnowed in a vessel and the big and heavy once, which sink down to the bottom by winnowing, are kept for sowing. Of course, from a strong grain comes a good crop, a weak one brings a bad. (Illud deinceps praecipiendum habeo, ut demessis segetibus iam in area futuro semini consularis. Nam quod ait CELSUS, ubi mediocris est fructus, optimum quamque spicam legere oportet, separatimque ex ea semen reponere; ex com rursus amplior messis provenerit, quidquid exteretur, capisterio expurgandum erit, ut sempre, quod mag- nitudinem ac pondus in imo subsederit, at semen reservandum. Nam id pluri prodest, quamvis celerius locis humidis, tamen etiam siccis frumenta degenerant, nisi cura talis adhibeatur, De re rustica II.)

VERGIL acknowledged:

I have seen how cereals run wild even given greatest care, if not every year the best grains are selected by hand. (Vidi ego lecta diu et multa spectata labore, degenerare tamen, nisi humana quotannis max- uma quaque manu legeret, sic omnia fatis in pejus ruere ac retro sublapsi referri, Georgica Lib. I.)

Thus, permanent reselection was known in order to prevent degeneration of seed. In present days, it is called maintenance breeding.

The best seed is the one-year-old one, less good the two-year-old, the three-year-old is the worst, and even older does not come up at all. For all crops the same rule is valid: what lies first on the threshing floor is kept for sowing because it is the best and the heaviest. There is no other kind of differentiation. Spikes sat seeds only with gaps are drawn away. The best seed is reddish and has the same color if it is bite through with teeth, bad is that grain, which is white inside. (C. PLINIUS: "Naturalis historica")

How precise the Romans recorded and worked proves the quotation of COLUMELLA in "De arboribus:"

In order to get good cuttings from good grape rootstocks, those rootstocks that show big, faultless, and ripen fruits are labeled with a mixture of vinegar and red chalk during harvest, which is not washed away by rain, and continues this procedure three or more years if the rootstock remains lasting good. Then one has sufficient evidence that the variety is excellent, and the quality and amount of berries is not incidentally influenced by a suitable vintage.

Those ancient approaches remind us on modern long-term and multilocation testing—a testing of breeder's strains and varieties over a more or less long period and on several geographically different sites in order to estimate the adaptive environmental response and/or performance stability.

2.5 ARABIC INFLUENCES ON WESTERN AGRICULTURE

The Arabs were a key influence in the exchange of crops between East and West. From India, they brought sugarcane, rice, spinach, artichokes, eggplants, orange, lemon, coconut, banana, and old world cotton; from Africa, they introduced the watermelon and sorghum. The watermelon originally came from Africa, but after domestication it thrived in hot climates in the Middle East and southern Europe.

It probably became common in European gardens and markets around 1600. Old watermelons, like the one in a still life painting of the Italian G. STANCHI, likely tasted pretty good. One thinks that the sugar content has been reasonably high, as melons were eaten fresh and occasionally fermented into wine. But they still looked a lot different (Figure 2.10). Modern breds show almost no seeds anymore. They are just fleshy. That fleshy interior is actually the watermelon's placenta, which holds the seeds. Before it was fully domesticated, that placenta lacked the high amounts of lycopene that give it the red color.

From the Middle East Arabs introduced the hard wheat (*Triticum durum*) and artichokes as well. The flour of hard wheat is known as semolina from the Arab "semoules" (the German word "Semmel" = "roll" in English, derives from the same Arab word). Having less gluten, it is not good for bread but can be used for a number of processed products including couscous and various pastas, such as spaghetti and ravioli that we now associated with Italian cuisine.

The land of the Arab countries was dry. Thus, the great contribution of Arab agriculture was the introduction of summer irrigation, which greatly intensified cultivation. The traditional Mediterranean agriculture was based on winter or spring production. This explains the incentive of Rome to conquer territory to import foodstuffs. The Romans managed the shortage of water by fallow, in which some land was not farmed to conserve moisture.

FIGURE 2.10 Painting of Giovanni Stanchi (1645–1672) "Watermelons, peaches, pears, and other fruit in a landscape," oil on canvas; the less domesticated watermelons showing, thick skin, strong chambering, and heavy seeds. (Courtesy of Christie's, New York. With Permission.)

When the Arabs[8] moved into Europe (they entered Spain in 800 AD and move as far north as southern France), they carried their irrigation technology with them allowing cultivation in the dry summers. The remains of the irrigation technology can be seen today in southern Portugal in the Algarve, where wells dot the countryside and primitive water-lifting devices, chain of pots, have only recently been electrified. Arabs introduced the Indian and African summer crops and employed fertilization including manure as well as ground bones, crop residues, ashes, and limestone. They introduced sugarcane and refined techniques for sugar manufacture.

2.6 MEDIEVAL AND RENAISSANCE AGRICULTURE IN EUROPE

The breakdown of the Roman Empire in the sixth century resulted in the destruction of the large cities but left the rural areas and organizations relatively intact. There was a decline in knowledge in the Western World and a period of regression. During the next 600 years, the center of gravity of intellectual thought shifted to the Moslem world, a byproduct of the invasion and conquest of Byzantium. Moslem and Jewish scholars who collected and translated the manuscripts from antiquity and developed scientific and technological schools of learning translated the manuscripts of antiquity from Greek to Arabic. Yet, vestiges of scholarship persisted in the monastery libraries where learning was kept alive in the West, although the influence of the Church was more interested in theology than natural history.

A feudal society developed, which involved the relation between land and the people who owned and worked it. There was an exchange of service for protection under a hereditary system. The lord of the manor who owned the land was served by the tenant farmers or vassals who offered homage, fealty, and owed a debt of labor.

The cultivation regime was rigidly prescribed. The arable land was divided into three fields: one sown in the autumn in wheat or rye; a second sown in the spring in barley, rye, oats, beans, or peas; and the third left fallow. The fields were laid out in strips distributed over the three fields, and without hedges or fences to separate one strip from another. About the eighth century, a 4-year cycle of rotation of fallow appeared. The annual plowing routine on 400 ha would be 100 ha plowed in the autumn and 100 in the spring, and 200 ha of fallow plowed in June. These three periods of plowing, over the year, could produce two crops on 200 ha, depending on the weather.

Agriculture was the principal source of wealth in Medieval Europe and so the owner of land had great power. In the typical system, the land was divided by use into traditional categories, such as cultivated fields for grains, pulses, and fodder; meadows for grazing cattle, sheep, and pigs; and forests, which provided timber and game, usually restricted to the lords entertainment. Annexing the manors or tenant houses were kitchen gardens that provided fruit, vegetables, and herbs. Typically, the unfenced "open" fields were divided into strips allocated between villagers such that each had a portion of good and poor ground. Each worked their own strips and in addition provided labor to the lord. Systems of crop rotation were developed, with crops one year and fallow every other or every third year. The basic crops were cereals: emmer and bread wheat, barley, rye, oats, and millet. The system became very traditional with little initiative left to individuals. Later, the church became a competing economic force as monasteries became large landholders.

"The Physica" of HILDEGARD of Bingen (1099–1179) was the first book in which a woman discusses plants in relation to medical properties and the earliest book on natural history in Germany. She was a mystic, prolific author, and abbess of a Benedictine convent. The "Physica," her only scientific work, was to have a great influence on the German botanists of the sixteenth century. In it, the presence of the beet (*Beta vulgaris*) was mentioned occurring for the time in Germany.

In many gardens inside of monasteries, all sorts of plants were grown. In 1170, Cistercian monks from Burgundy mentioned the "Borsdorfer Apple," possibly the oldest variety. In this way, also Cistercian monks distributed the pear tree during the twelfth century from Burgundy to

---

[8] The Arabs entered Spain in the eighth century AD (AD 711) and moved as far north as southern France. In 1492, the last Moors were expelled from Spain.

Bedfordshire (UK). They bred the first variety from a Burgundian rootstock. About 400 years later, in Italy, 232 varieties of pear were known. In 1254, the quince (*Cydonia oblonga*) was presumably introduced by the crusader ELEANOR of Castile and cultivated in the monk's gardens (cf. Section 3.9).

ALBERTUS MAGNUS (1193 or 1207–1280), German name: Albert von Bollstädt, was an early member of the Dominican Order of Preaching Friars that established scholarly houses in the European Centers of Learning.

He was responsible for the translations of ARISTOTELES and THEOPHRASTOS and influenced the revival of botanical and horticultural information based on the writing of antiquity in his work "De vegetabilibus libri" (1256). Most medieval writers thereafter drew inspiration from the Dioscoridian–Plinian tradition, although the description increasing drew on the delineation of living plants and efforts were made to reconcile ancient writings with native flora. Issues, such as inheritance and modificability, are discussed in the fifth book. The unification of one plant species with another and the conversion of one species in another are particularly treated. It appears again one of THEOPHRASTOS'S assumption that, for example, rye can be turned into wheat after the second or third year after sowing, as wheat can convert into spelt wheat and vice versa again into rye. The thesis of THEOPHRASTOS is believed until the modern times. THEOPHRASTUS noted the kinship of wild-olive with the cultivated olive, but his correspondents informed him that no amount of pruning and transplanting could transform *kotinos* into *olea*. Through lack of cultivation, he knew, some cultivated forms of olive, pear, or fig might run wild, but in the "rare" case where wild olive was spontaneously transformed to a fruit-bearing one, it was to be classed among portents. He noted that wild kinds, such as wild pears and the wild olive tended to bear more fruits than cultivated trees, though of inferior quality, but that if a wild olive was topped, it might bear a larger quantity of its inedible fruits. He also noted that the leaf buds were opposite.

A. MAGNUS saw the reason of conversion in the metabolism of soil, where plants take up the nutrients, that is, that potency of matters of one organism may be transferred to another. He believed that from stubs of oaks and beeches poplars and birches can develop as mushrooms and grasses can develop from rotten substances. Another type of conversion is believed by grafting of plants (▶ LYSENKO; cf. Chapter 3). For example, if plums are grafted onto willow seedless fruits will originate. Grapevine scions grafted on cherry, apple or pear rootstocks result in scions showing the same ripening time as the rootstock (Figure 2.11). The conversion of the wild into the crop plants he explained as influence of nutrition, soil preparation, environment, sowing time, and so on. He did not recognize heritable changes, although he accepted malformations in animals that are inherited.

By the fourteenth century, a trend toward more naturalistic drawings became apparent, suggesting the influence of the Renaissance in art and ideas (Figure 2.10). Plants were no longer slavishly copied from past manuscript but redrawn afresh, with artists taking advantage of the local flora. In addition to the idealistic presentations of plants, there are already very practical activities. Because plants always played a role in the diet, they also had to be cultivated. For this purpose, seed was needed. And this was already provided by dealers and merchants. The LORAS family from Lyon in France belonged to them. In 1440, they founded a business which, among other things, sold seeds of various vegetable and ornamental plants. This did the family more than 500 years, until now! Even a dealer of saffron is known from this time, Andreas WEISS (1426–1501) at Augsburg (Germany).

In the late Middle Ages, the concept of doctrine of signatures developed in which the inner qualities of plants were thought to be revealed by external signs. This was codified by mystical writers such as PARACELSUS (1493–1541) and G. PORTA, the author of "Phytognomonica" (1588). Thus, long-lived plants would lengthen a man's life, short-lived plants would abbreviate it. Yellow sap would cure jaundice; plants with rough surface would cure diseases that destroy the smoothness of kin. Plants that resembled butterflies would cure insect bites. The concept was that medical plants were stamped with a clear indication of their uses. A seventeenth century dispensary explains the concept:

> The powers of Hypericum are deduced as follows: I have oft declared how by the outward shapes and qualities of things we know their inward virtues which God hath put in them for the good of man. So in John's wort we take notice of the form of the leaves and flowers, the porosity of the leaves, the veins.

FIGURE 2.11 Picture of gardeners crafting fruit trees. (Courtesy of Woodcarving from "Ruralium Commodorum" by Petrus de Crescentiis, 1486. With Permission.)

1. The porosity roles in the leaves signify to us, that his herb helps both inward and outward holes and cuts in the skin. 2. The flowers of Saint John's wort, when they are putrefied they are like blood; which teaches us, that this herb is food for wounds, to close them and fill the up (PARACELSUS).

This herb plant John's wort is recently subjected to breeding because the hypericin is extensively used as a mood elevator and nerve calmer. In Germany in the year 2005, its sales are comparable to those of the conventional antidepressants, purely on the extensive clinical research and data on it.

The long period of medieval agriculture in Europe led to our present divisions of agriculture into agronomy, horticulture, and forestry. Agronomy became involved with open fields and meadows for the production of grain and fodder for animals. The kitchen gardens of tree fruits and vine, vegetables, ornamental, and herbs became the domain of horticulture. The forests for timber and game became a special purview of forestry. Initially, the differences between agronomy and horticulture were based on crops and intensity of production. This system breaks down in the tropics where it is never clear where agronomy ends and horticulture begins.

With the age of printing, there was a tremendous demand for books of agricultural works including farm management and vine culture. Consequently, agricultural technology of the period is available from the printed record. A particularly good example is "L'agriculture et maison rustique" of C. ESTIENNE and J. LIÉBAULT (Paris, France, 1564) that went through many printings starting from the late 1500s. It was translated into English by R. SURFET and published as the "Country Farm" in 1600 (London) and enlarged in 1616 and went through a series of editions. These books were a tremendous source book of information about crop improvement and it is clear that by the Renaissance equals and then surpasses that described by the Roman agricultural writers. But it took over a thousand years!

The literature of the Middle Ages and Renaissance, which developed in the fourteenth century, has made it possible to develop a complete history of horticulture and agronomy. It was a difficult time in Europe duc a change in climate, the little ice age, which reduced yields, the rise of disease and the increase in populations. There was a demand for new land. The discovery of America by C. COLUMBUS (1451–1506) was a part of those strategies. He discovered America in 1492 and describes red pepper and allspice. In his journal, he writes:

> We ran along the coast of the island, westward from the islet and found its length to be 12 leagues as far as a cape which I named Cabo Hermoso, at the western end. The island is beautiful, ...I believe that there are many herbs and many trees that are worth much in Europe for dyes and for medicines; but I do not know them and this causes me great sorrow. There are trees of a thousand sorts, and all have their several fruits; and I feel the unhappy man in the world not to know them, but I am well assured they are valuable.

Two of his soldiers reported about people on the island in Cuba eating flour from seeds of new sort of grass (probably maize!). By this time, about 200 differing varieties of flint maize had been developed by the American Indians. Dent maize appeared in northeastern Mexico. They have the advantage of a substantial amount of soft starch in their kernels surrounded by a thin outer layer of hard starch. Those maizes were introduced to Spain in 1520. In 1596, the monk Romano PANE, one of COLUMBUS' crew, for the first time, described tobacco and how the Indians smoke it. Since 1550, the tobacco plant is also grown in Spain.

In 1551, H. BOCK published the "Kräuterbuch" ("Book of herbs") in Basel (Switzerland) containing many description of horticultural and agricultural plants, such as broad bean, pea, white lupine (*Lupinus albus*), and clover (*Trifolium pratense*). Carolus CLUSIUS (1525–1609) or Charles de l'ESCLUSE (1526–1609), a French medical doctor and botanist as well as temporarily working at Leiden (the Netherlands), introduced the tulip to Holland. In addition, he discovered many plant species during his many travels through Spain, Austria, Hungary, and Bohemia. Some of them were tested and described as medical and ornamental crops, for example, in his books "Historia Stirpium per Hispanias" or "Rariorum Stirpium per Pannoniam, Austriam et vicinas provincias observatarum historia."

By the sixteenth century, population was increasing in Europe, and agricultural production was again expanding. The nature of agriculture and horticulture there and in other areas was to change considerably in succeeding centuries. Several reasons can be identified. Europe was cut off from Asia and the Middle East by an extension of Turkish power. New economic theories were being put into practice, directly affecting agriculture. Also, continued wars between England and France, within each of these countries, and in Germany consumed capital and human resources. Colonial agriculture was carried out not only to feed the colonists but also to produce cash crops and to supply food for the home country. This meant cultivation of such crops as sugar, cotton, tobacco, and tea, and stimulating more detailed studies of plant development.

At the end of sixteenth century, we receive the first comprehensive description of a heritable change that spontaneously occurred in a higher plant. The druggist and professor of botany at the University of Heidelberg (Germany), P. S. SPRENGER, found in his experimental garden in 1590 a new type of celandine (*Chelidonium majus*) with deeply lopped leaves. He called it *Chelidonium folio lanciniato*. The mutant form showed long-lasting stable inheritance without any segregation,

TABLE 2.2
Foundation of First Botanical Gardens in Europe

Year	Location	Country
1543	Pisa	Italy
1543/1544	Padua	
1545	Florence	
1567	Bologna	
1590	Leiden	Holland
~1550	Eichstätt	Germany
1579	Leipzig	
1597	Heidelberg	
1625	Altdorf	
1605	Giessen	
1629	Jena	
1593	Montpellier	France
1620	Strasbourg	
1626	Paris	
1600	Copenhagen	Denmark
1655	Uppsala	Sweden
1621	Oxford	England
1730	Kew	
1670	Edinburgh	Scotland

clearly shown by a later study over 40 years later ▶ MILLER (1768). The spontaneous mutation in *Linaria* with formation of peloria flowers that took LINNÈ to doubt on the unique creation of species was not the only example of higher plants known in eighteenth century. Similar reports are given by MARCHANT (1719) in *Mercurialis annua* with laciniate habit, or in *Fragaria vesca* with a simple oval leaf (DUCHNESNE 1766).

Something special is also the large collection of plants, each defined by low box hedging (*Buxus sempervirens*) at the German town Eichstätt (Table 2.2). All the plants here were illustrated in a remarkable book published in 1613—*Hortus Eystettensis*, a codex produced by Basilius BESLER (1561–1629). Commissioned by Prince-Bishop Johann Conrad von GEMMINGEN (1581–1612), the book illustrates a collection of plants that had been gathered together in the castle garden. This was not a collection of medical herbs or vegetables, but something altogether more modern, that is, ornamental and exotic plants have been included. Before Archbishop GEMMINGEN, who was appointed from 1595 and commissioned the physician and botanist Joachim CAMERARIUS (1534–1598) to expand the garden, Prince-Bishop Martin von SCHAUMBERG (1523–1590) left "newe gardens behind the castle." After his death, the Nuremberg pharmacist Basilius BESLER continued his work.

The sixteenth and seventeenth centuries saw a change in intellectual attitudes about the natural world, at the core of which was a curiosity that began to develop into what we called later scientific revolution.

2.7 PLANT BREEDING BY EXPERIENCE DURING THE SEVENTEENTH TO NINETEENTH CENTURIES

The scientific revolution resulting from the Renaissance and the Age of Enlightenment in Europe encouraged experimentation in agriculture as well as in other fields. Cereal yields were low, around 0.8–1 t/ha; higher were the yields of cabbage (about 25 t/ha). Therefore, fermented cabbage became a more popular diet in Central Europe than cereal products. Trial-and-error efforts in plant breeding produced improved crops. In addition, farmers and landowners began to correspond and network, driven by commercial pressure. Samuel HARTLIB (~1600–1662), a London publisher and great intellectual figure of the period, inspirited a German physician Johann Brun to introduce new apple varieties to England, as Francis LODWICK, a merchant, introduced black cherry cultivars from Flanders to England (KINGSBURY 2009).

After the first step of unconscious influencing and improvement of cropped plants to a conscious selection of the best, biggest, or most beautiful, what was proved as good under cultivation—at least partially—the way was free for breeding by experience or plant breeding as an "art" when only few people practiced it by their individual knowledge and experience. Particularly, in breeding ornamental plants, a breath of artistic interest determined the actions and success.

This form of breeding was devoid of every scientific basis, but brought up the basic types of modern crop plants over more or less long periods of selection.

An impressive example represents the wide variety of cabbages derived from a simple *Brassica* genotype. The gradual development can be followed by presentations in the old herb books. In Mexico, a second example can be given. On the daily market places, a wide range of at least 25 different varieties of common bean (*Phaseolus vulgaris*) is offered: white beans for meat meals, black beans are eaten with tortillas, some are only roasted or boiled, red beans a suitable for soups, and so on.

Over centuries, a complex culture of beans developed. On the one hand, it shows the habit of man to select for color, taste, constitution, nutritional value, or storage ability; on the other, it demonstrates how a high degree of genetic variability within a crop species is permanently maintained. This situation dramatically changed with modern plant breeding of the twentieth century where uniform and standardized plant products are required by food technology and international trade. Therefore, it became as major task of national genebanks of twenty-first century to maintain the genetic diversity of crop plants by collection, storage, and evaluation (cf. Chapter 4).

The foundation of botanical gardens during the sixteenth and seventeenth centuries did much in the way of advancing botany, horticulture, and agriculture (Table 2.2). They were at first appropriated chiefly to the cultivation of medical plants. This was especially the case at universities, where medical schools existed. The first botanical garden was established at Pisa in 1543.

Although there are rather few data about plant breeding from this epoch, the results demonstrate the progress of breeding even if steps forward were small. Most workers did not yet recognize the difference between genetic and environmental variation. More detailed references are available only from the last centuries, and only from countries where economic development mated with agricultural requirements was high. It is not surprising therefore that England played an outstanding role in this time.

In the 1670s, ▶ A. van LEEUWENHOEK (1632–1723, Holland) built the first microscopes with 250-fold magnification for plant and other studies, in 1665, ▶ R. HOOKE (1635–1703, England) described the cell as basic functional unit of organs, in 1675, ▶ M. MALPIGHI (1628–1694, Italy) conducted anatomical investigations in plants and repeated the observations of HOOKE. In 1676, T. MILLINGTON (England) saw anthers function as male organs, and, in 1670, ▶ GREW (1641–1712, England) suggested the function of pollen and ovules. In 1823, J. B. AMICI (1784–1860) noted pollen tubes in flower studies. However, long before them, C. PLINIUS already wrote that "the naturalists communicate that trees and herbs show both sexes" not explaining additional details. At that time, most researchers followed the thesis of ARISTOTLES that fertilization in plants is a sort of nutrition.

C. F. WOLFF (1733–1794), a German biologist, concluded from his studies on animals and plants, that nutrition and growth of plants are caused by an "essential force" ("vis essentialis") that converts a homogeneous, clear, and glassy matter into new organs not present ever before (with respect to an egg). His knowledge can be summarized as follows:

- All living beings show cell structures as a common trait.
- Plant and animal cells are basically similar.
- The morphological appearances (fibers, vessels, etc.) are modifications of cells.

The thought of constancy and variability are the two big biological phenomena, which were on his mind. He recognized two types of variabilities. The first is influenced by light, temperature, air, moisture, or nutrition. This variability caused by the environment, he called "variation." He knew that plants brought from St. Petersburg (Russia) to Siberia subsequently change their habit—and when returned to St. Petersburg, they took up the former shape. The idea of modification and its inheritability obviously was known to him:

> The variations are not inherited, and the outer shape itself and its internal structure would be constant if they would not be changed by random, not inheritable predicates.

Although GREW ("Anatomy of Plants," 1682) already explained that anthers "generate" the male substance of fertilization, in 1694, R. J. CAMERARIUS (1665–1721, Germany) published "De sexu plantarum epistola" summarizing all data on sexual knowledge and the first demonstrated sex in plants. He also suggested crossing as a method to get new types dealing with mulberry, ricinus, and maize. Systematic plant breeding was established at this point. By the puritan minister in the American colony of Massachusetts, C. MATHER (1663–1728, the USA), similar findings were described in 1716. He observed natural crossing in maize. Different races of "Indian corn" showing red or blue seeds with common yellow resulted in hybrids. He noted xenia. When ears of yellow maize are planted next to red and blue maize, they showed red and blue kernels in them during the first year. By this method, he was able to describe spontaneous outcrossing in neighboring rows. Supporting results on maize were later achieved in controlled experiments by P. DUDLEY (1724) and the American J. LOGAN (1736). In his book "Religio Philosophica," MATHER also reported about spontaneous hybrids between *Cucurbita pepo* var. *ovifera* and *C. pepo* var. *condensa* or *C. pepo* var. *maxima*. He noticed bitter fruits when *C. pepo* var. *ovifera* is pollinated with *C. pepo* var. *condensa*.

Other activities at America began in 1737 by W. PRINCE in New York, who offered raspberry plants for sale; in 1777, he offered 500 white mulberry trees, *Morus alba*, for sale. General OGLETHORPE in 1733 imported 500 white mulberry trees to Fort Frederica near Sea Island Georgia to suggest to the colonists, that there was an economical future for silk production. The President of the US Continental Congress, H. LAURENS, a native of Charleston, South Carolina, after the year 1755, introduced olives, limes, ginger, ever-bearing strawberry, red raspberry, and blue grapes from the South of France; he also introduced apples, pears, plums, the white Chassell's grape, which bore abundantly.

But T. FAIRCHILD (1667–1729, England)—a multidimensionally interested and talented gardener known for the introduction of exotic plants to England—created in 1717 the first proved artificial hybrid of carnation × sweet William (*Dianthus caryophyllus* × *D. barbata*), commonly known as "Fairchild's mule," so as Dutch breeders found the first hybrid in *Hyacinthus*. The carnation hybrid was sterile and was vegetatively propagated more than 100 years. It was the first interspecific hybrid documented.

E. LEHMANN (1916) mentioned a forerunner of ▶ J. G. KÖLREUTER (1733–1806, Germany) by reference of a Dutch botanist (MORREN [1858]—the Parisian postman N. GUYOT—who made first plant hybrids in *Ranunculus* and *Primula* based on his knowledge about sexual propagation, function of pollen grains, and so on. At that time, gardening became very fashionable among aristocrats (Figure 2.12).

KÖLREUTER was the first who did systematic studies on plant hybrids. He demonstrated in 1761 by using tobacco (*Nicotiana rustica* × *N. paniculata*, but also other hybrids of *Dianthus, Matthiola, Hyoscyamus, Verbascum, Hibiscus, Datura, Cucurbita, Aquilegia, Cheiranthus*) that hybrid offspring received traits from both parents (pollen and ovule transmit genetic information) and were intermediate in most traits. In the paper "Vorläufige Nachricht von einigen das Geschlecht der Pflanzen betreffenden Versuchen und Beobachtungen" (1763), he concluded that mother and father equally and specifically contribute to the character of the hybrid.

> For production of each plant in nature two equally liquid matters of different kind are needed for unification, which are predetermined by the creator of all things….

FIGURE 2.12 Reconstruction of a seed cabinet of gardeners during the eighteenth century, showing the high appreciation of their seed collection, modified after Christian Reichart (1685–1775), Erfurt (Nonne) 1788: "Anhang zu denen sechs Theilen des Land- und Garten-Schatzes." (Courtesy of Deutsches Gartenbaumuseum Erfurt, Erfurt, Germany, 2017. With Permission.)

He demonstrated the identity of reciprocal crosses and mentioned hybrid vigor (now called "heterosis," cf. Chapter 3), segregation of offspring (parental and nonparental types) from a hybrid.

Nevertheless, the hybridization experiments and the "change of species" by backcrossing were not recognized and even denied by KÖLREUTER'S contemporaries. As in other sciences, thus the history of genetics and plant breeding demonstrate that important discoveries were not recognized or ignored for long. The rejection of the works of CAMERARIUS, KÖLREUTER, and SPRENGEL was a big failure of an epoch, in which the dogma of invariability of species is still not overcome. SPRENGEL (1766–1833) also published in 1793 the role of insects in pollination of angiosperms and discussed the flower structure and color in relation to pollination and insect habits. He made the prerequisites for the development of plant genetics!

On the other hand, and unimpressed of the academic discussions, practical breeders, such as HOVEY (USA)—the "Father of the American Strawberry"—developed the first fruit cultivar of strawberry using controlled pollination in 1834.

Only by end of the eighteenth century, progressive English, French, and German plant breeders started systematic hybridization experiments in order to increase the crop yield and made experimental gardens for their observations. In addition to production of plants for consumption and trade, they carried out numerous crosses with the aim of new recombination of traits. In 1727, under the supervision of ▶ L. VILMORIN, the French company practiced the *pedigree method* for breeding sugarbeet. VILMORIN'S connection to the world of botany and the modern seeds industry stretches back to the mid-eighteenth century. In 1743, Madame Claude GEOFFREY, known as the "maitresse grainetière" in Paris, opened a boutique for plants and seeds on the city's Quai de la Mégisserie with her husband, Pierre d'ANDRIEUX, then the chief seed supplier and botanist for King LOUIS XV. Geoffrey and Andrieux's daughter later married Philippe Victoire de VILMORIN, who joined the business in 1775. The Vilmorin-Andrieux et Cie Seed Company extensively contributed to the development of plant-breeding knowledge and improved cultivars for over 260 years. In 1785, J. B. van Mons started systematic selection of horticultural plants. In his seed catalog of 1825 he listed 1050 cultivars (DECAISNE 1855). In 1825, LORAIN recognized the possibility of maize hybrids, and in 1830, "Red May," the first bread wheat cultivar selected in the USA, was released.

In 1779, ▶ T. A. KNIGHT (1759–1838)—President of the Horticultural Society London from 1811 to 1838—emphasized the practical aspects of hybrids and worked on improvement of frost resistance and high yield in different plants (grape, apple, pear, plum, garden pea and, probably even wheat) rather than on inheritance. However, he noted the advantage of outcrossing to produce new forms of crops, and that male and female parents contribute equally to the resulting hybrid, with segregation following in the next generation. In 1787, he made first wheat variety crosses in Europe to produce new varieties. He crossed domestic apples with the winter-hardy Siberian crab apple (*Malus baccata*) in order to improve the local Herefordshire fruit production. Because of the missing genetic knowledge, he came to the conclusion that varieties age, even when newly propagated. For example, budstock from a 100-year-old "Bridgewater Pippin" apple tree would be a 100 years too. Thus, as many new varieties have to be constantly created as possible. ▶ KNIGHT did not know yet that this decline is often associated with an accumulation of virus diseases.

His pea crosses from 1799 to 1823 are most important. He realized—as ▶ G. MENDEL much later—several varieties with distinct character, the self-fertility, the possibility of flower isolation, and simple cultivation. He carefully emasculated the flowers and used as control flowers without pollination. A luxuriance of hybrids was early observed. As first researcher, he found the dominance of the gray seed color over white. After backcrosses of white-seeded plants to the hybrid, he received segregating white-seeded and gray-seeded offspring, of course without calculating the numerical proportion.

Similar data were achieved by the Italian pomologist ▶ G. GALLESIO (1772–1839) using carnations. In his book, "Teoria della Riproduzione Vegetale" in 1816, he explained his latest observations about plant breeding:

> I fertilized white flowering carnations with pollen of red flowering plants and vice versa; the carnations I cultivated from the seeds showed flowers with mixed color.

> Therefore with their unification, which is not natural, arises an irregularity of their effects, and those soonly show the character of the one origin or the other, according to circumstances the one dominates.

MARTINI (1961) pointed to the fact that GALLESIO used for the first time the term "dominate," although ROBERTS (1929) noted in his book "Plant Hybridisation before Mendel" that ▶ A. SAGERET (1763–1851), a researcher and agronomist and member of the Société Royale et Centrale d'Agriculture de Paris, did it for the first time 1826, that is, about 10 years later!? SAGERET was mainly working with melons and studying the descent of traits through several generations. He discussed what came to be called segregation. He was the first used the term "dominant" for genetic characters, long before MENDEL. Another was Henri-Louis Duhamel du MONCEAU (1700–1782), who concluded that forms appear and disappear over the time. In his book, "La physique des arbres" (1778), he addressed the variability of cultivated plants, that is, most cultivated varieties are composites of older varieties brought about by mixing pollen.

Inspired by KÖLREUTER, ▶ W. HERBERT (1778–1847), an English vicar of Manchester and contemporary of ▶ T. A. KNIGHT—suggested that the character of winter hardiness is inherited. He first established the genus *Nerine* in 1820 too. HERBERT investigated the fertility of hybrids—no matter whether fertile or sterile—as a measure for the relationship between species and varieties. Although he recognized the importance of hybridization for horticulture and agriculture—as ▶ KNIGHT for improvement of fruit trees—he ignored the cross-ability as the measure for phylogenetic relationships:

> The only thing certain is, that we are ignorant of the origin of races; that God has revealed nothing to us on the subject; thereon; but we cannot obtain negative proof, that is, proof that two creatures or vegetables of the same family did not descend from one source. But can we prove the affirmative; and that is the use of hybridizing experiments, which I have invariably suggested; for if I can produce a fertile offspring between two plants that botanists have reckoned fundamentally distinct, I consider that I have shown them to be one kind, and indeed I am inclined to think that, if a well-formed and healthy offspring proceeds at all from their union, it would be rash, to hold them of distinct origin (Herbert 1847).

Later, however, G. GODRON used in his 1844–1863 experiments the character of sterility of interspecific hybrids as evidence that the parents were distinct species. But a German botanist, A. F. WIEGEMANN (1771–1853), a member of the "Kaiserliche Akademie der Naturforscher" (Leopoldina), already refuted several prejudices of his contemporaries in the paper "Über die Bastarderzeugung im Pflanzenreiche" (Wiegemann 1828). His thesis can be summarized as follows:

- There is hybridization in plant kingdom. It gives evidence for sexuality in plants. The influence of the male parent by transfer of the pollen to the stigma can be seen on several traits.
- The pollen substance is taken up by the secretion of the stigma and transferred through pistil to the carpel by mediation of the cell tissue.
- Some hybrids bear a slight resemblance to the father, some are intermediate. More frequently, the hybrids take after the mother.
- LINNÉ'S hypothesis is disproved that sexual organs of hybrids are from the mother and foliage and habit from the father.
- Hybrids are fertile if they derive from crosses between varieties and species.
- Individuals deriving from hybrids often strongly deviate from the characters of the mother or father and show heterogeneous habits.
- The closer are the parental plants, the easier is the production of hybrids. The best is the cross-ability between varieties, followed by species of one genus and plants of different genera.

Based on these studies, the Dutch Academy of Sciences at Haarlem in 1830 initiated a competition: "What teaches the experience concerning creation of new species and varieties by artificial fertilization of flowers by pollen, and which crop and ornamental plants can be produced and multiplied in this way?" The winner C. F. von GÄRTNER (1772–1850) received the award belated in 1837. Within the period of 1827–1849, he made the most extensive crossing experiments to date, with 10,000 crosses in 700 species from 80 genera. By analysis of more than 9000 experiments over a period of 7 years, he briefly reported to the Academy. Although his statements are similar to WIEGEMANN'S, he concluded that the origin of new varieties after hybridization just confirms the irrefutable border of species that cannot be overcome. As deserving his work was, it still has brought criticism to him 17 years later. ▶ G. MENDEL (1866) wrote in the introduction remarks to his famous paper "Versuche über Pflanzenhybriden" ("Experiments in Plant Hybridization") and in a letter to C. von NÄGELI that GÄRTNER'S experiments are difficult to explain because clear descriptions of trials and diagnosis of hybrids, are missing as well as the characterization of hybrid traits is perfunctory. With this paper, MENDEL was the first to explain phenotype ratios occurring in the progeny generations following a biparental hybridization. As genes were unknown to MENDEL, he hypothesized "elements" present in pollen and egg cells as causing the heritable differences between phenotypes. In the twentieth century, the concept of gene was developed, and the gene became known as the unit of inheritance, function, recombination, and mutation (see Chapter 3).

C. NAUDIN (1863) described in the paper "Noevelles recherches sur l'hybridité dans les végé-taux" the separation of essences in various proportions in germ cells and recombination in the off-spring of a cross when he used *Datura* hybrids for his rather complex experiments. He thought that the whole plant was a simple mosaic of the two parents.

Since the beginning of the nineteenth century, experiments with pea were en vogue. As the English gardener ▶ KNIGHT in 1789, also ▶ J. GOSS chose varieties of pea as subject for hybridization experiments. In 1822, he reported to the Secretary of the Horticultural Society of London about his crosses of "Blue Prussian" with "Dwarf Spanish" peas:

> … that these white seeds had produced some pods with all blue, some with all white, and many with both blue and white peas in the same pod …

Did he separately grow blue and white seeds then blue seeds produced only plants with blue seeds; however, the white seeds gave plants with white and blue seeds in the pods. In 1824, A. SETON mentioned a cross of the green-seeded pea "Dwarf Imperial" with a white-seeds variety resulting in:

> … they were all completely either of one color or if the other, none of them having an intermediate tint.

Both GOSS and SETON thus observed in their experiments the phenomenon of *dominance* and segregation as ▶ Knight. A few decades later (1866, 1872), the breeder ▶ T. LAXTON again carried out hybridizations with pea that confirmed the findings of ▶ KNIGHT and GOSS: dominance of one of the parental traits, for example, seed color or seed shape and segregation of the traits in the second generation. All segregants were separately harvested and checked in further generations. The result reflected the grouping of segregants with two associated features in the sense of MENDEL. LAXTON (1872) explained:

> I have noticed that a cross between a round white and a blue wrinkled pea, will in the third and fourth generations (second and third year produce) at times bring forth blue round, blue wrinkled, white round, and white wrinkled peas in the same pod, that the white round seeds when again sown, will produce only white round seeds, that the white wrinkled seeds will, up to the fourth or fifth generation, produce both blue and white wrinkled and round peas, that the blue round peas will produce blue wrinkled and round peas, but that the blue wrinkled peas will bear only blue wrinkled seeds.

As a breeder, LAXTON also knew that maximum variability can be expected in the third and fourth generation after crossing and impossible to get stable new varieties before the third and fourth generation, whereas a setback to parental traits may happen earlier. The French breeder L. de VILMORIN, the one who introduced the principle of individual *testing of progeny*, in 1860, stressed the value of self-pollination in breeding stable cultivars of wheat and sugarbeet ("Notes sur la création d'une nouvelle race de betterave et considerations sur l'hérédité des plantes," 1856) and confirmed LAXTON'S data in lupine (*Lupinus hirsutus*) in experiments between 1856 and 1860. His son, H. de VILMORIN (1843–1899), in particular, was interested in new combinations (*recombination*) of (formerly) associated characters. In 1878, he did the first extensive work with interspecific crosses in cereals, for example, *Triticum vulgare* × *T. polonicum* and *T. durum*.

Many breeders of the nineteenth century were aware of spontaneous variations and used them for the selection of new strains. ▶ K. RÜMKER (1889) mentioned in his book "Anleitung zur Getreidezüchtung auf wissenschaftlicher und praktischer Grundlage" some examples. Obviously, COUTEUR systematically selected deviating segregants from hybrid combinations. In 1841, R. HOPE described "fenton" wheat derived from a single plant (spontaneous mutation found in

1835). In 1873, the Scottish breeder ▶ P. SHIRREFF (1791–1876) gave most examples. He developed more than 11 varieties of oat and wheat based on the selection of spontaneous variants:

- "Mungowells wheat," based on one plant from 1819, stable after the fourth generation
- "Hopetoun oats," based on one plant from 1824
- "Hopetoun wheat," based on one plant from 1832
- "Shirreff oats," origin unknown
- "Shirreff's bearded red wheat" (by 1860 distributed for sale)
- "Shirreff's bearded white wheat" (by 1860 distributed for sale)
- "Pringles wheat" (by 1860 distributed for sale)
- "Early fellow oats" (by 1865 distributed for sale)
- "Fine fellow oats" (by 1865 distributed for sale)
- "Long fellow oats" (by 1865 distributed for sale)
- "Early Angus oats" (by 1865 distributed for sale)

SHIRREFF had first nursery plots, with 70 Scottish families by 1857. He suggested crossing parents with desirable characteristics to obtain progeny of value. Already in 1819, he observed in his wheat field a particularly vigorous type. He fertilized it, removed all surrounding plants, and harvested 63 spikes with 2473 grains. After propagation over many years, the offspring showed specific characters deviating from the original population. Nowadays, it would be called "mutant." In this way, he developed several other wheat varieties. However, SHIRREFF also mentioned that RANNBIRD from Basingstoke (UK) carried out first hybridizations in wheat. Such products were shown on the international fair of London in 1851, together with hybrids of MAUND (Worcester, UK). There were nevertheless still long doubts about the utilization of wheat hybrids, as the German KÖRNICKE claimed in 1875 (Landwirtschaftliches Journal, 1877, p. 217).

In rye, similar approaches were reported by MARTINI (1871), who tried to fix the three-flowered spikelets as BLOMEYER did it with the variety "Leipziger Roggen." WOLLNY (1885) was successful with breeding the "Igelroggen" (hedgehog rye) and the "Laxblättriger Roggen" (lax-leaf rye) based on single plant selection. In 1883, ▶ RIMPAU (1899) found in two-rowed barley and rye plants with branched spikes, a white-glumed, an awned and a compactum plant in a red-glumed awnless landrace of wheat. Of particular breeding value was the awnless "Anderbecker Hafer" (Anderbecker oats) of BESELER (Germany) originating from a single plant selection within the awned "Probstei Hafer." DRECHSLER found on the experimental field of Göttingen (Germany) an awned type of the square head wheat. METZGER (1841) reported 12 variants of maize. However, with few exceptions, the most of those selections could not be developed as varieties (cf. Section 7.1).

3 MENDEL's Contribution to Genetics and Breeding

The prevailing view of heredity during the middle of the nineteenth century assumed two gross misconceptions: the acceptance of a blending of hereditary factors and the heritability of acquired characters. Evidence for a particulate basis for inheritance, such as the reappearance of ancestral traits, was common knowledge but considered exceptional. In fact, all the *discoveries* attributed to MENDEL, such as the equivalence of reciprocal crosses, dominance, uniformity of hybrids, and segregation in the generation following hybridization, are gleanable from the pre-Mendelian literature.

The many achievements in sciences and agriculture by the middle of the nineteenth century, the numerous hybridization experiments (very often with pea varieties), and ▶ C. DARWIN's paper on *Origin of Species* demonstrating genetic variation, inbreeding, sterility, and differences in reciprocal crosses and others can be seen as fertile lap from which G. MENDEL (1822–1884, Figure 3.1), an Austrian Augustinian monk in the monastery of Brünn (now Brno, Czech Republic), received some of his inspiration.

DARWIN's explanation of evolution by natural selection became a well-established theory in the years following publication in 1859 despite any factual evidence to explain either the nature or the transmission of hereditary variation. DARWIN, aware that blending inheritance led to the disappearance of variation, relied on the inheritance of acquired characters to generate the variability essential to his theory. His clear but inaccurate formulation of a model of inheritance, which involved particles (*gemmules*) passing from somatic cells to reproductive cells, exposed the lack of any factual basis. The concept was merely a restatement of views dating from HIPPOCRATES in 400 BC and endlessly reformulated.

In his famous paper "Versuche über Pflanzen-Hybriden," G. MENDEL found that the plant's respective offspring retained the essential traits of the parents, and therefore were not influenced by the environment. This simple test gave birth to the idea of heredity. This understanding of a gene proved to be useful for plant breeding throughout many decades and contributed to significant breeding progress in major crop species. Later, new insights from molecular genetics brought considerable changes to the concept of the gene and subsequently to plant breeding: although a more modern imagination of a gene as a coding sequence is strongly challenged by discoveries, such as split genes, alternative splicing, or the significant finding of noncoding RNA, single nucleotides within a gene became the ultimate units of interest both for selection and for genetic modification (GAYON 2016).

The proof of the particulate nature of the *elements* was made possible by the nature of the plants and traits studied. The traits chosen were contrasting (e.g., yellow versus green cotyledon, tall versus dwarf plant) and constant (i.e., true breeding) after normal self-pollination in the original lines. In seven of the eight characters chosen for study, the hybrid trait resembled one of the parents; in one character, bloom date, the hybrid trait was intermediate. MENDEL called traits that pass into hybrid association *entirely* or *almost entirely unchanged* as dominating and the latent trait as recessive (recidivous) because the trait reappeared in subsequent crosses. He saw that the traits were inherited in certain numerical ratios. Thus, he then came up with the idea of dominance and segregation of genes and set out to test it in peas. The phenomenon of dominance was clearly not an essential part of the particulate nature of the genetic elements but was important to classify the progeny of hybrids. The disappearance of recessive traits in hybrids and their reappearance, unchanged, in subsequent generations was striking proof that the elements responsible for the traits (now called genes) were unaffected or contaminated in their transmission through generations.

FIGURE 3.1 Gregor MENDEL, about 1880. (Courtesy of MENDEL Museum of Masaryk University, Czech Republic. With Permission.)

When the hybrids of plants differing by a single trait were self-pollinated, three-fourths of the progeny displayed the dominating trait and one-fourth the recessive. In the next self-generation, progeny of plants displaying the recessive trait remained constant (nonsegregating), but those with the dominant trait produced one of two patterns of inheritance. One-third was constant, as in the original parent with the dominating trait, and two-thirds were segregating as in the hybrid. Thus, the 3:1 ratio in the first segregating generation (now called the F_2 generation) was broken down into a ratio of 1 (true breeding dominant):2 (segregating dominant as in the hybrid):1 (true breeding recessive). The explanation proposed was that theme parental plants had paired elements (e.g., *AA* or *aa*, respectively) and the hybrids of such a cross were of the constitution *Aa*.

It took seven years to cross and score the plants to the thousand to prove the laws of inheritance. From his studies, MENDEL derived certain basic laws of heredity:

- Hereditary factors do not combine, but are passed intact.
- Each member of the parental generation transmits only half of its hereditary factors to each offspring (with certain factors *dominant* over others).
- Different offspring of the same parents receive different sets of hereditary factors.

MENDEL, in his paper, spoke about the "law of combination of different characters" and talked about "the law of independent assortment." He implied that the segregation of factors occurred in the production of sex cells.

Further, elements were distributed to conceptualize a genetic theory that was to create a new biology. The standard approach to unravel the mysteries of heredity was to analyze the complexity of characters usually from wide crosses, a method that had failed for 2000 years and was to continue to fail even when applied by the combined talents of F. GALTON and K. PEARSON.

G. MENDEL succeeded because of his approach. His goal was grand, being no less than to obtain a *generally predictable law* of heredity. His previous crosses with ornamentals had indicated predictable patterns, and his assumption was that laws of heredity must be universal. He had reviewed the literature and noted:

> ... that of the numerous experiments, no one had been carried out to an extent or in a manner that would make it possible to determine the number of different forms in which hybrid progeny appear, permit classification of these forms in each generation with certainty, and ascertain their numerical relationships.

The experimental organism, the garden pea, was once again a perfect choice. MENDEL procured 34 cultivars from seedsmen, tested their uniformity over 2 years, and selected 22 for hybridization experiments. Results from preliminary crosses indicated that common traits were transmitted unchanged to progeny but that contrasting traits may form a new hybrid trait that changes in subsequent generations. A series of experiments followed traits carefully selected for discontinuity to permit definite and sharp classification rather than *more-or-less* distinctions. This was a key decision.

MENDEL restricted his attention to individual traits for each cross, avoiding the *noise* of extraneous characters. His analysis was quantitative and he displayed a mathematical sense in the analysis of data and the design of experiments. MENDEL had a clear feeling for probability and was not put off by large deviations in small samples. MENDEL was a meticulous researcher. The sheer mass of his data is impressive and his experiments build from the simple to the complex.

The immediate impact of MENDEL's paper, presented in 1866, was nil, although it was distributed to about 120 libraries throughout the world through the exchange list of the Brünn society and was available in England and the United States. The paper was listed in the Royal Society Catalogue of Scientific Papers for 1866 (England) and referred to without comment in a paper on beans by H. HOFFMANN in 1869. The only substantial reference was in the 590-page treatise on plant hybrids (Die Pflanzen-Mischlinge, ein Beitrag zur Biologie der Gewächse) by the American W. O. FOCKE in 1881. He also coined the word *xenia*. MENDEL's name is mentioned 17 times, but it is clear that FOCKE did not understand him. In the critical passage, he writes:

> MENDEL's numerous crossings gave results which were quite similar to those of KNIGHT but MENDEL believed he found constant numerical relationships between the types of the crosses.

Yet the period from 1866 to 1900, the classical period of cytology, the study of cells, was to establish the basic part of structural cell biology that put MENDEL's theoretical discovery of inferred genes into structures contained in each living cell. In 1866, E. HAECKEL (1834–1919) published his conclusion that the cell nucleus was responsible for heredity. Soon thereafter, the chromosomes, the physical framework for inheritance, became the focus of attention in mitosis, meiosis, and fertilization with speculation on their relation to heredity. Later, W. ROUX (1850–1924) postulated that each chromosome carried different hereditary determinants. But the experimental evidence was not reported until 1902, when T. BOVERI (1862–1915) announced that each of the 36 chromosomes of the sea urchin was necessary for normal development. The issue was cloudy because the details of the meiotic process were not well understood.

In 1891, the German plant breeder ▶ W. RIMPAU (1842–1903) published a paper in which he reported about spontaneous and induced hybrids, mainly in wheat and pea. He did not know MENDEL's paper but the book of W. O. FOCKE where MENDEL's pea crosses were pagewise quoted. RIMPAU described all his hybrids with intermediate characters considering the parental habits. However, he noticed the great heterogeneity in the hybrid progeny. He found complete setbacks toward the parents, combinations, several intermediate forms, and sometimes completely new traitsnot seen among the parents. Moreover, he described for the first time the self-sterility of rye—in one of the most famous commercial grain cultivars at that time, the "Schlanstedter Roggen" (Schlanstedt rye). He produced it through 20–25 years of meticulous mass selection. In 1883, he also selected from a two-rowed barley a plant with branched spikes, which could be bred true by constant selection. In 1877, he even discovered within a red-glumed and awnless landrace of wheat three new types of spikes:

1. spikes showing awns,
2. white-glumed plants, and
3. a compactum-type of spike linked with stiffy straw.

He was also the first producing in 1888 a fertile octoploid wheat–rye hybrid, which can be taken as the birth of the triticale research. RIMPAU already knew that in 1876 the English botanist and cytologist A. S. WILSON (whose famous work "The Cell," 1896, described chromosome behavior and speculated on their role in heredity) had presented stalks of two wheat–rye hybrid plants produced by him in 1875 to the Edinburgh Botanical Society. He also mentioned attempts on production of wheat–rye hybrids by BESTEHORN and by CARMAN (1882). Those hybrids showed complete sterility because of the polyhaploid chromosome status (ABDR).

A student of WILSON, W. S. SUTTON (1876–1916, England) soon after recognized in a 1902 paper that the association of paternal and material chromosomes is arranged in pairs and their subsequent separation during meiosis constituted the physical basis of Mendelian genetics. SUTTON wrote two of the most important papers in cytology but never received his Ph.D. At the end of the nineteenth century, also two Swedish and one American breeder came very close to the discovery of inheritance of single traits. A. ÅKERMAN and J. MACKEY (1948) mentioned that P. BOLIN and H. TEDIN reported in 1897 during Agricultural Congress in Stockholm (Sweden) about crossing experiments with barley, pea, and vetch cultivars started in 1890. They found no variation in F_1 but different combinations of parental features in F_2, which could be numerically predicted. In 1897, P. BOLIN reported on the Second Agricultural Congress:

Concerning the inheritance of typical characters in the second and following generations seems to exist regularity. The forms that occur namely represent all sorts of combinations of parental traits and thus can be calculated with mathematical precision.

In 1901, W. J. SPILLMAN reported during the 15th Annual Meeting of the Convent of American Agricultural Colleges and Experimental Stations about crosses between winter and spring wheat that he began in 1899. Similar to BOLIN, he wrote:

When these results were classified, they conformed the above suggestion; and if similar results are shown to follow the crossing of other groups of wheat, it seems entirely possible to predict, in the main, what types will result from crossing any two established varieties, and approximately the proportion of each type that will appear in the second generation.

SPILLMAN also found that the most frequent character in F_2 corresponds to the F_1 plant, and he knew that quantitative investigations are the key for the understanding of inheritance of single traits. H. F. ROBERTS (1929) reassembled SPILLMAN's experimental data and came to the following conclusion: The traits "length of spike," "awnless spike," "hirsute glumes," "glume color," and so on segregate in a 1:2:1 or 3:1 manner.

In a book published in 1885 by NÄGELI and PETER entitled "Die Hieracien Mitteleuropas," MENDEL's paper is cited out of context, lumped with his other inheritance paper on *Hieracium* species. The relation between MENDEL and NÄGELI is a shameful episode for academic science. NÄGELI (1817–1891) corresponded with MENDEL and received MENDEL's reprints, a reformulated explanation, and packets of seed of peas with notes by MENDEL, but he could not or would not understand the paper. His eternal punishment is that he may only be remembered for this fact.

There are other curious references. The Russian botanist I. F. SCHMALHAUSEN (1849–1894) appeared to have read and appreciated MENDEL's results. He is cited as a footnote to a literature review of 1874.

MENDEL received 40 reprints of his paper, of which three have been traced. One went to C. NÄGELI at München (Germany), one to A. K. von MARILAUN at Innsbruck (found after his death, uncut), and one turned up in the hands of M. W. BEIJERINCK, who sent it to ▶ H. de VRIES, undoubtedly the true source of VRIES' introduction to MENDEL. Thus, he knew MENDEL's paper before VRIES. He even claimed to be the rediscoverer of MENDEL's work, at least five years before VRIES (Itallie van Emden 1940). M. W. BEIJERINCK was also the first when, in 1888, he isolated the bacteria causing root nodulation in legumes. He called them *Bacterium radicicola*.

In this place should be remembered another man who was as close to the rediscovery of MENDEL's law of inheritance—W. BATESON (1861–1926). ▶ BATESON stated in his paper given in 1899 during an international conference of the Royal Horticultural Society, London:

What we first require is to know what happens when a variety is crossed with its nearest allies. If the result is to have a scientific value, it is almost absolutely necessary that the offspring of such crossing should then be examined statistically. It must be recorded how many of the offspring resembled each parent, and how many showed characters intermediate between those of the parents. If the parents differ in several characters, the offspring must be examined statistically, and marshaled, as it is called, in respect to each of those characters separately.... All that is really necessary is that some approximate numerical statement of the result should be kept.

He put therefore the same demand as 30 years before MENDEL that a statistical treatment of the progeny and a separation of individual traits are indispensable prerequisites for studies on inheritance. One year later, in a report for the "Evolution Committee of the Royal Society" he introduced the term "allelomorph" for the opposite characters of a pair of characters, the term "heterozygote" for nuclei that resemble the opposite "allelomorphs" (later abbreviated to "allele"), and the term "homozygote" for identical "allelomorphs." In 1909, he published "MENDEL's Principles of Heredity" explaining own experiments and of his students concerning heritable characters. As first, he translated MENDEL's paper into English, and presumably the most vehement promoter of genetics in England after 1900.

Finally, the pieces of the puzzle, however, quickly fit together only after the independent verification of MENDEL's result by the Dutch ▶ H. de VRIES (1848–1935), the German ▶ C. CORRENS (1864–1933)—a student of C. von NÄGELI (1817–1891), and the Austrian ▶ E. von TSCHERMAK-SEYSENEGG (1871–1962).

Yet, none of the rediscoverer's papers was in the class of MENDEL's paper in terms of either analysis or style. VRIES was vague on the role of dominance and CORRENS was convinced that the law of segregation could not be applied universally. TSCHERMAK's paper reported the 3:1 ratio in the first segregation generation but did not interpret the backcross of the hybrid to the recessive parent as a 1:1 ratio, casting doubt of his complete understanding at that time.

Conclusion: MENDEL's paper is a victory for human intellect, a beacon cutting through the fog of bewilderment and muddled thinking about heredity. The story of its origin, neglect, and so-called rediscovery has become a legend in biology. An obscure monk working alone in his garden discovers a great biological phenomenon, but the report is ignored for a third of a century only to be resurrected simultaneously by three scientists working independently.

The genetic revolution had a rapid impact on plant improvement. Although breeders had unconsciously been using many appropriate procedures via crossing and selection in the nineteenth century, the emerging science of genetics and, especially, the fusion of MENDELism and quantitative genetics put plant breeding on a firm theoretical basis.

The relation between genetics and post-Mendelian plant breeding is best exemplified by two routine breeding protocols. One is the extraction and recombination of inbreds combined with selection to produce heterozygous but homogeneous hybrids, whereby combinations are first disturbed to complete the final order. The other is backcross breeding, in which individual genes can be extracted and inserted with precision and predictability into new genetic backgrounds. The combination of backcross breeding to improve inbreds and hybrid breeding to capture heterosis is the basis of present-day maize and other hybrid crop improvement (cf. Section 3.4).

The success of the new science of plant breeding had a substantial impact on agriculture, horticulture, and forestry. Dramatic successes quickly followed: examples include hybrids and disease-resistant crops. A further spectacular example of plant breeding progress was the development of short-stemmed photoperiod-insensitive wheat and rice, the forerunners of the so-called "Green Revolution" for which ▶ N. BORLAUG (1914–2009) (Figure 3.9), a plant breeder with the Center for the Improvement of Maize and Wheat, Mexico (CIMMYT), was to receive the Nobel Prize for Peace in 1970. Short-statured rice cultivars developed under the guidance of T. T. CHANG at the International Rice Research Institute (IRRI), Los Baños, Philippines, became the basis of Green Revolution in rice production. The cultivars increased rice yields by more than 40% in the tropical Asia.

3.1   REDISCOVERY OF MENDEL'S LAWS—BEGINNING OF GENETIC RESEARCH

The Dutch botanist H. de VRIES (1848–1935) was always involved and fascinated about questions of the theory of the origin of species and the role mutation played in the evolution of plants. Besides that, he carried out crossing experiments with *Oenothera lamarckiana* × *O. brevistylis* in 1895, where he observed the uniformity of his crossing hybrids and the "dominance" of some prevailing characters. The detection of a citation of MENDEL's work in the book of FOCKE (1881) and the study of MENDEL's publication guided VRIES to work with peas. In his first report about the segregation of his pea hybrids, he did not mention MENDEL's name, but used the expressions "dominant" and "recessive." In the second, more precise paper, he confirmed MENDEL's result but concentrated again on his mutation theory. He was convinced that breeding efforts should concentrate on looking for spontaneous variations within population caused by "retrogressive" and "degressive" mutations. When he visited the Swedish breeding station Svalöf in 1901, the wealth of different forms was so overwhelming that he tried to convince the breeders that selection of "elementary units" within populations was the only method needed in plant breeding. Of course, this was not MENDEL's approach. So, in VRIES' textbook "Pflanzenzüchtung" (*Plant Breeding*, 1907) again he did not mention MENDEL's paper. His "mutation theory" fitted better into breeder's experiences. It became theoretical basis at the breeding station Svalöf. Later, TSCHERMAK visited Svalöf (1901) and presented to a new wheat breeder, ▶ H. NILSSON-EHLE, the idea that only MENDELism could provide "a new, rational basis" for breeding of new constant forms by hybridization. The combination of characters could proceed substantially more surely and simply than before. The ongoing discussion about these different opinions suggests how Svalöf became a testing ground for new ideas of heredity and evolution. However, the practical results achieved by the breeders with spontaneous mutations in all sorts of landraces were disappointing. In 1906, NILSSON-EHLE finally preferred hybridization as the most important contribution to breeding programs of Svalöf. In particular, the winter wheat breeding had run into trouble because all the new promising "elementary units," off-types, or races that had been selected within populations suffered from one or more serious weaknesses. If hybridization could be used to combine good properties and eliminate deleterious effects, progress could be made (NILSSON-EHLE 1908). At the same time (1909), the American ▶ E. M. EAST proposed the idea of multiple alleles at a locus that had consequences for several breeding programs.

NILSSON-EHLE used to refer always to the visit of TSCHERMAK as a main source of inspiration for his steps toward crossing methods. In a letter to TSCHERMAK from Cambridge (England) on August 14, 1909, he criticized strongly the former practices of selecting off-types within population to create better varieties:

> … than the spontaneous new form, which we specially tried to select and utilize in winter wheat in the years from 1989 to 1905 did not prove—even with extremely strong efforts—better results. The pedigree books show a terrible useless work in this respect; … the opinion, that the future breeding work must concentrate itself mainly in crossing work.

Nevertheless, H. de VRIES in 1900, independent of, but simultaneously with the biologists ▶ CORRENS and TSCHERMAK von Seysenegg, rediscovered MENDEL's historic paper on principles of heredity. He became particularly known by his mutation theory. New elementary species originated by what VRIES (1901) called "progressive mutation." This corresponded to the creation of a new sort of "pangene" (*pangenesis*). Within a species there could occur "retrogressive" and "degressive" mutations, which correspond to the modification of existing pangenes. Retrogression meant the disappearance of characters through inactivation of pangenes. Degression meant the reappearance of a character; the hereditary differences behaved differently in hybridization. Progressive mutations led to what he called "unisexual" hybrids, whereas the hybrids of retrogressive as well as degressive mutations were "bisexual." Only the latter type of hybrid was subject of MENDEL's law. Unisexual hybrids were constant and nonsegregating. VRIES' speculative explanation for this behavior was that with progressive mutations the new type of pangenes would be unpaired in the hybrid. There is no "antagonist," that is, a modified pangene of the same kind with which it could pair up. Therefore, progressive mutation was VRIES' main interest, not MENDELism. He also thought that individual selection is the only method needed in plant breeding. The production of new forms though hybridization was superfluous and mass selection could only create "local races." VRIES even gave currency to modern use of the term "mutation" and was a major inspiration for research on spontaneous change of hereditary factors. His book "Mutationstheorie" (*Mutation Theory*), which he published in two volumes (VRIES 1901, 1903), made a greater impact in some ways than the rediscovery of MENDEL's laws.

E. von TSCHERMAK-SEYSENEGG (1871–1962) was the second Austrian scientist after MENDEL who again detected the laws of inheritance by studying pea crossings and the first plant breeder who purposely applied the combination of genes as a scientific method to improve the agronomic characters and therefore the efficiency of cultivated plants in practical breeding. Taking his impressive scientific work with about 100 original papers, not only the manifold range of topics but also the originality of his experiments, observations, and theories can be noticed.

Among other important discoveries, he observed the phenomenon of xenia in seed pods of single F_1 plants when he made extended crosses with peas, whereby seed color and seed shape resemble the difference in parental characters at this early stage after hybridization. Xenia effects can then be used to demonstrate the segregation of the parental characters since the seed grains can be classified in alternative groups in the same year of production. Most sporogenous responses are not apparent after the cross until the hybrid seeds are planted and the characters become visible in the progeny plants only. However, in peas the difference in the color of the cotyledons (yellow versus green) or the shape of the seeds (round versus wrinkled[1]) of the crossing partners shows its F_2 segregation of the progenies already on their ripe F_1 plants. He planted this segregating seed lot individually to follow their behavior in the F_2 generation. These observations and the results of some backcrossing procedures in which parental characters appeared in a 1:1 segregation scheme when the hybrids were crossed again with their parental types formed the basis of his D.Sc. thesis in January 1900. In this publication, he demonstrated and discussed some of the results of his studies—similar data as already been achieved half a century before by ▶ MENDEL (1866). In March of the same year, ▶ VRIES (1900) published two similar papers about pea crossing and—coincidentally—C. CORRENS (1900) launched in April volume of the same periodical ("Berichte der Deutschen Botanischen Gesellschaft") a third paper about the same topic, but with special reference to MENDEL's experiments and figures from 1866.

C. E. CORRENS (1864–1934), the German botanist and geneticist, in 1900, independently but simultaneously rediscovered with the biologists E. TSCHERMAK von SEYSENEGG and H. de VRIES, G. MENDEL's historic paper: He entered the University of München (Germany) to study botany. C. NÄGELI, the botanist to whom MENDEL wrote to about his pea plant experiments, was no longer

---

[1] Molecular genetics provided data that the round versus wrinkled seed shape trait is due to an 800-bp insertion into the gene coding for a starch branching enzyme causing the accumulation of sugars in the wrinkled phenotype; the tall versus dwarf stem length trait is caused by a G to A nucleotide substitution in a gibberellic acid 3-oxidase 1 gene; and the yellow versus green cotyledon color originates from a 6-bp insertion to a stay-green gene prohibiting chlorophyll degradation at maturity.

lecturing at München. NÄGELI, however, knew CORRENS' parents and took an interest in him. NÄGELI was the one who encouraged CORRENS' interest in botany and advised him on his thesis subject. Before CORRENS was a tutor at the University of Tübingen (Germany) in 1892, he spent time in Graz (Austria), Leipzig, and Berlin where he studied anatomical and physiological issues. He began experimental work on inheritance of plants in 1893 at Tübingen (Germany). Ten years later, he accepted a call as professor of the University of Leipzig and in 1909 of München. He already knew about some of MENDEL's hawkweed plant experiments from NÄGELI. NÄGELI, however, never talked about MENDEL's pea plant results, so he was initially unaware of MENDEL's laws of heredity.

However, by 1900, when CORRENS submitted his own results for publication, the paper was called "Gregor MENDEL's Regel über das Verhalten der Nachkommenschaft der Bastarde" (G. MENDEL's Law Concerning the Behavior of the Progeny of Hybrids). CORRENS and VRIES were the ones who most clearly redefined MENDEL's laws. CORRENS was active in genetic research in Germany, and was modest enough to never have a problem with scientific creditor recognition. He believed that his other work was more important, and the rediscovery of MENDEL's laws only helped him with his other work. He was supposedly indignant that H. de VRIES did not mention G. MENDEL in his first printing. In 1913, CORRENS became the first director of the newly founded "Kaiser Wilhelm Institut für Biologie" in Berlin-Dahlem (Germany).

Several opinions have been launched about the so-called "independent rediscovery of the Mendelian laws by CORRENS, TSCHERMAK, and de VRIES in 1900." There are still some doubts about these "random publication events" within a few months in the year 1900. Some of the irregularities in the paper of TSCHERMAK (1900) were criticized in recent years (MONAGHAN and CORCOS 1986, 1987).

Why was MENDEL's paper, despite its clarity and incisiveness, ignored for 35 years? The best explanation is that MENDEL was ahead of his time and it took that long for the scientific community to catch up. Remarkably, MENDEL's paper was precytological and the cytological discoveries that were to provide a physical basis for heredity were published between 1882 and 1903. Nineteenth-century biology was not ready for MENDEL. Part of the reason is that science then, as now, is conservative. New ideas are absorbed with difficulty and old ones discarded only reluctantly. One paper is not enough. The human qualities that made MENDEL admirable as a person, modesty and reticence, worked against his receiving personal acclaim and fame during his lifetime.

Soon after the period of rediscovery, TSCHERMAK became fully aware that these fundamental principles of inheritance should be applied to achieve stable and uniform combinations of different characters of parental genotypes by crossings, individual selection, and separate testing of the progenies in all agricultural crops. So, he advocated the system of "combination breeding" instead of the only individual ear selection of phenotypically equal plant types within populations, which was very common in creating improved varieties at that time.

Due to his excellent crossing techniques and improved selection management, he started first with cereals. With various crossings of rye and wheat varieties of different origins, he tried to solve one of the most important problems cereal breeders were faced with in the dry areas of Lower Austria, Moravia, and Western Hungary, namely to combine earliness with high-yielding performance (TSCHERMAK 1901, 1906). The later performance of many of his varieties in barley and wheat perfectly showed the possibility to combine successfully even these negatively correlated characters.

3.2 SCIENTIFIC PLANT BREEDING WITH THE BEGINNING OF THE TWENTIETH CENTURY

Although the so-called breeder's view, that is, the breeder's experience, played an important role for centuries (actually up to the present time), the further progress was characterized by a more scientific penetration of the breeding process. It was the time, 1913, when the first volume of "Zeitschrift für Pflanzenzüchtung" (later *Journal of Plant Breeding*, 1971, and finally *Plant Breeding*, 1986) was published—being one of the first journals exclusively dealing with the upcoming, new discipline of plant breeding—by Paul Parey Scientific Publishers, Berlin (Germany). Leading scientists of that time became the journal's very first editors. C. FRUWIRTH (Vienna, Austria) was the first editor-in-chief who was assisted by L. KIESSLING (Weihenstephan, Germany), H. NIELSON-EHLE (Svalöf, Sweden), K. von RÜMKER (Berlin, Germany), and E. von TSCHERMAK (Vienna, Austria) (Figure 3.2).

Zeitschrift
für
Pflanzenzüchtung.

Zugleich Organ
der Gesellschaft zur Förderung deutscher Pflanzenzucht,
der
Österreichischen Gesellschaft für Pflanzenzüchtung
und des
Bayerischen Saatzuchtvereins.

Unter Mitwirkung
von

L. Kiessling,   H. Nilsson-Ehle,   K. v. Rümker,   E. v. Tschermak,
Weihenstephan   Svalöf   Berlin   Wien

herausgegeben
von

C. Fruwirth,
Wien.

Erster Band.
Mit 9 Tafeln und 38 Textabbildungen.

BERLIN
VERLAGSBUCHHANDLUNG PAUL PAREY
Verlag für Landwirtschaft, Gartenbau und Forstwesen
SW. 11, Hedemannstraße 10 u. 11
1913.

FIGURE 3.2 Facsimile of the very first title page of the journal *Zeitschrift für Pflanzenzüchtung* (later *Journal of Plant Breeding*, 1971, and finally *Plant Breeding*, 1986). (Courtesy of Library, New York Botanical Garden, New York.)

By the late nineteenth century, farmers still rely on advanced local landraces. However, commercial breeders, who established a dominating presence, for example, in Germany, wanted a single variety of each crop for the whole country—universal varieties ("Universalsorten"). Phillipe de VILMORIN (1872–1917) in France and W. A. BURPEE (1858–1915) in the United States had developed a pre-Mendelian system of plant breeding that was very successful. BURPEE was born in New Brunswick (Canada) and moved in 1888 to Fordhook near Doylestown (Pennsylvania), where his lettuce breed "Iceberg" and the "Stringless Green Pod Bean" (1894) became famous.

With the industrialization in Europe and the United States toward the end of the nineteenth century, the development of sciences, medicine, and agriculture paced along. It stimulated new products, new technologies, and cross innovations that also challenged plant breeding. The German philosopher, F. ENGELS (1820–1885), recognized already the meaning of the plant breeding as an important economy factor in his book "Anteil der Arbeit an der Menschwerdung des Affen" (*The Share of Labor at the Origin of Man Out of the Ape*):

> Through artificial breeding, plants and animals are changed under the hand of man in a manner that they are not to be recognized anymore.

Examples for that development are numerous: The increased demand for sugar led to the breeding of sugarbeet as a rather new crop plant and industrial crop. After the discovery of the German chemist A. S. MARGGRAF (1709–1782) that beets contain sugar in 1747, ▶ F. C. ACHARD (1753–1821), also a German pharmacist, was the first person who selected beet plants successfully for sugar production. He is considered the founder of the beet sugar industry. Later, he placed the selection work in the hands of the Silesian sugar manufacturer Moritz von KOPPY (1749–1814), and after him the son Georg Friedrich von KOPPY (1781–1864).

By crossing of the "Weisse Mangoldrübe" (Zuckerwurzel = sugar root) and the "Roter Mangold" (Rübenman's gold or Futterrübe = fodder beet), he was able to select genotypes showing high sugar content. It was the birth of a new crop plant. The first name of sugarbeet was "Weisse Schlesische Rübe." Around 1850, Louis VILMORIN (1816–1860) in France changed the selection method according to the sugar content instead of turnip morphology. He introduced the so-called "floating method," that is, cutting out a piece of turnip, dipping it in a concentrated salt and/or sugar solution, and determining the sugar content.

First cropping of sugarbeet started in central Germany around—Halberstadt—Kleinwanzleben (Figure 3.3), where heavy very fertile soils were available (it remained the center of sugarbeet production in Germany). Later, ▶ ACHARD founded the first sugarbeet factory of the world at Cunern (Silesian, Germany/Poland). Another company was founded in 1850 by the brothers ▶ Gustav Adolf

FIGURE 3.3 Postcard showing the first sugar factory at Kleinwanzleben (Germany). It was founded in 1838 by 19 farmers and craftsmen of the village Kleinwanzleben as "Zuckerfabrik Wallstab & Co." (Courtesy of KWS, Einbeck, Germany. With Permission.)

FIGURE 3.4 Monument to GUSTAV ADOLF DIPPE (1824–1890), at the gate of his former property at Quedlinburg, Neuer Weg 21. (Courtesy of Dr. M. STEIN, Quedlinburg Germany. With Permission.)

DIPPE (1824–1890) (Figure 3.4) and Christof Lorenz DIPPE at Quedlinburg (Germany). The company emerged from the small horticultural business of the formerly deceased father of founder Johann Martin DIPPE (1785–1836). Since 1714, the DIPPE family owned gardenland at the gates of the town of Quedlinburg. Christof Lorenz DIPPE left the company in 1863, but remained a shareholder. The most important product was sugarbeet seeds. Through new breeding and selection procedures, the breeds achieved a high yield and high sugar content. Toward the turn of the century from the nineteenth to the twentieth century, the ▶ DIPPE brothers covered one-sixth of the world's sugarbeet seeds. The company also produced vegetable and flower seeds. In 1890, 1800 workers and 120 gardeners were employed. An area of 2500 ha was cultivated. The company was considered the most versatile wholesaler in the world.

Fodder beets were already bred by W. von BORRIES in Eckendorf (Germany) from the beginning of the nineteenth century. In 1849, he grew 20 varieties in a nursery for morphological and yield comparison. In 1840, ▶ W. RIMPAU, Schlanstedt, and F. KNAUER (Gröbers, Germany) introduced the method of mother beet selection and its pedigree testing to speed up the breeding procedure for yield performance and sugar content of the so-called "Magdeburger Zuckerrübe." The method became even more efficient when, in 1862, M. C. RABBETHGE (1804–1902) and J. GIESEKE (1833–1881), the founders of the "KWS" international seed company in 1856, applied the light polarization of beet sap for screening and determination of sugar content (in 1811, discovered in France by F.-B. BIOT).

Other important inventions can be listed: In 1853, BULL (USA) produced "Concordi" grape by hybridizing European wine cultivars and wild grapes of New England. In 1855, R. L. VIRCHOW (1821–1902, Germany) established that the egg cell of one generation comes from the egg of the previous generation and demonstrated the "continuity of heredity." In 1858, W. HOFMEISTER (1824–1877, Germany) discovered the female gametophyte and the law of change of generations. BIDWELL (1867, USA) first introduced detasseling of maize in his breeding program. In 1875, the first hybrid oat ("Pringles Progress") was released in the United States. In the years 1875–1877,

STRASBURGER (Germany) gave the first adequate description and drawing of chromosomes and showed that nuclei arise only from nuclei. He suggested the terms "gamete" and "chromosome." In 1878–1881, Professor William J. BEAL (1833–1924) at Michigan State Agricultural College, USA, proposed crossing maize cultivars to increase the yield of commercial types. In 1879–1880, HORSFORD (USA) selected the first known hybrid cultivar of barley; in 1885–1887, ▶ A. WEISMANN (Germany) wrote 12 essays on heredity and evolution that became important in directing the trend of biology and thought. He summarized current knowledge and pointed out the broad implications of "continuity of germplasm," "equal inheritance from parents," and "nonheritability of acquired characters." In 1884, E. STRASBURGER (Germany) demonstrated fertilization and showed the fusion of the two nuclei to form the zygote, and in 1888, the reduction division in plants. In 1886, SCHINDEL (USA) developed the first wheat by hybridization "Fulcaster;" in 1890, FARRER (Australia) ran an extensive wheat breeding program using hybridization for selection of rust resistance. In the same year, ▶ W. M. HAYS (USA) used centgener[2] test and pedigree selection on wheat and oats after starting plant breeding in Minnesota in 1888. From the beginning, he used the individual-plant method of selection. "Improved Fife," "Minnesota 163," "Minnesota 169," and "Haynes Bluestem" were valuable new varieties of spring wheat selected by this method.

In 1891, KELLERMAN and ▶ SWINGLE did first counts of segregation for the starchy gene on maize ear. REID's "Yellow Dent Maize" gained the grand prize as "the world's most beautiful corn" at the World's Colombian Exposition in 1893 at Chicago. REID's maize became a major force in Midwestern agriculture and an important parent to modern hybrid maize.

In 1899, HOPKINS described the ear-to-row selection method (half-sib progeny testing). However, major advances in plant breeding followed the revelation of MENDEL's discovery. Breeders brought their new understanding of genetics to the traditional techniques of self-pollinating and cross-pollinating plants. Yields could be significantly increased by utilization of modern crossing and selection methods after 1900 (Figure 3.5). This trend is recognizable not only in Europe but also in America and Asia (Figure 1.1). Based on data from various world regions, it is estimated that between 1960 and 1980 around 20% of the yield gains in major cereals were directly attributed to improved seeds.

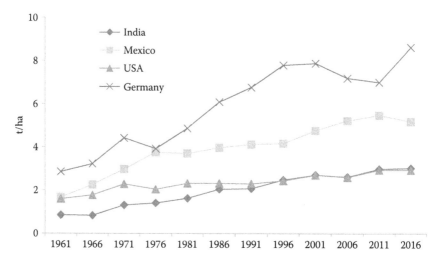

FIGURE 3.5 Development of yield increase in wheat since 1961 in some countries of the world significantly influenced by breeding approaches; data compiled from FAOSTAT. (From FAO, http://www.fao.org/faostat/en/#data/QC, 2017.)

---

[2] From testing plants grown 30.5 cm (1 ft) apart each way, ▶ W. M. HAYS went in 1888 to a centgener (his word signifying 100 offspring of a selected individual) system of 100 plants 10 cm (4 in.) apart planted by a machine he designed.

Since the end of the nineteenth century, plant breeding became scientific, complicated, expensive, and difficult. It was no longer possible for amateurs to breed crops. Plant breeding was institutionalized, usually in government-owned institutes or in government-financed universities (Table 3.1). In addition, several institutions arose: In 1886, the Swedish Society of Seed Breeding was founded in Svalöf by B. von NEERGARD. In 1888, a Seed Section was inaugurated in the German Agricultural Society by ▶ W. RIMPAU. In 1889, ▶ K. von RÜMKER held the first lecture on plant breeding at the University of Göttingen (Germany).

TABLE 3.1
Important International Breeding Institutions for Crop Improvement during 1800–1945

Institution and Localization	Country	Foundation	Breeders/Persons Involved
Institute of Agronomy and Plant Breeding (Vienna)	Austria	1872	F. HABERLANDT
Plant Breeding Station (Gross-Enzersdorf/ Vienna)	Austria	1903	E. von TSCHERMAK
Institute of Plant Breeding (Lednice)	Austria/Lichtenstein	1913	E. von TSCHERMAK
Centre de Recherches Agronomiques (Gembloux)	Belgium	1872	
Station de Selection du Boerenbond Belge (Héverlé)	Belgium	1925	A. G. DUMONT
Agricultural College/Ontario Agriculture & Experimental Union (Gelph)	Canada	1874/1878	C. A. ZAVITZ
Kaohsiung State Agricultural Experiment Station (Yuan)	China/Taiwan	1903	
Maison VILMORIN-Andrieux (Verriéres)	France	1815	M. H. VILMORIN, P. VILMORIN
Institut des Recherches Agronomiques (Noisy-le-Roi)	France	1921	E. TISSERAND, E. SCHRIBAUX
Centre National de Recherches Agronomiques (Versailles)	France	1923	E. ROUX
School of Forestry (Ilsenburg)	Germany	1764	H. D. V. ZANTHIER
Academy of Forestry (Waltershausen/ Dreissigacker)	Germany	1794/1801	J. M. BECHSTEIN
Institute of Forest Studies (Tharandt)	Germany	1811	H. COTTA
Seed Breeding Station (Emersleben/ Hadmersleben	Germany	1871	F. HEINE
Seed Breeding Station (Domaine Mahndorf/ Derenburg)	Germany	1878	F. HEINE
Bavarian Plant Breeding Institute (Weihenstephan/München)	Germany	1902	KRAUS
Institute of Plant Production and Breeding (Alferde/Breslau)	Germany	1872	F. W. BERKNER
Plant Breeding Institute (Halle/S.)	Germany	1863	J. KÜHN
Agricultural Experiment Station (Salzmünde)	Germany	1865	F. STOHMANN
Seed Association (Schwartau)	Germany	1901/1925	R. H. C. CARSTEN
Landessaatzuchtanstalt (Stuttgart/Hohenheim)	Germany	1905	C. FRUWIRTH
Plant Breeding Institute (Magyarovar)	Hungary	1909	E. GRABNER
Hungarian Royal Lowland Agricultural Institute (Szeged)	Hungary	1924	
Indian Agricultural Research Institute, "PUSA Institute" (New Delhi)	India	1905	A. HOWARD, F. J. F. SHAW

(Continued)

TABLE 3.1 (*Continued*)
Important International Breeding Institutions for Crop Improvement during 1800–1945

Institution and Localization	Country	Foundation	Breeders/Persons Involved
Paddy Breeding Station (Coimbatore), later associated with Tamil Nadu Agricultural University	India	1912	R. PARNELL
Institut di Genetica per la Cerealicoltura (Roma)	Italy	1919	N. STRAMPELLI
Stazione Sperimentale di Granicoltura (Rieti)	Italy	1907	N. STRAMPELLI
Institut di Allevamento per la Cerealicoltura (Bologna)	Italy	1920	F. TODARO
Agricultural Experiment Station Volcani Center (Ben Shemen)	Israel	1921	Y. E. VOLCANI
Agricultural Experiment Station (Sapporo)	Japan	1871	Under the advices of hired foreign experts
Agricultural Experiment Stations (Tokyo, Sendai, Kanazawa, Osaka, Hiroshima, Tokushima, Kumamoto)	Japan	1893	Governmental Edikt No. 18
State Plant Breeding Institute (Priekuļi)	Latvia	1913	
Agricultural Experiment Station (Ettelbrück)	Luxembourg	1884	C. ASCHMAN
Asgrow Vegetable Seeds Inc./Royal Sluis	the Netherlands	1865/1868	
Plant Breeding Institute (Wageningen)	the Netherlands	1876	L. BROEKEMA
Station de Recherches Agronomiques (Groningen)	the Netherlands	1889	
Institute for Breeding of Field Crops (Wageningen)	the Netherlands	1912	O. PITSCH
Central Agronomic Station (Bucharest)	Romania	1887	I. IONESCU, P. S. AURELIAN, C. SANDU-ALDEA
Agricultural Experiment Station (Sadilovsk)	Russia	1896	P. A. KOSTYCEV
Bureau of Applied Botany (St. Petersburg); now N. I. Vavilov Institute of Plant Industry, VIR	Russia	1894	A. F. BATALIN, R. E. REGEL
Plant Breeding Station (Moscow)	Russia	1903	D. L. RUDZINSKI
Agricultural Experiment Station (Kharkov)	Russia	1909	V. J. JUREV
College of Agriculture/Department of Genetics (Stellenbosch)	South Africa	1898	A. FISCHER
Elsenburg College of Agriculture (Elsenburg)	South Africa		J. H. NEETHLING
Plant Breeding Station (Svalöf)	Sweden	1886	N. H. NILSSON-EHLE
Weibullsholm's Plant Breeding Station (Landskrona)	Sweden	1870/1886	W. WEIBULL
Experimental Station (Rothamsted)	UK	1837, 1841	J. B. LAWES, on his farm
SUTTON & Sons Ltd. (Reading)	UK	~1850	
Carter Ltd. (London)	UK	1883	Carter brothers
John Innes Horticultural Institution (Merten/London)	UK	1910	J. INNES and W. BATESON
Plant Breeding Institute (Cambridge)	UK	1912	R. H. BIFFEN, F. ENGLEDOW
Institute of Forest Breeding and Genetics (Albany)	USA	1925	J. G. EDDY
Hi-Bred Seed Co. (present Hi-Bred Int.)	USA	1926	H. A. WALLACE
Ferry-Morse Seed Co.	USA	1930	D. M. FERRY and C. C. MORSE

But the first scientific "School of Forestry" worldwide was already founded in 1764 by the head forester of the counts of STOLBERG-WERNIGERODE, Hans Dietrich von ZANTHIER (1717–1778), at Ilsenburg (Germany). He initiated the first restocking program. In this approach, we can discern a source of the word "sustainability" so much in vogue today. This institution was followed by the University of Jena (Germany) at Zillbach (Thuringia), established by ▶ H. COTTA in 1786.

Breeders always tried numerous strategies to capitalize on the insights into heredity. The art of recognizing valuable traits and incorporating them into future generations is very important in plant breeding. Breeders have traditionally scrutinized their fields and traveled to foreign countries searching for individual plants that exhibit desirable traits. Such traits occasionally arise spontaneously through a process of mutation, but the natural rate of mutation is too slow and unreliable to produce all the plants that breeders would like to see.

3.2.1 BREEDING BY SELECTION

Practical breeders have for generations worked without knowledge of scientific principles of inheritance. In many cases, they achieved valuable results. The chief method of the progressive breeder has been, and continues to be, selection or the choice of those plants that approach most nearly to the type desired. Such selection has been based almost wholly on the appearance of single individuals or group of individuals, without regard to ancestry, progeny, or degree of relationship of the individuals selected. Van MONS (Belgium), ▶ KNIGHT (UK), and COOPER (USA) all wrote in the eighteenth and nineteenth centuries about improvements achieved by selecting superior types from their landraces.

J. le COUTEUR (1794–1875), a farmer of the Isle of Jersey, in 1843 noted the diversity of types in his wheat fields. He developed the concept of progeny test individual plant selection in cereals. Professor LAGASCA, from the University of Madrid (Spain), visited COUTEUR and pointed out numerous differences in plant type occurring in his wheat field. Selections were made and the progenies tested. Some proved superior to the commercial variety and were of more uniform habit of growth; other selections were of little value. The variety "Bellevue de Talevera," one of these selections, was of commercial importance for many years. Precise and scientific in his work, he was very conscientious in naming his invention. He was talking of replacing a "variety" by a "pure sort." He further stated it clearly that a "pure sort" was obtained from a single grain or ear. COUTEUR also was apparently the first to set down clearly the value of selecting individual plants in the improvement of autogamous small grains, although P. SHIRREFF (UK) had earlier used the same method in inbreeding some extensively grown varieties of wheat and oats. Like COUTEUR, he proceeded on the assumption that the selected single plants would breed true. New varieties produced by this means were grown extensively.

Since the introduction of pedigree breeding method, however, ancestry has become to some extent one of the criteria of selection, although in general practice of farms and early-specialized seed producers the old method prevails.

As a means for changing the character of a population, selection has usually been applied to characters, which show quantitative variation. In connection with such characters, it is necessary to recourse, since single recessive traits may be fixed in the population. With variable traits, such as size, shape, quality, or others, in which the extremes do not usually breed true, the breeder is forced to depend on the slower process of selecting individuals, with slight variation in the desired direction. The breeder has found that some characters yield readily to selection, and that progress is rapid up to a certain point, after which it becomes ineffective. On the other hand, the same characters may show no change under the same method. Similar systems of selection can give different results.

3.2.2 CROSS AND COMBINATION BREEDING

In 2016, the Indian rice breeder Nekkanti Subba RAO was honored for the development of the so-called "Miracle rice" (Figure 3.6). It is thought that "IR8" saved many millions of lives and transformed the lives of hundreds of millions of people. As a 29-year-old farmer, he discovered the

FIGURE 3.6 Nekkanti Subba RAO, rice breeder from Andhra Pradesh (India) on his field in 1967, showing the high-yielding variety "IR8." (Courtesy of IRRI, Los Baños, Philippines. With Permission.)

"IR8" variety's extraordinary properties on his small farm in the southeast state of Andhra Pradesh in 1967. Its yield was 10 tons per hectare in comparison to 1½ tons as common before. According to some studies, "IR8" yields in optimal conditions could be as much as 10 times that of traditional varieties.

In 1962, Dr. PETER JENNINGS made 38 crosses of various varieties at the newly established IRRI, Los Baños, Philippines, in 1960. The eighth was between a Chinese dwarf variety known as "Dee-geo-woo-gen" (DGWG) and a tall variety from Indonesia, "Peta." The eighth cross was promising, but only 130 seeds were produced. Nevertheless, 130 seeds became the famous "IR8" variety with outstanding results. There was never any instance in the history of the world where rice yields doubled in one step. Usually there was a 1% or 2% yield increase every year. Besides CIMMYT's semi-dwarf wheats, another story of "Green Revolution" actually began in India, moved to the Philippines, and then throughout Southeast Asia. This example of cross-breeding was so successful and so many farmers switched to it that "IR8" became the sole variety grown in some areas of Asia, reducing biodiversity and risking a catastrophe if a pest or disease hit the crop. It also led to a big increase in fertilizer use, leading to problems with pollution.

In all combination methods in breeding, it is common to cross more or less related genotypes to produce new combinations of traits. The theoretical basis is the application of Mendelian laws. Most crosses are carried out within species, that is, between varieties and strains. After a decade-long struggle between proponents and rejecting practitioners as well as scientists of cross-breeding as a useful means to improve the varieties, the "First International Conference on Hybridization and Crossbreeding" took place in 1899 at London (UK). Maxwell T. MASTERS (1833–1907) as a fellow of the Royal Society (London) defended the meeting in his introductory address (WILKS 1900):

Many worthy people objected to the production of hybrids on the grounds that it was an impious inter-ference with the laws of nature... The best answer to this prejudice was supplied by Dean Herbert,[3] whose orthodoxy was beyond suspicion... He succeeded in raising... many hybrid narcissi, such as he had seen wild in the Pyrenees, means of artificial cross-breeding. If such forms could exist in nature, there can be no improperly in producing them by the art of the gardener.

The history of barley breeding in North America shows that clear observations and targeted combi-nations were already practiced before and after 1900. ▶ W. M. HAYS was one of the four pioneers of barley improvement in the United States and Canada. Plot tests were reported from Wisconsin as early as 1871. Many types of barley, mostly of European origin, were brought to Wisconsin to be tested in comparison with "Manchuria" and "Oderbrucher."[4]

After many years of experience, R. A. MOORE decided that "Oderbrucher" was the barley best suited to his state. He improved it first by elite, or mass, and later by pure line selections (see foot-note 4). During much of the work, he was aided by L. A. STONE. In 1908, MOORE and STONE released "Wisconsin Pedigree No. 5" and "Wisconsin Pedigree No. 6." They were good variet-ies and are still being grown.

Barleys of hybrid origin were being tested in field plots as early as 1904. C. A. ZAVITZ at Guelph (Canada) had much to do with the dissemination and improvement of the "Manchuria" and "Oderbrucher" barleys. His variety "O.A.C. 21" was grown on a large acreage in Canada and is one of the important barleys in North America. C. E. SAUNDERS of Ottawa (Canada) led the way in the production of "hybrid" barleys (by crosses). As early as 1893 he was testing varieties of hybrid origin in field plots. Several varieties were foreign genotypes. "Club Mariout" was brought from Egypt in 1904.

Several varieties were foreign genotypes. "Club Mariout" was brought from Egypt in 1904. The two-rowed barley "Hannchen" is a selection of Svalöf (Sweden). Besides the selections from "Manchuria" and "Oderbrucher," there are four others of importance, such as "Atlas," "Trebi," "Horn," and "Tennessee Winter." "Atlas" is a selection of the old "Coast" variety.

It is similar to "Coast," but rather lighter in color than most strains of the original Spanish barley. "Trebi" is a selection from mixed barley imported from the hill country south of the Black Sea. It was widely grown in the Rocky Mountain region and in the Prairie States of the northern United States and adjacent Canada. "Horn," a two-rowed selection from European barley, was seeded in Montana and adjacent areas. The awns of most barley are stiff and heavily toothed. Such barleys were unpleasant to handle.

In 1840, a variety was introduced indirectly from Nepal, which has a hood in place of the awn. Its appeal to farmers was immediate and lasting. The same approach of the family team, William SAUNDERS (1863–1914) and his son Dr. Charles SAUNDERS (1867–1937), was applied in wheat. Gathering seeds from all over the world, crossings and selection, they succeeded with new varieties.

---

[3] The Hon. WILLIAM HERBERT (1778–1847) was a British botanist, botanical illustrator, poet, and clergyman.

[4] In the American breeding literature, it is spelled "Oderbrucker." The "Oderbrucher" barleys are known as six-rowed, rough-awned varieties with white kernels. They are particularly prized by maltsters in the upper Mississippi Valley of the United States. They yield well and are stiff-strawed and moderately tolerant to summer heat and humidity. Oderbrucher selections were grown in the United States on 40,960 ha in 1935. Oderbrucher was originally a variety identical or similar to the "Manchuria." As with the latter variety, it consisted of a large number of strains, both blue and white. A report in the old U.S. government records of 1865 states: "This variety is grown very extensively on the low, formerly swampy lands of the valley of Oder (Germany)," although these were drained during the reign of King FREDERICK the Great. In German, the word for the swampland is *Bruch,* so the barleys from there are called Oderbrucher. Evidently, by an error of translation or typing in the United States, this became Oderbrucker. An import by the federal government of the United States apparently never reached the farmers. In 1889, however, the Ontario Agricultural College at Guelph received this barley from Germany and later sent it to theWisconsin Agricultural Experiment Station (USA). It was widely distributed by the Wisconsin Station and most of the improvement was made there. In 1908, R. A. MOORE and A. L. STONE of Madison released Wisconsin Pedigree No. 5 and Wisconsin Pedigree No. 6. Both of these were selections of Oderbrucher.

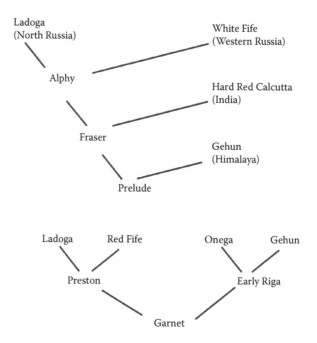

FIGURE 3.7 Breeding for frost-resistant Canadian wheats by combining Russian and Asian landraces. (Redrawn after CLARK, J. A. and BAYLES, B. B., *Bull. USDA*, 459, 1942.)

Their early-ripening varieties were mostly based on Russian wheats. One was "Ladoga" originating from Ladoga lake region near St. Petersburg (Figure 3.7).

F. HORSFORD at Charlotte made the first known barley hybrid selection in the United States, in 1879 or 1880. The "Horsford" variety and others originating in the same way have long been marketed under various names such as "Beardless," "Success," or "Success Beardless." Among the early breeders to employ "Nepal" in hybrids was R. WITHECOMBE. One of the more important projects for the production of a hooded barley started at the Tennessee Agricultural Experiment Station by C. A. MOOERS in 1905. One of the most interesting hybrid barley was "Arlington Awnless," developed by B. DERR in 1909. From the progeny of a cross of "Tennessee Winter" on "Black Arabian," DERR isolated an awnless variety. "Lion," smooth-awned black barley, was another introduction from Russia in 1911. Actually, it was reselected at the Michigan Agricultural Experiment Station and their variety, "Michigan Black." Two other examples of successful combination breeding for ecological adaptation in wheat are given in Figure 3.7.

By the imposing development of in vitro techniques during the past five decades, also intergeneric crosses became feasible on a large scale. However, they are mainly used in pre-breeding programs. So, crosses can be classified as intervarietal, interspecific, and intergeneric. As long as there is sufficient genetic variability within varieties and strains of species, breeders will prefer this type of intervarietal crosses to restrict recombination of the target genotype. For example, when SCHALLER and WIEBE (1952) tried to improve resistance to net blotch disease (*Helminthosporium teres*) in barley, they screened more than 4500 varieties and strains. Among them, 75 were resistant. From them, only a few were suitable for the crossing program. Depending on the trait and its genetic control, a cross is only once required or multiple times in the breeding cycle.

A final example of successful cross and combination breeding was the development of semidwarf varieties that revolutionized the world of wheat breeding. The story began in 1935 when the Japanese Gonjiro INAZUKA (1971) crossed a semi-dwarf Japanese wheat landrace with two American cultivars resulting in an improved semi-dwarf variety, known as "Norin 10" (Figure 3.8). Unlike other varieties, which stood taller than 150 cm, the *Rht1* and *Rht2* genes present in "Norin 10" reduced its height to 60–110 cm.

In the late 1940s, the breeder Orville VOGEL (1907–1991) at Washington State University, Pullman (USA), used "Norin 10" to help produce high-yielding, semi-dwarf winter wheat varieties. Eventually, his varieties ended up in the hands of ▶ Norman BORLAUG (Figure 3.9), who was working to develop rust-resistant wheat in Mexico.

In 1953, N. BORLAUG started crosses of semi-dwarf varieties of Pullman with Mexican varieties. The result was a new type of spring wheat: short and stiff-strawed varieties that tillered profusely, produced more grain per head, and were less likely to lodge. After a series of crosses and re-crosses, the semi-dwarf Mexican wheat progeny began to be distributed nationally, and within seven years, average wheat yields in Mexico had doubled. BORLAUG named two of the most successful varieties "Sonora 64" and "Lerma Rojo 64," and it was these two varieties that

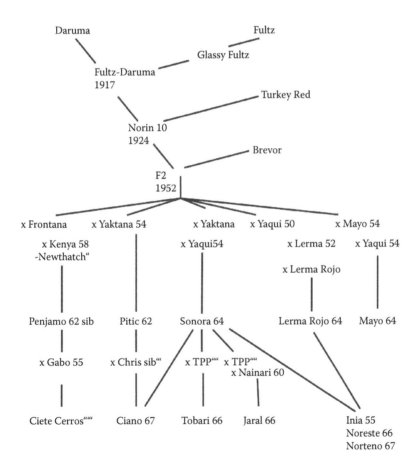

FIGURE 3.8 Pedigrees of the Japanese wheat variety "Norin 10" (after REITZ and SALMON 1968) as well as early semi-dwarf wheat varieties of CIMMYT. (Redrawn after LUMPKIN, T.A., *Proc. 12th Int. Wheat Genet. Symp.*, 13–20, 2015. With Permission.)

FIGURE 3.9 Norman BORLAUG (1919–2009) recording plant data at experimental wheat plots during the 1950s. (Courtesy of CIMMYT Archives, Mexico. With Permission.)

led to the so-called Green Revolution in India, Pakistan, and other countries. This international exchange of ideas and germplasm—starting with genetic resources from Japan—ultimately saved hundreds of millions of people from starvation.[5]

BORLAUG toured India and Pakistan as part of a team of FAO-Rockefeller Foundation scientists in 1960 when he visited the region for the first time. After that he agreed to provide training for regional breeders and in response Manzoor A. Bajwa became the first Pakistani to arrive in Mexico in 1961. While working alongside the team in Obregon (Mexico), Bajwa identified a medium-to-hard white grain line with a high gluten content that was ideal for making chapatti. The new variety also showed promising resistance to rust and powdery mildew. To mark this collaboration, the line was named "MexiPak"—to indicate that the line selection took place in Mexico by a Pakistani breeder. To accelerate the potential of CIMMYT wheat, in 1967 Pakistan imported about 42,000 tons of semi-dwarf varieties from Mexico—including 40,000 tons of "Indus-66," 1,500 tons of "MexiPak 65," 200 tons of "Sonora 64," and 20 tons of "Inia 66." At the time, this was the largest seed purchase in the history of agriculture (Anonymous 1989). In Pakistan, the name "MexiPak" became synonymous with the successes of the Green Revolution.

BORLAUG returned to the region in 1963, at the invitation of a 38-year-old Indian wheat cytogeneticist, M. S. SWAMINATHAN, before he had sent a few seeds of his high-yielding,

---

[5] In the early 1960s, South Asia was facing mass starvation and extreme food insecurity. To combat this challenge, scientists and governments in the region began assessing the value of the Mexican semi-dwarf wheat varieties for their countries. Trials in India and Pakistan were convincing and demonstrated high yields that offered the potential for a dramatic breakthrough in wheat production.

disease-resistant semi-dwarf wheat varieties to India to test their resistance to local rust strains. SWAMINATHAN, who like many wheat breeders of the time was interested in testing the semi-dwarf wheat varieties, immediately grasped their potential for Indian agriculture.

Wheat yield improvement in both India and Pakistan was unlike anything seen before. In just four years between 1963 and 1967, India's wheat harvest doubled to 20 million tons and the nation went from dependence on wheat imports to self-sufficiency. The trend has continued to recent times. In 2012, Indian farmers harvested about 95 million tons of wheat. A similar effect was experienced in Pakistan. Between 1965 and 1970, wheat production in Pakistan increased from 4.6 million tons to 6.7 million tons (Anonymous 1989). In 2012, Pakistani farmers harvested over 25 million tons of wheat (Figure 3.5).

Cross-breeding and combination breeding is still current in the twenty-first century. The program "MAGIC" (UK) impressively demonstrates this. It runs since 2015 by the National Institute of Agricultural Botany (NIAB, founded 1919) and University College London under the patronage of the Biotechnology and Biological Sciences Research Council (BBSRC). It has generated the UK's only Multiparent Advanced Generation Inter-Cross wheat populations, known as "MAGIC," by crossing multiple parent lines over multiple generations. "MAGIC" mixes up the genes from different parents far more than traditional crossing programs, creating greater diversity in the subsequent generations and the ability to investigate many more traits within a single population. It is a new approach in wheat research, with the potential to allow rapid advances in our understanding of how we can improve yield. The wheat resource used in this project will be developed and made available to the global wheat research community.

The research expects to uncover valuable diversity captured by the 16 wheat parent lines, that is, a selection of varieties grown in the United Kingdom from the 1940s through to the present day. Individuals within the population will carry unique combinations of genes, from both modern and heritage varieties. Reshuffling the genetic information in historical and more modern varieties provides a new way to mine potentially favorable combinations, including genetic information that may have been lost in the process of wheat improvement over time (Anonymous 2015).

3.2.3 Pure Lines and Improvement of Self-Pollinated Crops

With beginning of the twentieth century, the effect of selection has been carefully investigated. In 1903, the Danish botanist ▶ W. L. JOHANNSEN published some results on line selection considering the weights of individual seeds of the "Princess" bean. Because of the autogamous flowering habit of bean, outcrossing could be widely excluded from interpretation of trait variability. He found that plants grown from the lightest and from the heaviest beans of the same mother plant produce seeds of the same average weight; that is, selection among the seeds of same plant will not be effective. His explanation of the apparently contradictory behavior was quite simple. There is no hereditary variation, which may be fixed by selection, and selection is, therefore, ineffective. He distinguished two classes of variation:

* genotypic variation, or differences in the genotype or hereditary constitution, and
* phenotypic variation, caused by the combined action of many nonheritable factors.

Thus, selection is effective only when applied to the first class. He called the group of descendants a "pure line" when the selection among the progeny of a single genetically pure self-fertilized individual is ineffective. At present, the term is more broadly defined as a group of individuals all of whom have the same genetically homozygous genotype for one or more loci. The other conclusion of JOHANNSEN's work is that effectiveness of selection depends on the presence of genetic variability. The pure-line theory has led to a general conception of the way in which selection accomplishes its results. According to this explanation, selection merely sorts out and isolates the genetic factors responsible for the trait selected, and does not itself create anything new. Thus, breeding includes both searching for and creating of new variability and selection of the most suitable individuals.

3.2.4 POSITIVE AND NEGATIVE MASS SELECTION

The basic way of selection is to choose for breeding those individuals that vary in the desired characters, usually by their visible or measurable traits. This is the method used by the maize, barley, rye, wheat, or other crop grower who at harvest time selects the best spikes from the whole yield and rears his next year's crop from such seed. Alexander LIVINGSTON (1821–1898) of Ohio (USA) as a successful producer of tomatoes thought hybridization and crossings not as the best idea for crop improvement. For him, selection was all. He grew whole fields of tomatoes and endlessly trawled them for development of new varieties.

Almost at the same time, L. de VILMORIN of France recognized the basic importance of separate sowing of individually selected plants. He reported to the "Societé Industrielle" at Angers (France) in 1856 considering his rapeseed breeding:

> Only when using single individuals, the heritable fitness of plants can be determined.

When this principle of selection is kept for many generations, then a pedigree appears, that is, the scheme of pedigree breeding (cf. Section 3.2.5). This positive mass selection has been followed by the breeder for many years, and although slow and uncertain it has led to the origin of many improved plant varieties. The opposite way would be to discard all individuals showing negative habit and to propagate the remaining, leading to negative mass selection. L. de VILMORIN's company aimed at producing a variety that was stable and predictable enough to be called a final product. HALLETT's slow and gradual selection was a continual process, that is, an ongoing crop improvement. It was also adopted by many German breeders. They worked with a variant of mass selection. They produced "elite" strains. Every year they selected the best plants to grow on them for the next year's selection. This became known as the "German method."

The efficiency of positive mass selection was demonstrated in allogamous maize. Already in 1896, selection for high and low protein and high and low oil content of maize kernels was begun at the Illinois Agricultural Experiment Station (USA). It was initiated at the University of Illinois in by C. G. HOPKINS. As can be seen from Figure 3.10, ▶ E. M. EAST and ▶ D. F. JONES (1920) were able to demonstrate after 23 generations that positive and negative mass selection has plainly been effective in altering the chemical composition of the kernel. After almost 100 years, in one of the longest experiments ever, U.S. maize breeders have continuously selected to change the oil composition. Typically, mean oil concentration was estimated in about 60 ears, and seeds from only 12 were selected to propagate the next generation. The change after 100 generations in oil concentration was almost continuous and substantial: from a base of about 5%, the high oil-producing line now has about 20% oil, and the low oil-producing line has almost none. Kernel protein concentration gave similar responses, except that the low line reached a plateau at about 5% protein. Most changes occurred in the earlier generations (JANICK 2004). By basically the same procedure, new crop cultivars were developed from either primitive wild forms or landraces, as can be seen from Table 3.2.

This method is generally effective because most of the characteristics on which it is employed are apparently dependent on a large number of inherited factors that may be slowly sorted out and accumulated by subsequent selection. For example, in the United States, the number of wheat varieties grown from 126 in 1919 to >500 in 2005 was mainly achieved by methods mentioned earlier. Where such genetic differences do not exist or cannot be screened by mass method, of course, the selection is ineffective. The most rapid results accomplished by selection are those obtained in the differentiation of pure lines from mixed population of self-fertile plants.

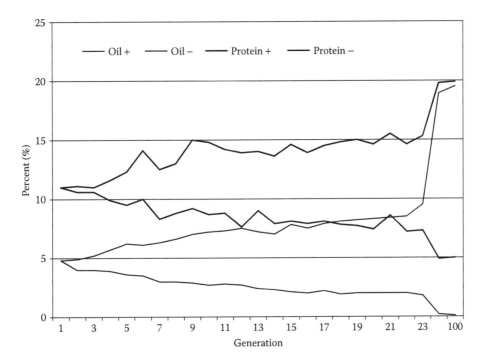

FIGURE 3.10 The results of selection for high and low protein as well as oil content of maize at the Illinois Agricultural Experiment Station during 1896–1919 and recent data. (Compiled and redrawn from EAST, E.M. and JONES, D.F., *Genetics*, 5, 543–610, 1920; Janick, J., *Plant Breed. Rev.*, 24, 1–384, 2004. With Permission.)

TABLE 3.2
Examples of Crop Plant Development from Wild Relatives

Primitive Form	Landrace	Cultivated Forms	Advanced Varieties[a]
Wild wheat, *T. boeoticum*		Wheat, einkorn, *T. monococcum*	
Wild wheat, *T. dicoccoides*		Wheat, *emmer*, *T. turgidum*	
Wild wheat, *T. araraticum*		Wheat, Armenian, *T. timopheevi*	
Wild wheat, *Triticum* species	Hulled hexaploid wheat, *Triticum spelta* ▶	Hulless wheat, *Triticum aestivum*, "**Breston**" (1896, USA)	"Mercia" (UK)
Wild barley, *Hordeum spontaneum* ▶	Hulled barley, *Hordeum vulgare*, "**Hanna**" (1875, Austria)	Hulless barley	"Weihenstephaner Hilte" (Germany)
Wild oats from Persia, *Avena sterilis, A. fatua* ▶	Hulled oats, *Avena sativa*, "**Hopetoun** oats" (1864, UK)	Hulless oats	"TSCHERMAK's Frühhafer" (1931, Austria)
Wild rye, *Secale vavilovii* ▶	Hulless rye[a], *S. cereale* "**Probsteier**" (Austria) ▶	"Petkuser Roggen" (1881, Germany) ▶	"Danae" (1962, Germany)

▶ indicates lineage.
[a] Advanced varieties derived from forms marked by bold letters.

3.2.5 PEDIGREE SELECTION

Already before the rediscovery of Mendelian laws, several crop varieties were developed by this approach. The English wheat breeder, Major F. F. HALLET from Brighton, developed in 1857–1874 the method of pedigree selection in wheat, oats, and barley, believing, apparently, that acquired characters were inherited and that improvement induced by favorable growing conditions would be transmitted to the progeny. He raised his plants, therefore, under the most favorable cultural conditions, selecting the best seed on the best-developed spike of the more vigorous plants, replanting, and following the same plan of selection in subsequent years. In this way, the wheat varieties "Golden Drop," "Victoria White," the barley variety "Chevallier barley," and the oat varieties "Pedigree White Canadian" and "Pedigree Black Tatarian" were developed. TAYLOR, a farmer from Yorkshire (UK), around 1860, found in a stand of "Victoria White" a mutant showing strong and short straw and a shared spike (RUSSEL 1966). It was the origin of the famous "Square Head Wheat," which came via Denmark (1874) to Germany (called "Dickkopf-Weizen"). One of the first advance winter wheat varieties "Rimpaus früher Bastard" (Germany) was based on the same method. It was derived from the cross-combination "Früher amerikanischer Landweizen" × "Squarehead" from England and was marketed in 1888.

Pedigree selection by utilizing the knowledge of early scientific genetics is known as the progeny test. In adopting the progeny test, breeders have recognized that permanent improvement must be improvement of the genotype, and the problem was, therefore, to develop some method of estimating the genotype itself. The best evidence as to the genotypic constitution of an individual is obtained from a study of its progeny. Among plants, seeds from single individuals are separately grown and the progenies compared. The most uniformly satisfactory progenies are saved for further selection and the others are abandoned.

In barley, the nestor of the Moravian plant breeders E. von PROSKOWETZ tried since 1875 to improve the old landraces of barley "Hanna" by single spike selection at his agricultural estate Kvasice and Tlumacova without success. In 1904, E. TSCHERMAK convinced him to change to individual selection and progeny testing and could find particularly early- and high-yielding lines of malting barley. The so-called "Kwassitzer Original Hanna Pedigree" became the mother of a wide range of malting barleys in Europe because of its high grain quality, earliness, yield stability, and high adaptability. In Germany, many years after the release of this variety, 14 sister varieties ("Mittlauer Hanna," "Eglfinger Hado," "Mahndorfer Hanna," "Heines Hanna," "Oppiner Hanna," "Weihenstephaner Hilte," "Braunes Hanna," "Dippes Hanna," "Selchower Landgerste," "Heines Haisa," "Criewener 403," "Rimpau's Hanna," "Mettes Hanna") demonstrated this early selection triumph, which was similarly successful in Sweden ("Svalöf's Hannchenkorn," "Svalöf's Hannchengerste 2") (WUNDERLICH 1951).

Sometimes the selected strains within the pedigree were called "genealogical lines," not to be mistaken for "pure lines." In allogamous plants, it cannot be a genealogical line when open pollination occurs. There are no true pedigrees, but maternal pedigrees without knowledge about the paternal contribution. Therefore, the terms "mother-pedigree breeding" or "maternal polyandry method" were used, as, for instance, in sugarbeet breeding (Section 3.2). In maize, between 1905 and 1907 WILLIAMS (Ohio, USA) developed the remnant seed-testing plan for maize breeding.

3.2.6 BULK SELECTION

When breeders want to avoid multiple single plant selections and its separate progeny testing, then they have to start selection in later generations. The portion of homozygous individuals in the segregating cross population is already comparably high. It was the American ▶ H. S. JENNINGS in 1912 who already found that selfing reduces heterozygosity by one-half each generation. Usually, from F_3 generation until the beginning of the selection the cross progenies are propagated as mixed populations in bigger field plots. Basically, the method of *bulk breeding* was first practiced in 1908 by H. NILSSON-EHLE (Sweden). It was the same year when he proposed the multiple factor explanation for inheritance of color in wheat pericarp.

In 1920, he introduced the method to the breeding station Åkarp for improvement of winter hardiness in wheat. Therefore, the method was sometimes called "Åkarp method." Later, the German geneticist ▶ E. BAUR strongly promoted the method because it was easy to handle even for large numbers of cross populations, it was particularly applicable for selection of winter hardiness, disease resistance, and insect resistance, and it was comparably cheap. With numerous modifications, the method is used in many breeding programs all over the world and for many crops until now. As disadvantage, breeders have to cope with yearly changes of environments and with lower selection efficiency. H. V. HARLAN and MARTINI (1938) as well as C. A. SUNESON and WIEBE (1942) demonstrated that just an accumulation of a certain genotype in a given population does not guarantee good expression of its traits in a population. However, partial bulk systems and composite crosses can minimize such doubts as demonstrated in a long-term experiment with barley (SUNESON 1956).

3.2.7 BACKCROSS BREEDING

It is a system of breeding whereby recurrent backcrosses are made to one of the parents of a hybrid, accompanied by selection for a specific character(s). In 1922, HARLAND and POPE described the backcross breeding technique for small grains. HARLAND also introduced later the mass-pedigree system (PARTHASARATHY and RAJAN 1953). ▶ F. N. BRIGGS is credited with the perfection of the backcross breeding method. He demonstrated the method with improvement of numerous wheat and barley varieties (BRIGGS 1938). It is a modification of the bulk method introduced by the Canadian J. B. HARRINGTON (1937).

By backcrossing, the characters of the donor parent are accumulated and the genes of the recipient are superseded. It is often applied when a single character has to be transferred to a, for example, high-yielding and adapted recurrent parent (see Figure 3.11). The desired character of the donor and the genotype of the recurrent parent must be recovered in the segregating populations. The advantages are the stepwise improvement of cultivar, the handling of multiple generations per year, small population sizes, and extensive yield tests are not necessary. Moreover, the results of backcrossing are predictable and need little testing efforts. However, the method is ineffective for traits with low heritability. Another disadvantage is the restricted chance of recombination.

Although the method can be classified as conservative, this is the method for producing genetic engineered crop varieties nowadays. The molecular engineered gene and/or trait will be introduced either in a suitable recipient genotype or in a high-yielding and highly adapted variety and then backcrossed when needed.

When parallel backcrosses are used for variety improvement, it is sometimes called "convergence breeding." It was preferentially applied in resistance breeding. But it was also successfully used for breeding monocarpic sugarbeet. First monocarpic sugarbeet were described by BORDONOS (1940). As interesting reminiscence, it should be added that in 1794 ▶ J. G. KÖLREUTER proposed backcrosses, however, with the aim to "change species" of his tobacco hybrids (cf. Section 2.7).

3.2.8 SINGLE-SEED DESCENT

Single-seed descent (SSD) methods (single-seed, single-hill, multiple-seed) are easy ways to maintain populations during inbreeding. Natural selection cannot influence the population, unless genotypes differ in their ability to produce viable seeds. The artificial selection is based on the phenotype of individual plants, not on the progeny performance. It was first proposed by ▶ GOULDEN (1941) and modified by BRIM (1966). It guarantees rapid generation advance by handling of a large number of crosses and speeds up homozygosity within the population. It requires little field space, but greenhouse capacity for winter annual crops. It is particularly suitable for selecting characters with low heritability. A first application of the SSD method was reported in oats by J. E. GRAFIUS (1965).

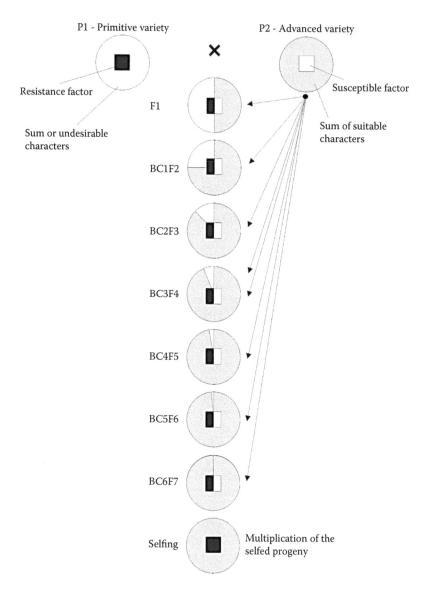

FIGURE 3.11 General scheme of sixfold backcrossing of a disease-resistant landrace to an advanced variety susceptible to disease for improvement of the advanced material. (Courtesy of R. Schlegel. With Permission.)

3.2.9 NEAR-ISOGENIC LINES AS A BREEDING TOOL

Near-isogenic lines (NILs) are not fully isogenic. For example, in maize, two distinct composites of F_3 lines develop from a single cross, one consisting of lines homozygous recessive and the other consisting of lines homozygous dominant for a certain gene (i.e., there is same genetic background), however, differing only in being homozygous dominant versus recessive for the genes. In wheat, near-isogenic lines were produced for different *Rht* genes (reduced height) causing different straw length.

The procedure involves generation of a series of NILs by sequentially replacing segments of an elite line (the recipient genome that is targeted for improvement) with corresponding segments from the donor genome. The objective is to generate a set of NILs containing, collectively, the complete genome of the donor source, with each NIL containing a different chromosomal segment from the donor. Marker-facilitated backcrossing, followed by marker-facilitated selfing to fix introgressed segments, is used to monitor the transfer of the targeted segments from the donor and to recover the recipient genotype in the remainder of the genome. The number of backcrosses required depends on the number of evaluations that can be made in the marker laboratory. As few as two backcross generations and one selfing generation will suffice if the laboratory resources are adequate to handle the required number of plant samples.

In maize, the NILs are crossed to an appropriate tester(s) to create hybrid testcross progenies that are evaluated in replicated field trials (with appropriate checks) for the desired traits. The NILs would be tested *per se* for crops, such as soybean and wheat. The superior performing testcrosses will be presumed to have received donor segments that contain favorable quantitative trait loci (QTLs). Thus, QTLs are mapped by function, which should be an excellent criterion for QTL detection. The breeding scheme not only creates enhanced elite lines that are essentially identical to the original line, but also provides for the identification and mapping of QTLs as a fringe benefit with no additional cost. The scheme is based on having a reasonably good marker map with distinct alternate alleles in the donor and recipient lines.

This breeding strategy should be an excellent procedure for tapping into the potential of exotic germplasm. BROWN et al. (1989) used a somewhat similar marker-facilitated backcrossing scheme to transfer isozyme-marked segments from wild barley (*Hordeum spontaneum*) into an elite barley (*H. vulgare* L.) cultivar. Each of the 84 NILs was then made homozygous for a single isozyme-marked segment with two selfing generations. After evaluating these lines, *per se*, in the field, they also concluded that this was a useful approach for identifying QTLs for improving yield in divergent germplasm.

A major advantage of this NIL approach is that once a favorable QTL has been identified, it is already fixed in the elite recipient line and the breeding work is essentially completed. Also, because only a small segment of the genome of the recipient line has been modified, the enhanced line is nearly identical to the original line and the amount of field testing required is minimal. In addition, lines with favorable QTL alleles can be easily maintained and then used for pyramiding several favorable QTL alleles into a single line. A possible disadvantage of this approach is that favorable epistatic complexes between QTLs may not be identified. However, there is little experimental evidence documenting the occurrence of such epistatic interactions (Table 3.7).

3.2.10 Polycross Method

It is generally impractical to make large numbers of diallel combinations or to obtain sufficient seed from such matings for adequate tests. For this, the polycross method was evolved independently by FRANDSEN (1940) in Denmark, by TYSDAL et al. (1942) in the United States, and by WELLENSIEK (1947) in the Netherlands. TYSDAL proposed the term "polycross" for the progeny from seed of a line that was subject to outcrossing with other selected lines growing in the same nursery. Selected lines or clones are planted in a field isolated from other material of the same crop plant and with several replications of the clones or seed progenies to be tested. By clonally dividing the provisionally selected plants and placing them throughout an isolated crossing block, one achieved a more nearly random pollination. Then, if the seed from the various plants representing a clone in the crossing nursery is bulked, an adequate supply for testing is available and represents a more nearly random sample of the various possible cross-combinations than that from individual clones, which are pollinated largely by their neighbors. The polycross seed of each clone is grown out in replicated nurseries at one or more locations, and after the progenies are evaluated for their general characteristics, the general combining ability of the clones may be determined. In the FRANDSEN technique, grass clones were divided into 100 cuttings, and 20 plots of 5 cuttings each are randomly placed in the various replications of the polycross nursery. The method is particularly suitable for cross-fertilized crops sensitive to inbreeding. The main reason is the self-incompatibility. E. M. EAST and A. J. MANGELSDORF (1925) first discovered the gametophytic system of self-incompatibility in *Nicotiana sanderae*. The sporophytic system of self-incompatibility was first discovered in *Crepis foetida* by HUGHES and BABCOCK (1950).

3.2.11 SHUTTLE BREEDING

In 1962, ▶ N. BORLAUG (Figure 3.9) developed high-yielding dwarf strains of wheat while working at the ROCKEFELLER-financed CIMMYT (Agricultural Station in Mexico City, Mexico). Along with improvements in rice productivity at the IRRI at Los Baños (Philippines) and other crops at yet more agricultural stations, the so-called "Green Revolution" came into being. During that time, a program was developed by BORLAUG that shuttles seed between two (or more) locations to be grown at each. It is now a proven contributor to success of CIMMYT breeding programs (CURTIS et al. 2002). Shuttle breeding allows two breeding cycles per year instead of one, for example, a winter cycle in the northern desert of Sonora and a summer crop in the central Mexican highlands. This not only fast-forwards selection, but also exposes test varieties to radically different day lengths, temperatures, altitudes, and diseases. Resulting plants are broadly adapted. They grow well in numerous environments. The method is broadly used in recent breeding programs of internationally operating companies. They provide best prerequisites for this sort of seed exchange and seed testing under different environments.

3.2.12 EVOLUTIONARY BREEDING

Like the fungi, the insects, and the weeds, recent crops also have the ability to evolve and to adapt to changes. The advantages of exploiting this ability were first understood by Coit SUNESON (1956), an American agronomist, who, in 1956, proposed an evolutionary breeding method even though the same idea is implicit in a more than 100-year-old publication by Herbert WEBBER (1908).

The method consists in planting in farmers' field mixtures—so-called evolutionary populations—of very many different genotypes of the same crop, preferably, but not necessarily, using early segregating generations. These populations (one for each crop) will be planted and harvested year after year, and due to the natural crossing, the genetic composition of the population that is harvested is never the same as the genetic composition of the population that was planted as a result of differences in individual's fitness. In other words, and in the presence of directional selection, the population will evolve to become progressively better adapted to the environment in which it is grown. As the climatic conditions vary from one year to the next, the genetic makeup of the population will fluctuate, but if the tendency is toward hotter and drier climatic conditions, as expected in view of climate change, the genotypes better adapted to those conditions will become progressively more frequent.

While the base population is evolving, breeders or farmers can practice artificial selection, with specific modalities depending on the crop and on the objective, thus deriving a flow of continuously better adapted improved varieties. Thus, evolutionary–participatory plant breeding reconciles agro-biodiversity, sustainable production increases, and adaptation to climate change (CECCARELLI 2009). Evolutionary–participatory plant breeding assumes that, while the population evolves, it maintains sufficient genetic diversity for evolution to proceed. However, it can accommodate the injection of novel genetic diversity any time it is required. It is also possible and indeed desirable for farmers to share the seed of the population with other farmers in other locations affected by different stresses or different combinations of stresses.

One of the critical issues is whether this methodology can be applied for self-pollinated crops, which include some of the most important food crops, such as wheat, rice, barley, or millets. This issue has been addressed, indirectly, through an experiment conducted by MORRAN et al. (2009), who used experimental evolution to test the hypothesis that outcrossing organisms are able to adapt more rapidly to environmental changes than self-fertilizing organisms. The experiment suggests that even low outcrossing rates (e.g., <0.05), comparable with those observed in self-pollinated crops, such as barley, wheat, and rice, allow adaptation to stressful conditions. Interestingly, the outcrossing rate increased in the course of the experiments, offering a partial solution to the problem of linkage drag. Linkage drag is a drawback in evolutionary breeding, particularly in self-pollinated species. This can be overcome by increasing recombination through, for example, the use of male sterility when available.

Normally, the performance of elite breeding populations is negatively affected when exotic germplasm is introduced. Sewall WRIGHT's shifting balance theory (WRIGHT 1931) suggests that crossing between a locally adapted genotype and a distant genotype will result in inferior progeny in the local environment. Elite breeding populations are well adapted to the target agricultural environment, but exotic germplasm is not adapted to the same environment. To avoid the negative effects of linkage drag in wide crossing, breeders have used multiple backcrosses to reconstruct the performance of the original elite parent and have focused on introgression of major genes from the exotic parent. Incorporation of beneficial minor alleles for quantitative traits from exotic parents requires strict adherence to quantitative genetics theory, because the value of such alleles may not be known until the incorporation is almost complete. Three cycles of crossing and selection were necessary to find useful improvements in the yield in six-row barley after introgression of alleles from two-row barley. This suggests that yield benefits can be obtained from incorporation of relatively small doses of donor alleles while maintaining favorable gene combinations in the adapted population.

Evolutionary plant breeding has been put into practice recently, and hard data show that it actually works (CECCARELLI 2014). Similar concepts are used in population breeding and have been applied with success to the management of mixtures in relation to resistance to diseases and insects.

Evolutionary plant breeding is, therefore, much more dynamic than either conventional or GM varieties, in providing farmers with a continuous flow of novel varieties, and appears as the most dynamic and cheap way of adapting crops to a moving target, such as climate change and its collateral effects on pests. Evolutionary populations of different crops are currently grown by farmers in Jordan, France, Ethiopia, Iran, and Italy on cereal crops (maize, barley, and bread and durum wheat), grain legumes (common bean), and horticultural crops (tomato and summer squash). Research activities on evolutionary populations are ongoing in a number of European countries.

Evolutionary plant breeding has the potential of making farmers independent of the seed market, which is now, globally, in the hands of large corporations. The world's top three corporations control 53% of the world's commercial seed market; the top 10 control 76%, whereas small farmers overwhelmingly rely on seeds that they save from their own crops and which they donate, exchange, or sell.

The industrial type of agriculture, of which GM crops are the most recent aspect, leads to an extension of monocultures, to a significant loss of agrobiodiversity, and to accelerated soil erosion. One of its most potentially devastating impacts is its contribution to increased greenhouse gas emissions, which amounts to 30%–32% of the total man-made greenhouse gas emissions attributable to food systems.

3.3 RESISTANCE BREEDING AS A PERMANENT CHALLENGE

The research and breeding for resistance to fungal, bacterial, and viral pathogens is constantly moving in the field of interactions between the host plant, the pathogens, and the environment. All those areas of activity must be taken into account for successful breeding. Neglecting one of these components can lead to serious misjudgments and loss of yield. Major resistance genes are of great importance in breeding. However, resistance based on this sort of genes can be short-lived under agricultural conditions due to the pathogen adaptation. Therefore, there is a growing need to develop more sustainable uses of major resistance genes. The understanding of the molecular interactions of host and pathogen molecules has to be improved and should result in a more sustainable use of the precious resources of major resistance genes.

3.3.1 RESISTANCE TO PESTS

The annual economic losses by plant diseases and pests are tremendous if all cultivated plants are considered throughout the world. At present, the world spends more than US$ 15 billion annually

on crop protection chemicals. Despite this, preharvest crop losses due to pests and diseases are estimated at 25%. In food crops alone, these losses are enough to feed about one billion people.

Many breeding programs center on recombination of disease resistance with other important
agronomic characters. Commonly, by subsequent domestication of plants natural resistance genes
got lost or new pathogen races developed under the selection pressure of large-scale crop cultivation. Breeders became aware of the problem from the early beginning. According to COONS (1953),

> … the first and fundamental principle of breeding for disease resistance … is … that where host and
> parasite are long associated, and then in the evolutionary process resistant forms are developed by
> natural selection…

The principle has been brought into focus on numerous occasions, when parasites have been introduced into new environments where only susceptible host plants were growing. Breeding for resistance to late blight disease in potato started in 1850. The German J. F. KLOTSCH in Berlin crossed
Solanum tuberosum with the wild species *S. demissum* to improve the resistance. Breeding for
resistance to wilt diseases, for example, *Verticillium* species, has been an accepted practice since
the pioneering work of the American W. A. ORTON with watermelons and cotton in 1900, and that
of H. L. BOLLEY (1865–1956) with flax in 1901. In 1899, ORTON already selected resistant cotton plants utilizing natural and artificial infection. In England, ▶ BIFFEN (1905) carried out first
inheritance studies on disease resistance to yellow rust in wheat. He found that stripe rust resistance
was due to a single gene. Although he misinterpreted incomplete dominance, the work initiates a
century of resistance gene use for disease control.

In 1915, JONES and GILMAN released a *Fusarium*-resistant cabbage. HAYES et al. (1920)
successfully transferred stem rust resistance from *Triticum durum* to bread wheat, *T. aestivum*,
in the famous cross of "Marquis" × "Lumillo" as it was similarly repeated in 1930 by ▶ E. S.
McFADDEN. Resistance to stem rust was transferred from *Triticum turgidum* species *dicoccum*
("Yaroslav") into hexaploid wheat producing the variety "Hope." It was later determined that stem
rust resistance in "Hope" was largely controlled by a single gene located on the short arm of
chromosome 3B. The resistance gene was named *Sr2* and assumed to have come from *Triticum
turgidum*, though the original tetraploid accession was lost and could not be confirmed to carry the
gene. The cultivar "Hope" was used in Mexico during the 1940s as the donor for developing the
stem rust-resistant wheat cultivar "Yaqui 48." Since then, the *Sr2* gene has been employed widely
by CIMMYT's global wheat improvement program in Mexico, and germplasm exchange from
CIMMYT has distributed it to many wheat production regions of the world. The gene has provided
durable, broad-spectrum rust resistance effective against all isolates of *Puccinia graminis* worldwide for more than 50 years. On the basis of its past performance, *Sr2* has been described as one
of the most important disease resistance genes deployed in modern plant breeding.

In 1921, H. K. HAYES and E. C. STAKMAN suggested different races of stem rust that have to
be considered in resistance breeding.

Before 1900, breeders looked at inherited characteristics as expressed on a quantitative basis,
that is, genetically inherited characteristics could be present in varying degrees. For example, a
susceptible plant crossed with a resistant plant would produce progeny with every degree of susceptibility, ranging from the one extreme of minimum resistance to the other extreme of maximum
resistance.

However, the recognition of Mendelian laws in 1900 drastically changed the approach. Now
breeders looked at inherited characteristics as expressed on a qualitative basis. The recognition of
MENDEL's laws marked the beginning of a conflict between the Mendelians and the biometricians
on whether genetically inherited characteristics are based on one single gene or on polygenes. After
the description of ▶ BIFFEN that wheat resistance can be controlled by a single gene, suddenly the
Mendelian school of breeders discovered that genetically controlled characteristics that were of economic importance indeed exist. An explosion of research followed. As a result, resistance of many

plant diseases was shown to be controlled by single genes. This was the resistance that is now called "vertical resistance." A differentiation of horizontal and vertical resistance was first introduced by PLANCK (1963).

It was also shown that the quantitative characteristics valued by the biometricians did not conflict with genetic laws. Instead of being controlled by single genes, these quantitative characteristics were controlled by many genes. By the time the conflict with the biometricians was resolved, the Mendelians, focusing on single-gene inheritance, were in complete control of plant breeding.

The single-gene resistances have both advantages and disadvantages. The advantages are complete protection against the parasite in question and compatibility with breeding for wide climatic adaptation. These characteristics are attractive to large, centralized breeding institutes since these institutes target large areas and thus broad adaptability in their breeding material. The main disadvantage of vertical resistance is its temporary nature, since it breaks down to new strains of the parasite. Other disadvantages include a loss of horizontal resistance while breeding for vertical resistance and the fact that single genes for resistance cannot always be found. Thus, it has appeared impossible to breed for vertical resistance to some species of crop parasites, including many of the insect pests of crops. Moreover, vertical resistance has been misused in agriculture. Breeders have employed vertical resistance in uniform crop varieties, in which every host individual in one cultivar has the resistant gene.

H. H. FLOR, in 1940, discovered the gene-for-gene relationship while working on flax rust (*Melampsora lini*) in Illinois (USA). He showed that the inheritance of both resistances in the host and parasite ability to cause disease is controlled by pairs of matching genes. One is a plant gene called the resistance I gene. The other is a parasite gene called the avirulence (*Avr*) gene. Plants producing a specific *R* gene product are resistant toward a pathogen that produces the corresponding *Avr* gene product (FLOR 1971). Gene-for-gene relationships are a widespread and very important aspect of plant disease resistance. Later, Clayton O. PERSON was the first scientist to study plant pathosystem ratios rather than genetics ratios in host–parasite systems. In doing so, he discovered the differential interaction that is common to all gene-for-gene relationships and that is now known as the Person differential interaction (PERSON 1959).

Most resistance genes are autosomal dominant but there are some, most notably the *mlo* gene in barley, in which monogenic resistance is conferred by recessive alleles. Gene *mlo* protects barley against nearly all pathovars of powdery mildew.

Scientific acceptance of horizontal resistance began slowly but many plant pathologists still doubt the value of horizontal resistance, and some even its very existence. The reasons are that horizontal resistance is often difficult to measure. It is dependent on environmental effects and is often partial.

During the 1960s, many breeders began to doubt the profitability of breeding for vertical resistance. The commercial life of most vertically resistant cultivars was too short to justify the amount of breeding work. This attitude, combined with the development of improved crop protection chemicals, and the investments of chemical industries in breeding, led to a gradual abandonment of resistance breeding in favor of crop protection by chemicals.This strategy changed during the recent years. Breeding for (horizontal) resistance can be cheaper than crop protection by chemicals and it spares the environment. Commonly, it is not necessary to find a good source of resistance, as when breeding for vertical resistance. Transgressive segregation within a population of susceptible plants will usually accumulate all the horizontal resistance needed. Should it not, merely widening the original genetic base will probably suffice. In addition, recurrent mass selection is applied. This means that about thousand original parents, high-quality modern cultivars but also landraces, are cross-pollinated in all combinations.

Recurrent selection is designed to concentrate favorable genes scattered among a number of individuals. It is performed by repeated selection in each generation among the progeny produced by matings *inter se* of the selected individuals of the previous generation. In practice, plants from a population are selfed and, after the yield of the selfed seeds, the progenies of the phenotypically best individuals are grown in the second year. The best progenies are then crossed in as many combinations as possible and the seeds received thereby are grown in the third year as a population. Within the already improved population, selection and selfing can be carried out again. With this population, a second cycle of recurrent selection can be started. In 1920, E. M. EAST and ▶ D. F. JONES gave initial idea about recurrent selection. M. T. JENKINS (1940) described the procedure in detail.

The progeny should total some thousands of individuals that are screened for resistance by being cultivated without any crop protection chemicals. The majority of this early screening population dies, and the parasites do most of the work of screening. The survivors become the parents of the next generation. This process is repeated until enough horizontal resistance is accumulated. Usually, a maximum of about 10–15 generations of recurrent mass selection will produce high levels of horizontal resistance to all locally important parasites. The recurrent mass selection must be performed in the area of future cultivation, during the season of future cultivation, and according to the farming system of future cultivation. This may produce new cultivars that are in balance with the local agroecosystem.

In the historical context of breeding, the wide field of plant pathology and resistance breeding can only be stressed. As already noted at the beginning, resistance breeding and approaches that provide new source of resistant plants against hundreds of fungal diseases and pests are a continuous subject of research and practical breeding work, including the genetic engineering. In 1995, the first maize variety was registered carrying the insecticidal *Bt* gene of *Bacillus thuringiensis* providing resistance against the European maize borer, *Ostrinia nubilalis*. It was another mileSTONE in resistance breeding.

3.3.2   RESISTANCE TO ENVIRONMENTAL STRESS

3.3.2.1   Salt Stress

Salt stress is an important abiotic stressor affecting crop growth and productivity. Of the 20% (12.78 billion ha) of the terrestrial earth's surface available as agricultural land, 50% is estimated by the United Nations Environment Program to be salinized to the level that crops growing on it will be salt-stressed. Increased soil salinity[6] has profound effects on seed germination and germinating seedlings as they are frequently confronted with much higher salinities than vigorously growing plants, because germination usually occurs in surface soils, the site of greatest soluble salt accumulation. With the recent climatic change, salt stress became a global thread. In 2016, more than 2000 ha get lost just by salting of irrigation.

Some crops, such as cotton, barley, safflower, or sugarbeet, can be grown in relatively saline soils. Others, including beans and maize, wheat, and rice, can be grown only in nonsaline soils. Morphological symptoms are:

white leaf tips, tip burning	less florets per plant
leaf browning and death	less seed weight
stunted plant growth	modified flowering duration
low tillering	leaf rolling
floret sterility	white leaf blotches
low harvest index	poor root growth, etc.

---

[6] The term "salinity" includes all the problems due to salts present in the soil, while in strict term these soils are categorized into two types: sodic (or alkali) and saline; however, the third type of salt-affected soils are also found and referred to as saline–sodic soils. Sodic soils are dominated by excess sodium on exchange complex and a high concentration of carbonate/bicarbonate anions. Such soils have high pH (>8.5) with a high sodium absorption ratio and poor soil structure. Saline soils are again dominated by sodium cations with electrical conductivity more than 4 dS/m, but the dominant anions are usually soluble chloride and sulfate. pH and SAR values of these soils are much lower than those in sodic soils. Saline–sodic soils are also called saline–alkali soils.

FIGURE 3.12 Differences in salinity tolerance among perennial ryegrass (*Lolium perenne*) genotypes under field testing. (Courtesy of Stacy A. BONOS, Rutgers University, Newark, NJ. With Permission.)

It is intriguing to speculate that a sensitive crop plant might be genetically altered to withstand high salinities. Breeders have considered this approach for many years, but research along these lines has been neglected in favor of other problems. Instead, management options have been used to alleviate saline conditions and, during reclamation of salt-affected soils, farmers have limited their choice of crops to the more tolerant species (Figure 3.12).

However, salt-tolerant species cannot substitute for good management practices that prevent salt accumulation in the soil. Such species can be valuable for cropping saline soils that are undergoing gradual reclamation or that cannot be fully reclaimed because of limited natural or economic resources.

Tolerant species can also be useful where good quality water is not available. Consideration could be given to developing crops that are productive when grown with saline irrigation waters, such as brackish underground well water, drain water, or even diluted sea water. It may be possible to improve the genetic tolerance of crops to salinity and thereby increase the productivity of marginal lands. Salt-affected fields are extremely variable in salinity. In different portions of the field, the productivity of a crop can vary from zero to normal, but the total yield is decreased.

The plant breeding approach to producing salt-tolerant crops has two limitations: limited sources of genes for salinity tolerance and lack of rapid, precise evaluation methods, at least in the beginning. Genetic resources for breeding have typically been selected from the best materials within the crop species, but there are many sources of tolerance in wild relatives of crop plants. There are examples of genetic variability with the crop species. EPSTEIN and RAINS (1987) have identified salt-tolerant barley, triticale, wheat, and tomatoes by growing them in salinized solution-tank cultures throughout their entire life cycles. A plant must germinate under high salinity (50%–75% the salinity of seawater) and grow to maturity to be considered salt tolerant, for example, "Kharchia" wheat from India. After evaluation, the selected materials in a number of field sites with natural salinity gradients demonstrated that some of the best screening entries did quite well in the field, but others did not. Some of the standard salt-sensitive, high-yielding varieties produced so well at low salinity that their productivity in a variably salt-affected field would be greater than that of a lower yielding, salt-tolerant type (Figure 3.12). The goal now is to produce varieties with both high yields and salt tolerance.

An early example of gene introgression is the wild tomato, *Lycopersicon cheesmanii*, collected in the Galapagos Islands by ▶ C. M. RICK (Davis, California). After hybridization with cultivated tomato, *L. esculentum*, it appeared to be an excellent source of tolerance to salinity. Recently, tomatoes have been genetically engineered enabling them to grow in salty water. These tomatoes take up salts from the soil and store them primarily in leaves. Similar results were achieved in bread wheat, when genes of the tall wheatgrass, *Elytrigia pontica*, were transferred to wheat.

As for resistance against pests (Section 3.3.1), the breeding methodology for resistance against salinity is diverse. Almost all the conventional breeding methods have been followed for the development of the salt-tolerant materials, that is, introduction, selection, hybridization, mutation, and shuttle breeding approach. For example, most of the initial salt-tolerant rice varieties "Damodar," "Dasal," and "Getu" were the pure line selections from the local traditional cultivars prevailing in the Sunderban areas in West Bengal, India. Later on, other salt-tolerant varietal series were developed through recombination breeding. The genotypes are grouped into different categories based on the physiological mechanism for salt tolerance. Crosses are made between the parents/donors possessing contrasting physiological traits such as tissue tolerance, Na^+ exclusion, K^+ uptake, and Cl^- exclusion to pyramid the genes governing or contributing for salinity tolerance into one agronomically superior background. Recent strategies to supplement the conventional breeding program are the marker-assisted selection (Section 4.2.1), if possible in early generations, and marker-assisted backcrossing.

Most of the traits of economic importance, such as yield and stresses, are controlled by polygenes and considerably influenced by environment and g × e interaction for their expression. These traits are most difficult to breed conventionally and using nonconventional techniques such as marker-assisted selection as well. Molecular markers are of great importance to plant breeding if they are used to aid selection for quantitative traits. Nowadays microsatellites or simple sequence repeat (SSR) markers are the first choice, whereas single-nucleotide polymorphisms (SNPs) and others are going to be the most preferred markers of the future. After completing saturated marker maps including physical localizations, tracing of polygenic traits became much easier as any time before. The limited phenotyping accuracy of the QTL is now the limiting factor. QTLs do not have direct phenotypic variation; hence, their chromosomal location is typically inferred by calculating the likelihood of odds (LOD) value. The chromosomal location with maximum LOD value has the likelihood for the QTLs of the trait but there is always a good possibility that the QTL is not located precisely at the maximum likelihood position. This precision could be increased by increasing size of mapping population.

In rice, a major gene for salt tolerance has been mapped to chromosome 7 using RFLP markers. Random amplified polymorphic DNA markers were also shown to be linked to salt tolerance: using the same population six novel clones were found that showed insignificant homology to any of the existing expressed sequenced tagged (EST) database and were differently regulated in salt-tolerant ("CSR27" and "Pokkali") and salt-sensitive ("PUSA Basmati 1") rice varieties under salt stress. Several QTLs for salinity tolerance have also been identified. Others detected one microsatellite marker, *Xrm223*, associated with salt tolerance at vegetative stage. Based on RFLP markers, QTLs for Na and K uptake as well as Na:K ratio were identified. Major QTLs were found on chromosomes 6, 4, 1, and 9 (SINGH 2016).

Major research programs on marker-assisted selection at IRRI are concentrated on introgression *Sub1* and *Saltol* QTLs for submergence and salinity tolerance, respectively, into the improved germplasm or adapted varieties. The major QTL for salinity tolerance, that is, *Saltol*, is also being transferred to the improved background as well as adapted rice varieties. *Saltol* is located on chromosome 1 and linked with SSR markers such as *Xrm8094*, *Xrm493*, and *Xrm3412*.

3.3.2.2 Drought Tolerance

The northern part of the Badia desert in the northeast of Jordan is a fascinating but inhospitable place: extinct volcanoes and millions of basalt sponges cover the ground, and the climate is extremely hot and dry. However, people lived here 6000 years ago. During excavations, not only wallings and simple residential buildings, but also terrace gardens were identified. They were irrigated artificially by means of ingenious irrigation systems with the little rainwater, for example, to cultivate grain.

This is long before the 5000-year-old Megacity Uruk in Mesopotamia had irrigation systems. There are numerous testimonies of distance trade and luxury production, with which social elite stood out from the majority of the inhabitants. Their right to existence drew this upper class from the ability to organize the ever-more complex irrigation of the fields and fields to ensure the distribution of the yields.

Recently, one discovered the oldest town in America, the 4600-year-old "Holy City of Caral" near Lima. The ancient irrigation system, further developed from the Inca to perfection, and adopted by the present Indian peasants, is the archaic foundation of the Andean agriculture. Without this ingenious invention, the archaeologists of Caral concluded, the early civilization, which was originally settled by the sea, could scarcely have survived the desolate line of death between the coast and the mountains.

Qingcheng Shan (China) is one of the origins of Daoism. In 143 AD, during the Eastern Han Dynasty, Grand Master Zhang DAOLING came to the Qingcheng Shan to spread Daoism. Together with the jagged, mountainous Qingcheng Shan, 17 kilometers away, an irrigation system has been a UNESCO World Heritage Site since 2000. The Dujiangyan irrigation system is located near the city of Dujiangyan, which is 60 km northwest of the city center of Chengdu. It is a 2300-year-old backwater system that controls the Min river and draws large amounts of water to irrigate the Red Basin. The facilities were built from 256 BC to 251 BC under the Qin administrator Li BING and his son.

All these early evidence of irrigation systems show that people were always faced with water scarcity and drought. It was certainly not only the technical achievements of those times, but also the plants that were selected for the respective conditions and were most likely bred. Drought becomes a challenge for modern breeding. During periods of severe drought, these losses can be much higher and can potentially result in complete crop failure. Obviously, drought is currently the leading threat to the world's food security. Various climate models from the Intergovernmental Panel on Climate CHANGe (IPCC) suggest that the average global surface temperatures in 2100 will be 1.1°C–6.4°C higher than they were in 2000, potentially resulting in more extreme environmental conditions than we find in today's droughts. Since 1880 there has been no month that was as hot as July 2016. The NASA reported that July 2016 was about 0.84°C warmer than the global average of all July months from 1950 to 1980.

However, drought in agriculture is not new. In rice, growers in the northeastern Indian state of Jharkhand have known for long time how to maximize crop yields during droughts: they simply cultivate their seed beds during the hot, dry summer months and transplant the seedlings at the start of the rainy season in July. However, this method cannot be applied always and for all crops. Crops in many regions face both water shortage and high temperatures—stresses that trigger different physiological responses that can compound each other's effects. The average annual yield loss of maize attributable to drought is approximately 15%. Soybean is considered the most drought-sensitive plant because drought may reduce soybean yield by approximately 40%.

John BOYER (1982) already estimated that up to 69% of yield losses of wheat and maize in North America were potentially attributable to abiotic stresses—those arising from environmental conditions such as heat and water deprivation. Bolstering crop defenses against such stresses has been a priority for decades. Today's grain plants are dramatically more bountiful than their ancestors, and maintain higher yields in hostile conditions. Since the 1950s, annual yields of maize have increased by 60–100 kilograms per hectare under both water-scarce and normal conditions.

This steady improvement prevented the recent drought worldwide from becoming a full-blown catastrophe. Any further rise in temperature threatens productivity and raises costs for staple crops in many regions. This environment of heightened risk and uncertainty has injected more urgency into the hunt for drought-tolerant plants. Unfortunately, these strategies are increasingly ineffective in the face of highly erratic rainfall and drought patterns, and the results have been devastating (EISENSTEIN 2013). In most critical regions of the world, a lot of wheat and maize as staple crops are growing at the limits of where they like to be, temperature-wise.

The adjustment the flowering time was the biggest factor in improving, for example, wheat yield. Heat and drought are deadliest during a plant's flowering phase, when reproduction is occurring. Many crops have evolved mechanisms to accelerate flowering before the dry season arrives. Breeders have exploited this feature to generate early varieties of crops through traditional cross-breeding. In rice, the variety "Sahbhagi Dhan," which is now cultivated throughout South Asia, is a result of such a conventional breeding approach. Until 2015, more than 50 new drought-tolerant varieties and hybrids have been developed and released for dissemination by private seed companies, national agencies, and nongovernmental organizations (Figure 3.13). African farmers now grow many of those varieties, which yield 20%–50% more than others under drought, on hundreds of thousands of hectares. Nevertheless, this way of progress is slow. It needs the most recent breeding methodology to gain more drought-tolerant varieties.

SHINOZAKI and YAMAGUCHI-SHINOZAKI (2007) indicated that genes involved in these responses can be grouped in two main classes: single function genes and regulatory genes based on their biological function. The single function genes encode enzymes associated with the accumulation of osmolytes, proteins, and enzymes scavenging oxygen radicals, proteins associated with the uptake and transport of water and ions, and proteins involved in lipid biosynthesis. The regulatory genes are involved in signaling cascades and transcriptional or posttranscriptional regulation of gene expression such as transcription factors, protein kinases, protein phosphatases, and proteinases. Meanwhile, regulatory proteins have been proven to play crucial roles in the responses of plants to drought stress conditions. Consequently, the modification of the expression of a regulatory gene is more efficacious and is probably to be widely used in genetically modified crops with abiotic stress tolerance.

FIGURE 3.13 Cob of a strain of drought-tolerant maize bred for Africa (left) as compared to the standard cultivars (right). (Courtesy of P. LOWE, CIMMYT, Mexico. With Permission.)

Rajeev VARSHNEY and colleagues at the International Crops Research Institute for the Semi-Arid Tropics (ICRISAT) in Andhra Pradesh, India, identified a genomic region in the chickpea that contains QTLs for several drought tolerance-related traits. A combination of two approaches, namely QTL analysis and gene enrichment analysis, was used to identify candidate genes in a QTL-hotspot region for drought tolerance present on the Ca4 pseudomolecule. In the first approach, a high-density bin map was developed using 53,223 SNPs identified in the recombinant inbred line (RIL) population of ICC 4958 (drought-tolerant) and ICC 1882 (drought-sensitive) cross. QTL analysis using recombination bins as markers along with the phenotyping data for 17 drought tolerance-related traits obtained over one to five seasons and one to five locations split the QTL-hotspot region into two subregions, namely "QTL-hotspot region a" (15 genes) and "QTL-hotspot region b" (11 genes). In the second approach, gene enrichment analysis using significant marker trait associations based on SNPs from the Ca4 pseudomolecule with the above-mentioned phenotyping data and the candidate genes from the refined QTL-hotspot region showed enrichment for 23 genes. Twelve genes were found common in both approaches. Functional validation using quantitative real-time PCR (qRT-PCR) indicated four promising candidate genes having functional implications on the effect of QTL-hotspot region (KALE et al. 2015). In praxi, they found a gain of 10%–20% higher yield in an already drought-tolerant chickpea variety.

Australian researchers describe an alternative approach of drought tolerance improvement in wheat and rice. Although intracellular signaling during oxidative stress is complex, with organelle-to-nucleus retrograde communication pathways ill-defined or incomplete, they discovered an enzyme that senses adverse drought and sunlight conditions, and how it works from atomic to overall plant levels. It is the 3′-phosphoadenosine 5′-phosphate (PAP) phosphatase SAL1 as a conserved oxidative stress sensor in chloroplasts. The sensor in plant leaves is constantly sensing the state of its environment in terms of water and light levels. It is able to sense when conditions become unfavorable, such as during extreme drought stress, by changing itself into a form with altered shape and activity. This sets off an alarm in the plant, telling it to respond to drought by making beneficial chemical compounds, for instance. But in the field, this can occur too late and the plant would have suffered damage already. If one can get the alarm to go off at the first signs of water deficit, one can help the plant survive severe droughts. Within a few years, PETER Mabbitt hopes to identify potential compounds for a chemical spray that will rescue crop yields by a triggering enzyme (CHAN et al. 2016).

The newer approach of creating genetically modified crops has so far delivered little in the way of improved drought resistance. It took until 2009 for the first drought-tolerant GM crop. It was maize "MON 87460," registered as "DroughtGard™" and developed by Monsanto Company (USA). In 2013, it was first planted in the United States. Its acreage increased 5.5-fold from 50,000 ha in 2013 to 275,000 ha in 2014—though even the U.S. Department of Agriculture admitted that it was no more effective than non-GM varieties. Hybrid seed sold under this trademark combined a novel transgenic trait (based on the bacterial cold shock protein B, i.e., *cspB* gene) with the best of Monsanto's conventional breeding program. This sort of genes act as RNA chaperones, which help to maintain normal physiological performance during stress events by binding and unfolding tangled RNA molecules so that they can function normally.

Another way of engineering drought tolerance is by taking genes from plants that are naturally drought tolerant and introducing them to crops. The resurrection plant (*Xerophyta viscosa*), a native of dry regions of southernmost Africa, possesses a gene for a unique protein in its cell membrane. Experiments have shown that plants given this gene are less prone to stress from drought and excess salinity. COSTA et al. (2016) demonstrated that those genes coding for antioxidant activities and polyphenolic profiles are often present in highly adapted crop landraces, but got suppressed in advanced breeding stains. As in maize, a targeted and specific re-expression may increase desiccation tolerance, that is, the ability of certain organisms to survive severe dehydration (see Figure 3.13).

Drought tolerance has to be considered as a very complex system, requiring a systems-based approach that combines agronomics (e.g., less tilling), breeding, and biotechnology. Worldwide climate change remains an acute threat to agricultural productivity, pushing more than 100 million people back into poverty by 2030, as reported by the World Bank Group. To contribute to global food security to some extent in the years to come, there are currently worldwide efforts to develop new crop varieties with tolerance to drought not only by conventional breeding, but also by using the tools of modern biotechnology. Each program has to be adjusted to the specific local conditions, otherwise it will not work.

3.4 HYBRID BREEDING

In 1877, a German gardener, C. KLEINERT from Gruschen near Schlichtingshausen (Poznan, today Poland), reported about rye crosses between the "Swedish Snow rye" and "Correns rye." He found better tillering and grain formation in F_1 as compared to "Probsteier" population rye. Obviously, he demonstrated for the first time heterosis effects in rye, long before the phenomenon was discovered in maize and other crop plants (RIMPAU 1883). The "Russett Burbank" hybrid potato (*Solanum tuberosum*) variety was launched in 1923 and the first hybrid maize (*Zea mays*) variety was not released until 1933.

A very basic question—Why make hybrids? Why not bring all the desirable traits in one plant itself and happily self-pollinate and propagate?

The very simple answer is that it cannot be done—or at least there is no record of a successful attempt so far. Breeders have been trying to understand the phenomenon of hybrid vigor for over a century. It depends at least partially on heterozygosity and therefore cannot be fully recapitulated in a stable inbred line. F_1 hybrids usually express higher and more consistent yield than the parents. No matter which interesting traits you can stack into an inbred line, you will lose this heterosis effect.

A conspirative answer is that there are obvious commercial benefits to marketing hybrid seeds instead of inbreds. Plant breeding costs money. It has to be paid either through government agencies by reluctant tax payers or by private investors who need a payback. Payback can be more easily achieved with hybrids than with self-pollinating varieties that farmers can propagate in their fields year after year without losing the value.

Hybrid breeding is the development of varieties that are the direct product of a cross between two parents. Hybrid varieties are increasingly grown worldwide. In maize, about 65% of the area is covered with hybrid material and a yield advantage of about 15%, in sunflower 60% and 50%, in rapeseed 50% and 12%, in sorghum 48% and 40%, in rice 12% and 30%, and in rye 5% and 20%, respectively.

The term hybrid breeding is used in narrow sense to describe breeding strategies utilizing effects of hybrid vigor, often observed in the F_1 hybrids between plants of two varieties, species, or higher order. In 1916, ▶ G. H. SHULL introduced the term "heterosis" to replace "heterozygosis" describing the phenomenon of luxuriance of hybrids or hybrid vigor; that is, progeny of diverse (inbred) genotypes exhibit greater biomass, speed of development, and fertility than the better of the two parents (see KÖLREUTER, SPRENGEL, KNIGHT, GÄRTNER, NAUDIN, MENDEL, DARWIN, FOCKE, WEBER, SANBORN [1890], or JOHNSON [1891]).

In general, the degree of heterosis increases as the "genetic disparity" of the parents becomes greater. A research project on maize was initiated in 1895 by DAVENPORT (1908) and HOLDEN, both students of ▶ W. J. BEAL (who in 1871 reported 10%–15% higher yields in maize hybrids from variety crosses; RICHEY, 1920), at the University of Illinois. Since 1900, ▶ E. M. EAST (1908), the cofounder of the heterosis conception, was appointed at the department. Although still not confessed generally, the important effects of heterosis were discovered already. However, they required a causal explanation. ▶ G. H. SHULL (1908) started his experiments with maize at Cold Spring Harbor (USA) in 1904, without special consideration of heterosis. Independently,

EAST began inbreeding and crossing experiments, which were continued till 1911. He published his work on inbreeding in 1908. The segregation of homozygotes and heterozygotes in inbred lines can be calculated by the binom $1 + (2^r - 1)^n$, where r is the number of inbred generations and n is the number of genes involved.

Before EAST, SHAMEL (1905) reported yields of maize lines inbred for three generations and of hybrids made by crossing inbred lines. In the years 1908–1910, ▶ SHULL realized that the yield of hybrids succeeded not only the parental inbred lines but also the original stock by 22% (ratio = 29:100:122). With first genetic interpretation of the results, the first step toward the various theories of heterosis was made. ▶ SHULL (1948) has summarized them as follows:

- The entrance of a sperm into a foreign cytoplasmic environment may in some cases produce an initial favorable reaction, which may manifest itself in the F_1 and not be repeated or not repeated in the same degree in F_2 and subsequent generations (SHULL 1912).
- Protoplasmic heterogeneity may favor increased metabolic activity.
- Linked dominant favorable factors may confer some advantages in heterozygotes as compared with homozygotes. The dominance theory was first outlined by BRUCE (1910), KEEBLE and PELLEW (1910), and expanded by JONES (1917) to include linkage. JONES developed the first commercial hybrid of maize. In this hypothesis, heterosis results from cumulative action of many dominant favorable genes contributing to vigor. The biggest heterosis effect should occur when the maximum number of loci with dominant favorable alleles are present.
- Not the accumulation of favorable dominants, but the rare occurrence of an unfavorable recessive induces heterosis (JONES 1945).
- A sensitization of a nonmutated parent gene by its mutated allele, or vice versa, of the nature of an anaphylaxis produces the heterosis effect (CASTLE 1946).

Although the dominance hypothesis was the most favored theory of heterosis, evidence has accumulated, which indicates that it does not take adequate account of various manifestations of heterosis. In 1908, C. B. DAVENPORT (1908) first proposed dominance hypothesis of heterosis. Much later, after experimental results E. M. EAST (1936) supported the dominance hypothesis.

Including the molecular studies of recent time, heterosis is concluded as a complex phenomenon of dominance actions and regulatory gene allelic interactions (SONG and MESSING 2003, BIRCHLER et al. 2003). It seems to be a matter of optimization of genetic, physiological, and environmental influences that contribute to the final phenomenon of heterosis, particularly when the breeding application is considered. Fixation of heterosis effects in breeding is feasible in several ways:

- If vegetative propagation of the plants is combinable with heterosis,
- If apomixis is associated with heterosis effects,
- If heterosis is caused by dominant alleles for a given trait or bases on complementary gene action, segregating in homozygous manner,
- If allopolyploid genome constitutions are present,
- If both alleles causing the heterosis effect are available as pseudoalleles,
- If complex heterozygosity or inversion heterozygosity of chromosomes permanently produces heterozygosity of certain genes, or
- If heterosis bases on favorable genotype–plasmotype interactions.

In nature, additional mechanisms of maintaining heterozygosity are described, such as heterostyly, protandry, protogyny, or sexual differentiation. Thus, not all crop plants are suitable for hybrid breeding. The genetic prerequisites, the economical expenditure, and the expected profit must get always against each other balanced. In maize, the hybrid seed production could be highly optimized by mechanical emasculation (detasseling) and stripe cropping of the parental inbreds.

Controlled experiments with varietal hybrids were begun at the Michigan Agricultural Experiment Station prior to 1880 and were carried out until the 1920s. In general, those crosses yielded in excess of the parents. Flint[7] and flour types crossed to dent types generally gave the must sizable yield increase.

Later the ear-to-row selection was the first method to be put to rather extensive use. The selected ears were grown out in ear rows the next year, and ears were again selected from the highest-yielding rows. Thus, the principle of progeny testing became an integral part of the breeding system, even though only the female parentage was under control and plants in a population were of heterogeneous nature, which was not reproducible.

Self-fertilization by detasseling the monoecious maize became the standard practice of fixing/isolating the genotypes. Together with their utilization in producing hybrid combinations, the basis for hybrid maize industry was formed. In 1926, Pioneer Hi-bred Corn Co. in Johnston, Iowa (USA) was the first seed company to organize maize breeding—now Pioneer Hi-Bred International, Inc. is one of the largest seed companies in the world.

After the first experimental four-way hybrid "Burr" × "Leaming" (1917), JONES (1920) suggested to use double crosses as a practical way of avoiding low-yielding inbred lines as seed parent in commercial seed production. Subsequently, several modifications of hybrid seed production were developed, which now are standard worldwide (Table 3.3).

The inbred lines used in the beginning of hybrid development were extracted from open-pollinated varieties. More productive inbreds were derived later from controlled hybrid combinations. Pedigree systems, topcrossing, backcrossing, or convergent crosses (RICHEY 1927) have been used in improvement of inbred lines.

The value of the method, which is equivalent to double backcrossing, is that it furnishes a plan for the improvement of each of two inbred lines that combine well in a single cross without modifying the yielding ability of the single cross. Multiple convergence, a modification of convergent improvement as presented by RICHEY (1946), implies the convergence of several small streams of germplasm as tributaries to one larger stream for the improvement of inbred lines and their use in crosses. HAYES and JOHNSON (1939) and HAYES et al. (1946) have shown that selection for improvement of combining ability is possible. In 1932, JENKINS and BRUNSON used the topcross method to give comparative tests for combining ability. One year earlier, ▶ RHOADES (1931) discovered cytoplasmic male sterility (CMS) in maize. In tobacco it was found in 1932 (EAST), in radish and sunflower 1968 (OGURA, LECLERCQ), and rice 1969 (SHINJYO). In maize, CMS is a maternally inherited trait that makes the plant produce sterile pollen, enabling the production of hybrids and removing the need for detasseling. Similar mutants were then found in many crops, even leading to innovative systems for hybrid seed production. Rye (*Secale cereale*) is one of them. Hybrid seed production in rye became a success story in Europe during the past 10 years (SCHLEGEL 2016).

Recently, in rapeseed nuclear male sterility is created through genetic engineering of anther-specific genes, for example, *Bcp1*. It was shown that antisense RNA inactivation of this unique gene leads to nuclear male sterility in *Arabidopsis* and *Brassica*.

Since the 1930s, when maize breeders started the commercial development of double-cross hybrids, hybrid breeding was also followed by the extensive utilization of single crop hybrids since the 1960s. For example, hybrid rice was a particular breeding contribution to reduce hunger in the world. It represents an extraordinary achievement since wisdom had considered the incorporation

---

[7] Dent and flint maizes differ in many traits. By about 1800, North American colonists from Europe found that hybrids between the two types could be advantageous: in particular, flints provide alleles for earlier maturity and dents for higher yield. Between 1800 and 1920, farmers in Canada and the United States, principally from flint × dent crosses, had developed over 500 open-pollinated varieties of maize. According to ▶ R. W. ALLARD (1999), the allogamous variety "REID yellow dent" became the most widely adapted and popular maize in the United States during the late 1880s. The farmers R. and J. REID moved from Ohio to Illinois in 1846 and brought dent maize with them. Probably, by spontaneous crosses with Indian flint maize surrounding the dent stands a sort of a hybrid variety was established.

TABLE 3.3

Basic Methods of Selection Adapted from More than 100 Years' Breeding Experience

Selection for Varieties					Maintenance Selection	
Line selection	Selection of populations		Hybrid breeding	Clonal varieties	Single spike selection	
Pedigree method	Mass selection		Variety hybrid (population 1 × population 2 or variety 1 × variety 2)		Mass selection	
	Positive	Negative	Topcross hybrid [line or single cross × variety; A × population or (A × B) × population]		Positive	Negative
Bulk method	Single-plant selection method (Illinois method)		Single hybrid (line 1 × line 2 or A × B)		Single-plant selection method	
Partial bulk method	Single-plant selection method with progeny testing (ear-to-row selection or Ohio method)		Modified single hybrid [sister lines crosses × line; (A' × A) × B]		Single-plant selection method with progeny testing	
Pedigree trial method	Synthetic varieties		Double modified single hybrid [sister line crosses × sister line crosses; (A' × A) × (B' × B)]		Recurrent selection	
Single seed descent method (SSD)			Three-way hybrid [single hybrid ×line; (A × B) × C]			
Backcrossing method			Modified three-way hybrid [single hybrid × sister line crosses; (A × B) × (C' × C)]			
			Double hybrid [single cross × single cross; (A × B) × (C × D)]			

of heterosis impractical in self-pollinated crops. It was successfully developed and released after the discovery of male-sterile cytoplasm at Hainan Island (China) in 1970 (LI and YUAN 2000).

The three-line system uses a male sterile or *A* line, a maintainer or *B* line, and a restorer or *R* line.

The first hybrid rice combinations were put into commercial production in China in 1976. The area under hybrid rice increased from 2.1 million ha in 1977 to >32 million ha in 2005. Recently, hybrid rice has maximum yielded >18 t/ha (average 7.5 t/ha) compared with 5.4 t/ha for conventional cultivars, a yield increase of about 30%. Hybrid rice technology has now expanded to India and other Asian countries.

On the basis of a comparison of 36 widely grown hybrids adapted to central Iowa and released at intervals from 1934 to 1991, DUVICK (1997) reported that the increase in maize grain yield during that time span averaged nearly 74 kg/ha per year (Figures 1.1, 3.5, and 3.10). Hybrid comparisons were based on side-by-side trials, so all of the gain could be attributed to genetic improvement. Earlier studies indicated that genetic improvement usually accounted for about one-half of the total yield increase, with the remainder attributed to changes in cultural practices such as increased rates of mineral fertilizers and the use of herbicides for weed control and pesticides for control of insects and diseases. It is suggested that the increased grain yielding ability of these widely successful hybrids was due primarily to improved tolerance of abiotic and biotic stresses, coupled with the maintenance of the ability to maximize yield per plant under non-stress growing conditions.

Another crop that was adjusted to hybrid seed production was sorghum. During the 1940s and 1950s, the inbred method as in maize was successfully applied. Hybrid onions came to commercial production in 1952; asparagus, strawberries, squashes, tomatoes, sunflower, and so on followed it. In 1970, ▶ C. T. PATEL developed world's first cotton hybrid for commercial cultivation in India; in 1991, ICRISAT presented first pigeon pea hybrid "ICPH 8."

Hybrid seed can be produced in not only cross-pollinated crops but also self-fertile plants, such as wheat. Because of self-pollination, creating hybrid varieties is extremely labor intensive; the high cost of hybrid wheat seed relative to its moderate benefits has kept farmers from adopting them widely, despite nearly 90 years of effort. Heterosis or hybrid vigor occurs in common wheat, but it is difficult to produce seed of hybrid cultivars on a commercial scale as is done with maize. Wheat flowers are perfect in the botanical sense, meaning they have both male and female parts, and normally self-pollinate. It was DeKalb Co. (USA) that launched the first hybrid wheat in 1974 (EDWARDS 2001). Commercial hybrid wheat seed has been produced during the 1980s using chemical hybridizing agents, plant growth regulators that selectively interfere with pollen development, or naturally occurring cytoplasmic male sterility systems (since the 1960s). Hybrid wheat has been a limited commercial success in Europe (particularly France), the United States, and South Africa. The price for wheat was very low, and the cost for the hybrid seed was too high at the time. However, today there is a better handle on the genes, availability of genomic tools, and better prices. Optimistic breeders expect first hybrid varieties of the next generation within five years (ZHANG 2016).

3.4.1 Synthetics

The idea of synthetic varieties is old. It was first mentioned by HAYES and GARBER (1919). Instead of hybrid varieties, they proposed synthetic populations in maize. Successful production of synthetics was already demonstrated during 1940s (HAYES et al. 1946).

SPRAGUE and JENKINS (1943) described synthetic varieties as populations that are derived from more than four lines after open pollination. They can be grown not only as F_1 population but also further. S_1 lines of high combining ability may be combined into *synthetic varieties* for fringe-area production and as a stopgap improvement measure as proposed by LONNQUIST and MCGILL (1956). On the basis of testcross progeny performance, a number of lines can be selected to be included in the synthetic variety (Figure 3.14). The synthetic variety is generally formed by compositing equal amounts of seed of all possible single crosses between the selected S_1 lines. In subsequent generations, mass selection may be practiced to maintain the synthetic variety. In other cereals than maize, multiline varieties can be superior over pure line varieties as demonstrated by ▶ JENSEN (1965), particularly in spite of their better adaptability against environmental stress and pathogens. N. F. Jensen (1952) first suggested the use of multilines in oats; in 1953, ▶ N. E. BORLAUG first outlined the method of developing multilines in wheat. For example, the intensity of leaf rust in multilines can significantly affect at all stages of rust development. A reduction between 32% and 89% over the average of components can be expected.

3.5 MUTATION BREEDING

The utilization of modified genotypes was already known in ancient China. In a Chinese book "Lulan" published in 300 BC, first documentation of mutant selection in plant breeding was described, namely maturity and other traits of cereals (HUANG and LIANG 1980). In 1590, a spontaneous plant mutant, that is, the "incisa" mutant of greater celandine (*Chelidonium majus*), was reported.

Meanwhile, more than 3235 new mutant varieties have been bred and used by millions of farmers, which has significantly contributed to world food security (ANONYMOUS 2017). The economic

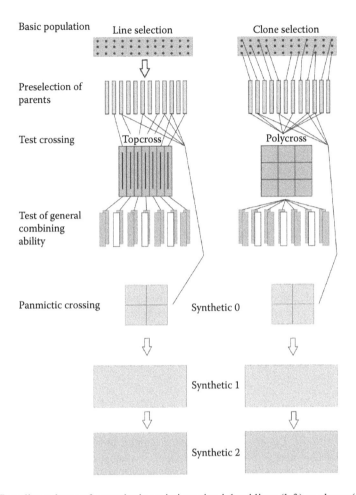

FIGURE 3.14   Breeding schemes for synthetic varieties using inbred lines (left) or clones (right).

contribution of the selected mutant varieties of rice, barley, cotton, groundnut, pulses, sunflower, rapeseed, Japanese pear, and many others is obvious. More than 3250 mutant varieties—including numerous crops, ornamentals, and trees—have officially been released for commercial use in more than 210 plant species from over 70 countries. A lot more mutants have also served as tools for gene discovery and functional analysis.

Particular impact had been shown by mutants, for example, durum wheat varieties that have occupied about 90% of the cultivating area of Bulgaria since the 1980s; in Italy it was the variety "Creso." In Pakistan, the hexaploid wheat variety "Kiran 95" stabilized the wheat production. In China, the following mutant varieties have cumulatively been grown on acreage of more than 10 million ha: the rice varieties "Yuanfengzao," "Zhefu 802," and "Yangdao 6," the bread wheat variety "Yangmai 156," and the cotton variety "Lumian 1." In Pakistan, the cotton variety "NIAB 78" was planted on over 80% of the cultivation area. The rice variety "Camago" occupied 30% of the cultivated area of Costa Rica. Most rice varieties of Japan carry the *sd1* mutant allele from the variety "Reimei." "Calrose 76" was the *sd1* donor for more than 10 successful varieties of the United States. The barley varieties "Diamant" or "Golden Promise" are widely grown in Europe. In India, the "TAU 1" mutant of blackgram has occupied 95% of the blackgram acreage in the state of Maharashtra and the groundnut varieties of the "TG" series (e.g., TG24 and TG37) cover 40% of the groundnut acreage. Japanese pear cultivation was rescued from extinction by the development of disease-resistant mutant varieties "Gold Nijisseiki" and its derivatives. "Star Ruby" and "Rio Red" are the two most important commercial grapefruit varieties of the United States (FORSTER AND SHU 2012).

Recent genomic approaches including TILLING (Targeting Induced Local Lesions In Genome) have enabled screening of mutations at the molecular level. Reports indicate that chemically induced mutations are mostly nucleotide substitutions, but that mutation frequencies fluctuate among plant species ranging from one DNA lesion per 300 kbp in *Arabidopsis* to one DNA lesion per 30 kbp in bread wheat, which reciprocate with an increase in ploidy levels. The majority (>95%) of chemically induced DNA lesions are silent or missense mutations. Mutations induced by physical mutagens seem to be more diverse, including single-nucleotide substitutions, insertions, inversions, and translocations, although short deletions (<15 bp) are relatively more predominant. The proportion of complex mutations (translocation, inversions, etc.) may increase with an increase in the linear energy transfer of radiations (NAWAZ and SHU 2014).

Radiation breeding was widely used in the developing world, thanks largely to the atomic agency's efforts. Beneficiaries have included Bangladesh, Brazil, China, Costa Rica, Egypt, Ghana, India, Indonesia, Japan, Kenya, Nigeria, Pakistan, Peru, Sri Lanka, Sudan, Thailand, and Vietnam. Politically, the method is one of many *quid pro quos* the agency, an arm of the United Nations in Vienna, offers client states. Its own agenda is to inspect ostensibly peaceful atomic installations in an effort to find and deter secret work on nuclear weapons. Some scientists believe that radiation breeding is not going to solve the world food crisis but it will help. Modern breeders are using every tool they can get. By promoting crop flexibility, it could help feed billions of added mouths despite shrinking land and water, rising oil and fertilizer costs, increasing soil exhaustion, growing resistance of insects to pesticides, and looming climate change. Globally, food prices are already rising fast.

3.5.1 INDUCED MUTATION BY MUTAGENS

DNA is stable but not static. Occasionally mutations happen in DNA, changing the sequence of base pairs. A *mutation* can have serious consequences if it changes the synthesis or function of an important protein. However, most mutations have no immediate consequences for the organism, as they take place in noncoding sequences, either within or outside a gene. Such harmless, neutral, mutations form the basis for most of the differences between individual genomes.

One of the rediscoverer of MENDEL's laws, ▶ H. de VRIES significantly stimulated the mutation research. In his book "Mutationstheorie" (1901), he summarized his observation on spontaneous mutations. Spontaneous mutations of diverse kind have been observed in plants. Nevertheless, there is no evidence that the mutation rate increases as a result of cultivation, apart from exceptional circumstances such as unequal crossing-over in hybrids heterozygous for small-scale structural changes in the chromosomes of maize or selection in favor of unstable, "ever-sporting" genotypes in garden pea. The Italian Petrus de CRECENTIA describes the oldest gene mutant in pea in the thirteenth century. It was a pea with white flowers and white seeds—without doubt a recessive mutant, from which modern vegetable pea varieties were derived.

Ancient peas found in Hissarlyk (Asia Minor) dated back to 2000 BC, Egyptian findings, and peas from the Hallstatt times (750–400 BC) show gray and small seeds and probably purple flowers—a pea type similar to current field peas. Another plant mutant was described by SPRENGER in 1590 for leaf shape in *Chelidonium majus* (see Section 2.6). The first mutant variety seemed to be the so-called "emperor rice"—a mutant accidentally found by Chinese Emperor KHANG-HI (1662–1723) during a walk along the rice fields. It was characterized by earliness and high quality, and therefore preferred at the imperial court. For long it was the only variety that could be grown north of the Great Wall. Because of the earliness, it sometimes brought two harvests in southern regions.

Famous were NILSSON-EHLE's chlorophyll mutations in wheat and rye (1914), the seed color and fatoid mutations in oats (1907–1921), and the speltoid mutations in wheat (1917–1927). TAMMES (1924) noticed in crosses of wild linseed (*Linum angustifolium*) × cultivated linseed

(*L. usitatissimum*) that all important wild and cultivated characteristics of linseed are controlled by the same genes, but by different alleles. Thus, the recessive alleles of the cultivated linseed were derived from their dominant wild alleles by mutations. It means that changes or improvements from wild to crops can be stepwise produced by single induced mutations.

In coffee, a variety of economical important cultivars based on a gene mutant with strong *pleiotropic* action, such as "caturra," "maragogipe," "mocca," or "bourbon," arose from primitive forms (*Coffea arabica* var. *abyssinica*). The spike brittleness of wild barley (*Hordeum spontaneum*) is based on two dominant genes *Bt1* and *Bt2*. Cultivated barleys with the genetic constitution *Bt1 Bt1 bt2 bt2* or *bt1 bt1 Bt2 Bt2* show less tough rachis than *bt1 bt1 bt2 bt2* genotypes. Therefore, the elimination of brittleness in cultivated barley required two mutations. Many other examples demonstrate the importance of gene mutations for crop evolution and breeding.

A first comprehensive review on spontaneous and induced mutations was presented from MATSUURA (1933). He also mentioned that on south Japanese islands a giant radish ("Sakurajima Mammoth Globe") was bred weighing >30 kg, of course under optimal subtropical growing conditions. The Russian ▶ N. I. VAVILOV called this radish "the world champion opus of breeding."

Early studies of GAGER (1908), later by PETRY (1922), or STEIN (1922) failed, however, to demonstrate clearly the *mutagenic action* of radiation, although effects on chromosomes were described. Such a demonstration was not forthcoming until the careful experiments of ▶ MULLER (1927) on *Drosophila*, Lewis J. Stadler of the University of Missouri (1928) on barley and maize, and GOODSPEED and OLSON (1928) on tobacco. When STADLER used X-rays to zap barley seeds, it resulted in plants that were white, yellow, pale yellow, and some had white stripes—nothing of any practical value.

These experiments showed behind a doubt that ionizing radiation could greatly increase the mutation frequency of genes, under certain conditions up to 10,000 times that of controls. The so-called *mutation breeding* accelerated after 1945, the end of World War II, when the techniques of the nuclear age became widely available. Plants were exposed to gamma rays, protons, neutrons, and alpha and beta particles to see if these would induce useful mutations. Critical dosages for dormant seeds ranged between 7500 r in *Cannabis sativa, Pisum sativum*, or *Glycine max* and 50,000 r in *Linum usitatissimum* (GUSTAFSSON 1947). Credit for realizing the potentiality of radiation as a tool for plant breeding at an early date must go to two Russians. In 1931, DELAUNAY published the results of induced mutations in wheat. By 1934, he selected a number of valuable mutations. SAPEHIN's (1935) results led him to state that artificial mutations were becoming valuable as a method in plant breeding. Theoretical aspects of utilizing mutations in German plant breeding programs were investigated by ▶ STUBBE, SCHICK, ▶ KUCKUCK, ▶ LEIN, GAUL, ▶ FREISLEBEN, and others (▶ HOFFMANN 1951).

After the discovery of the first chemical (ethylorethan), acting as mutagen in *Oenothera* by OEHLKERS (1943), a second wave of mutation research spilled through the world. Different chemicals were identified as mutagens, such as sodium azide, ethyl methanesulfonate, DMSO, MMS, and MNH. Examples of crop varieties that were produced via mutation breeding are given in Table 3.4. Despite those mutations, which are considered as so-called factorial, gene, or point mutations, mutations in a broad sense also include chromosome, genome, and plasmon or plastidome mutations.

Mutation breeding was particularly popular around the world during the 1970s. Although interest has waned somewhat in recent years, occasional varieties continue to be produced using these methods. For example, the new herbicide-resistant wheat variety "Above" was developed using exposure to sodium azide. Mutation breeding efforts continue around the world today. Modern mutation research is focused on *site-directed mutagenesis*, that is, introducing specific base pair mutations into a gene to make a protein that differs slightly in its structure from the native type. Computer-aided site-directed mutagenesis is already used in protein engineering. For example, major pollen allergen genes of ryegrass and rice were recently identified and cloned in Australia. These genes are being used to develop new diagnostic reagents and allergy therapeutics by site-directed mutagenesis

TABLE 3.4

A Small Selection of Crop Varieties of Different Genera and Species Bred by Induced Mutation Programs and Application of Different Mutagens

Legumes	*Cajanus cajan* (pigeon pea)	"Co 3," "Co 5," India, 1977, 1984
	Cicer arietinum (chickpea)	"Binasola-2," "Binasola-3," "Binasola-4," Bangladesh, 1998, 2001, 2001; "BGM 547," India, 2005; "CM 2000," "CM 72," "CM-2008," "CM-88," "CM-98," Pakistan, 2000, 1983, 2008, 1994, 1998
	Glycine max (soybean)	"Cerag No.1," Algeria, 1979; "Bisser," "Boriana," Bulgaria, 1984, 1981; "Anji 2," "Beinong 103," China, 1989, 2009; "Botula I," "Botula II," Congo, 1972, 1972; "Birsa-1," India, 1983; "Akita-midori," "Aomori-Fukumaru," "Aomori-Toyomaru," Japan, 2002, 2001, 2001; "Bangsa-Kong," Korea, 1985; "Albisoara," "Amelina," "Clavera," Moldava, 2010, 2010, 2010; "Dioskuriye," Russia, 1980; "Aida," Slovakia, 1984; "Arkadiya Odesskaya," "Chudo Gruzii 74," Russia, 1986, 1974; "Chiang Mai 5," "Doi kham," Thailand, 2006, 1986
	Lathyrus sativus (grasspea)	"Binakhesari-1," Bangladesh, 2001; "Bogdan," Moldova, 2008
	Lens culinaris (lentil)	"Binamasur-1," "Binamasur-2," "Binamasur-3," Bangladesh, 2001, 2005, 2005; "Djudje," Bulgaria, 2000; "Aurie," Moldova, 2008
	Pisum sativum (marrow pea)	"Caoyuan 10," China, 1980; "Agra," "Bosman," "Diament," Poland, 1994, 1989, 1989; "Bitug," Russia, 1990
	Phaseolus vulgaris	"Campeiro," "CAP-1070," Brazil, 2003, 1986; "Centralia," Canada, 1988; "Alfa," Slovakia, 1972; "CIAT 899," Tunisia, 2007; "Albion," "Black Magic," "Blackhawk," "C-20," "Domino," USA, ?, 1987, 1990, 1982, 1987
	Vicia faba (broad bean)	"Babylon," Iraq, 1994; "Bronto," "Dino," Poland, 1989, 1987; "Chabanskii," Russia, 1985
	Vigna angularis (azuki bean)	"Beni-nambu," Japan, 1978
	Vigna mungo (black gram)	"Binamash-1," Bangladesh, 1994; "Co 4," India, 1978
	Vigna radiata (mungbean)	"Binamoog-1," "Binamoog-2," "Binamoog-3," "Binamoog-4," "Binamoog-5," "Binamoog-6," "Binamoog-7," Bangladesh, 1992, 1994, 1997, 1997, 1998, 2005, 2005; "BM 4," "Co 4," "Dhauli," India, 1992, 1982, 1979; "Camar," Indonesia, 1987; "AEM-96," Pakistan, 1998; "Chai Nat 72," Thailand, 1999
	Vigna unguiculata (cowbean)	"COCP 702," "Co 5," "Cowpea-88," India, 2002, 1984, 1990
Cereals	*Amaranthus* ssp. (amarant)	"Centenario," Peru, 2006
	Avena sativa (oat)	"Dolphin," Australia, 1984; "AC Ronald," Canada, 2000; "Belozernii," Russia, 1979; "Alamo-X," "Bates," "Bay," "Belle," "Bob," "Centennial," USA, 1961, 1977, 1995, 1977, 1987
	Fagopyrum sagittatum (buckwheat)	"Dewa-kaori," Japan, 1996 "Aelita," "Aromat," "Chernoplodnaya," Russia, 1978, 1985, 1980

(Continued)

TABLE 3.4 (Continued)
A Small Selection of Crop Varieties of Different Genera and Species Bred by Induced Mutation Programs and Application of Different Mutagens

Cereals (Continued)		
	Hordeum vulgare (barley)	"Alpina," "Amalia," "Berolina," "Berta," "Carmen," Austria, 1995, 1988, 1982, 1982, 1986; "Dinky," Belgium, 1987; "Diana," Bulgaria, 1983; "AC-Albright," "AC-Stacey," "Atlanta," Canada, 1993, 1995, 1977; "7938," China, 1984; "Ametyst," "Atlas," "Diamant," Czech Republic, 1972, 1976, 1965; "Alf," "Alis," "Anker," "Anna Abed," "Carnen," "Camir," "Canor," "Canut," "Carula," "Catrin," Denmark, 1978, 1985, 1986, 1979, 1989, 1985, 1985, 1988, 1989, 1985; "Anni," Estonia, 1993; "Balder J.," Finland, 1960; "Baraka," "Betina," "Cargine," France, 1986, 1970, 1986; "Acclaim," "Alexis," "Allasch," "Alondra," "Amazone," "Amei," "Annabell," "Arena," "Beate," "Bitrana," "Brenda," "Chariot," "Cheri," "Comtesse," "Consista," "Cork," "Cruiser," "Defla," "Dera," "Derkado," Germany, 1984, 1986, 1963, 1994, 1986, 1966, 1999, 1983, 1984, 1990, 1995, 1994, 1987, 1988, 1979, 1995, 2001, 1984, 1984, 1987, 1982, 1987; "BH-75," "DL-253," India, 1983, 1981; "Amil," "Baraka," Iraq, 1994, 1994; "Amagi Nijo 1," Japan, 1971; "Centenario," Peru, 2006; "BPI Ri-10," "BPI-121-407," Philippines, 1983, 1971; "Accord," "Araraty 7," "Bastion," "BIOS-1," "Debut," "Dobrynia-3," Russia, 1987, 1983, 1992, 1993, 1982, 2001; "Bonus," "Denar," "Diabas," Slovakia, 1984, 1969, 1977; "Ariel," Sweden, 1988; "Akdeniz M-Q-54," Turkey, 1998; "Ayr," "Bacchus," "Beauty," "Blenheim," "Camargue," "Carnival," "Corgi," "Corniche," "Cromarty," "Donan," UK, 1986, 1981, 1983, 1987, 1986, 1981, 1985, 1985, 1983, 1983; "Badjory," Ukraine, 1999; "Advance," "Blazer," "Bonneville 70," "Boyer," "Dawn," USA, 1979, 1974, 1969, 1975, 1975
	Oryza sativa (rice)	"Binadhan 4," "Binadhan 5," "Binadhan 6," "Binadhan-7," "Binadhan-9," "Binasail," Bangladesh, 1998, 1998, 2007, 2012, 1987; "1870," "202," "240," "Iiyou 3027," "Iiyou623," "Iiyou673," "Iiyou802," "Iiyou 949," "Iiyouhang 2," "652," "69–280," "7404," "7738," "Aifu 9," "Ailiutiaohong," "Baifengyou 48," "Baofu 766," "B-fu 1," "CHANGGwanxian," "CHANGyouzao 1," "Chenzao 5," "Chiyou 65," "Chiyou S162," "Dongting 3," China, 1984, 1973, 1980, 2000, 2007, 2008, 1996, 2001, 2007, 1979, 1969, 1977, 1980, 1966, 1989, 2007, 1988, 1982, 1992, 1995, 1979, 2006, 2005, 1976; "Camago-8," Costa Rica, 1996 "Arlatan," "Calendal," "Delta," France, 1979, 1979, 1970; "ADT-41," "Anashwara," "Au-1," "Biraj," "CNM 20," "CNM 25," "CNM 31," "CNM 6 (Lakshmi)," "CRM 2007-1," "CRM 49," "CRM 51," "CRM 53," "Dhanu," "Dharitri," India, 1994, 2006, 1976, 1982, 1980, 1979, 1978, 1982, 2006, 1999, 1999, 2000, 1989; "Atomita 1," "Atomita 2," "Atomita 3," "Atomita 4," "Bestari," "Cilosari," "Danau atas," "Diah Suci," Indonesia, 1982, 1983, 1990, 1991, 2008, 1996, 1988, 2003; "Amber-Baghdad," "Amber-Furat," "Amber-Manathera," Iraq, 1994, 1995, 1995; "Aichinokaori," "Aichi-no-kaori SBL," "Akane-fuji," "Akichikara," "Aki-geshiki," "Akigumo," "Akihikari," "Akineiro," "Aki-no-sei," "Aki-roman," "Akisayaka," "Aneko-mochi," "Asahi-no-yume," "Asamurasaki," "Asa-tsuyu," "Awaminori," "Aya," "Banbanzai," "Bekoaoba," "Benigoromo," "Benika," "Benisarasa," "Chiho-no-kaori," "Chiyo-no-mochi," "Churahikari," "Chuukan-bohon Nou-13," "Chuukan-bohon Nou-14," "Daichi-no-kaze," "Daisenminori," "Dewahikari," "Dewanosato," "Dewasansan," "Doman-naka," "Dontokoi," Japan, 1988, 2003, 1997, 1987, 2000, 2004, 1976, 2005, 2000, 1996, 2002, 1992, 2000, 1998, 2001, 2002, 1992, 2001, 2005, 2002, 2004, 2002, 2003, 1991, 1991, 2002, 1988, 1996, 2004, 1997, 1993, 1997; "Azmil Mutant," "Bengawan Mutant," Philippines, 1976, 1984; "Dalris 11," Russia, 1988; "A201," "Aromatic se," "Calamylow-201," "Calhikari-201," "Calmati-201," "Calmochi 201," "Calmochi 202," "Calmochi-101," "Calpearl," "Calrose 76," "Dellmont." USA, 1997, 2004, 2006, 2001, 2001, 1979, 1981, 1986, 1981, 1977, 1993; "6B," "A20," "CM1," "CM6," "DB 250," "DB-2," "DCM-1," Vietnam, 1986, 1990, 1994, 1999, 1987, 1988

(Continued)

TABLE 3.4 (Continued)
A Small Selection of Crop Varieties of Different Genera and Species Bred by Induced Mutation Programs and Application of Different Mutagens

Cereals		
(Continued)	Panicum miliaceum (millet)	"Cheget," Russia, 1993
	Secale cereale (rye)	"Donar," Germany, 1981
	Sesamum indicum (sesame)	"Binatil-1," "Bangladesh, 2004; "Cairo White 8," Egypt, 1992; "Babil," Iraq 1992; "Ansanggae," Korea, 1984; "ANK-S2," Sri Lanka, 1995; "Birkan," Turkey, 2011
	Setaria sp. (millet)	"Angu 221," "CHANGwei 74," "CHANGwei 75," "Chigu 4," China, 1978, 1974, 1975, 1987
	Sorghum bicolor (sorghum)	"Co 21," India, 1977; "Djeman," "Djemanin," Mali, 1998, 1998
	Sorghum sudanense (sorghum)	"Donetskaya 5," Russia, 1982
	Triticum aestivum (bread wheat)	"BR4," Brazil, 1979; "Altimir 67," Bulgaria, 1979; "092," "1161," "352," "503," "62–8," "62–10," "77–L15," "78A," "Baichun 5," "CHANGwei 19," "CHANGwei 20," "CHANGwei 51503;" "Chuanfu 1," "Chuanfu 2," "Chuanfu 3," "Chuanfu 4," "Chuanfu 5," China, 1966, 1966, 1983, 1975, 1985, 1985, 1983, 1986, 2000, 1978, 179, 1983, 1982, 1989, 1989, 1993, 2002; "Carolina," Chile, 1981; "Claudia (=Mv8)," Italy, 1979; "Akebono-mochi," Japan, 2000; "Bajio Plus," "Centauro," Mexico, ?,?; "Darkhan-106," "Darkhan-35," "Darkhan-49," Mongolia, 2004, 1994, 1997; "Bakhtawar-92," Pakistan, 1994; "Albidum 12;" "Belchanka," "Birlik," "Deda," "Dnestryanka," Russia, 1984, 1992, 1989, 1983, 1989; "Above," USA, ?
	Triticum turgidum ssp. durum	"Arpad," "Attila," Austria, 1987, 1980; "Beloslava," Bulgaria, 1997; "Cargidurox," France, 1981; "Augusto," "Castel del Monte," "Castelfusano," "Castelnuovo," "Castelporziano," "Creso," Italy, 1976, 1969, 1969, 1971, 1968, 1974
	Zea mays (corn)	"CHANGdan 3," China, 1985; "CR MB/3b," Congo, 1972: "De 2205 SC," Hungary, 1987; "Collectivnii 100 TV (H)," "Collectivnii 210 (H)," "Collectivnii 210 ATV," "Collectivnii 225 MV (H)," "Collectivnii 244 MV (H)," "Collectivnii 95 M (H)," "Collectivnii 100SV," Russia, 1988, 1982, 1984, 1994, 1986, 1992, 1988; "CE 200," "CE 268," "CE 330," Slovakia, 1979, 1979, 1979
Vegetables	Abelmoschus esculentus (okra)	"Anjitha," India, 2006
	Allium cepa (onion)	"Brunette," "Compas," Netherlands, 1973, 1970
	Brassica campestris (cabbage)	"Binasarisha-6," Bangladesh, 2002
	Brassica pekinensis (Chinese cabbage)	"Baicai 9," China, 1978
	Brassica juncea (oriental mustard)	"Agrani," Bangladesh, 1991
	Colocasia esculenta (taro)	"Chiba-maru," Japan, 2007
	Capsicum annuum (green pepper)	"Albena," Bulgaria, 1976
	Cucumis sativus (cucumber)	"Altaj," Russia, 1981
	Curcuma domestica (turmeric)	"BSR 1," "Co 1," India, 1986, 1983

(Continued)

TABLE 3.4 (Continued)

A Small Selection of Crop Varieties of Different Genera and Species Bred by Induced Mutation Programs and Application of Different Mutagens

Vegetables (Continued)		
Dolichos lablab (hyacinth bean)	"Co 10," India, 1983	
Lactuca sativa	"Blush," USA, 1992	
Lycium chinense (boxthorn)	"Bullo," "Cheondae," "Cheongyang," Korea, 2001, 2001, 1998	
Lycopersicon esculentum (tomato)	"Bahar," "Binatomato-2," "Binatomato-3," Bangladesh, 1992, 1997, 1997; "Co.3," India, 1977	
Nelumbo nucifera (lotus)	"Dandinyuge," "Dianezhuang," China, 1997, 1983	
Phaseolus vulgaris (common brean)	"AC Hensall," "AC Skipper," Canada, 1997, 1996; "Arapaho," USA, 1995	
Solanum melongena (eggplant)	"Daijiro," Japan, 2004	
Solanum tuberosa (potato)	"Desital," Italy, 1987	
Ipomoea batatas	"91-C3-15," China, 1999	
Wasabia japonica (wasabi)	"Amagi-nishiki," Japan 1998	
Oil crops	*Arachis hypogaea* (groundnut)	"Colorado Irradiado," Argentina, 1972; "Binachinabadam-1," "Binachinabadam-2," "Binachinabadam-3," Bangladesh, 2000, 2000, 2008; "78961," "8130," "CHANGhua 4," China, 1988, 1988, 1972; "BG 1," "BG 2," "Co 2," India, 1979, 1979, 1984; "ANK-G1(Tissa)," Sri Lanka, 1995; "B 5000," Vietnam, 1985
	Brassica napus	"Binasarisha-5," "Binasharisha-3," "Binasharisha-4," Bangladesh, 2002, 1997, 1997; "Abasin-95," Pakistan, 1995
	Linum usitatissimum (flax)	"Baltyuchai," Russia, 1991
	Papaver somniferum (opium poppy)	"BC-28/9/4 (Vivek)," India, 1992
	Helianthus annuus	2
	Sinapis alba	5
	Ricinus communis (castor bean)	"Aruna," India, 1969
	Cymbopogon winterianus (citronella)	"Bhanumati (OJC-11)," "Bibhuti (OJC-5)," India, 1987, 1987
	Olea europaea	1

(Continued)

TABLE 3.4 (Continued)

A Small Selection of Crop Varieties of Different Genera and Species Bred by Induced Mutation Programs and Application of Different Mutagens

Ornamental plants	
Achimenes sp. (achimenes)	"Compact Arnold," "Cupido," Netherlands, 1971, 1973
Alstroemeria sp. (alstroemeria)	"Audino," "Chimbotina," Germany, 1979, 1981; "Appelbloesem," "Atlas," "Canaria," "Capitol," Netherlands, 1980, 1984, 1970, 1977
Antirrhinum sp. (snapdragon)	"Antirrhinum Juliva," Germany, 1961; "Bright Butterflies," USA, 1966
Begonia sp. (begonia)	"Big-Cross," Japan, 1976; "Aphrodite Joy," "Aphrodite Peach," "Aphrodite Twinkles," USA, 1974, 1974, 1974
Bougainvillea sp.	"Abhimanyu," "Arjuna," India, 2010, 1976
Canna indica (canna lilies)	"Caixiao," "Caixui," China, 1986, 1986; "Cream Prapanpong," Thailand, 1999
Chrysanthemum sp.	"Copper Marcon," "Dark Red Marconi," "Dark Torino," "Belgium, 1985, 1985, 1985; "Angshoujingshi," "Baiyunyong," "Chongyangshaoyao," "Chuntao," "China, 1989, 1989, 1991; "Babette Gelb," "Bronce Kalinka," "Dark Gaby," "Dark Mario," Germany, 1988, 1987, 1988, 1983; "Agnisikha," "Alankar," "Anamika," "Aruna," "Asha," "Ashankit," "Basant," "Basanti," "Colchi Bahar," "Cosmonaut," India, 1987, 1982, 1975, 1969, 1975, 1974, 1975, 1979, 1985, 1984; "Amason," "Arajin," "Baiogiku rainbow orange," "Baiogiku rainbow peach," "Baiogiku rainbow pink," "Baiogiku rainbow red," "Baiogiku rainbow white," "Baiogiku rainbow yellow," "Dipu Sei Roza," Japan, 1998, 2006, 1985, 1985, 1985, 1985, 1985, 1997; "ARTIpurple," Korea, 2011; "Amber Boston," "Apricot Deholta," "Apricot Impala," "Blue Redemine," "Blue Redemine," "Blue Star," "Blue Winner," "Bright Lameet," "Bright Star," "Bright Westland," "Bronze Byoux," "Bronze Charmette," "Bronze Clinspy," "Bronze Miros," "Bronze Redemine," "Bronze Star," "Bronze Westland," "Bronze Winner," "Cherry Deholta," "Coral Refla," "Coral Winner," "Cream Clingo," "Cream Deholta," "Cream Impala," "Danny Boy," "Danny's Cape," "Danny's Pearl," "Dark Charmette," "Dark Deep Tuneful," "Dark Lymon," "Dark Miros," "Dark Oriette," "Dark Westland," "Dark/Royal Rendez-Vous," Netherlands, 1978, 1983, 1984, 1984, 1977, 1975, 1978, 1977, 1976, 1985, 1976, 1978, 1979, 1986, 1977, 1976, 1975, 1985, 1986, 1975, 1979, 1985, 1984, 1973, 1973, 1976, 1969, 1985, 1979, 1976, 1976, 1986; "Dalekaya zoezda," Russia, 1976
Chrysanthemum morifolium	"Batik," India, 1994
Cordyline fruticosa (ti plant)	"Afable," "Medina," Philippines, 2009, 2005
Cymbidium sp. (boat orchids)	"Cocktail Dress," Japan, 1997; "Dong-i," Korea, 2005
Dendranthema grandiflorum (chrysanthemum)	"Cristiane," Brazil, 1995; "ARTIqueen," Korea, 2011
Dahlia sp.	"Adagio," "Allegro," "Altamira," "Amalfi," "Annibal," France, 1970, 1970, 1970, 1970, 1970; "Bichitra," "Black Beauty," India, 1978, 1978; "Autumn Harmony," Netherlands, 1967
Dianthus caryophyllus	"Bonitas," "Dione," Germany, 1985, 1977; "Boh-red," "Dark Pink Vital Ion," Japan, 2005, 2005; "Accent," "Cerise Kortina," Netherlands, 1982, 1985; "Chaichoompon," Thailand, 1983

(Continued)

TABLE 3.4 (Continued)

A Small Selection of Crop Varieties of Different Genera and Species Bred by Induced Mutation Programs and Application of Different Mutagens

Ornamental plants (Continued)		
Euphorbia fulgens (scarlet plume)	"Albora," Netherlands, 1976	
Forsythia x intermedia (forsythia)	"Courtadic," "Courtalyn," France, 1984, 1984	
Gypsophyla elegans (annual baby's-breath)	"Ayami," "Buranka," Japan, 1999, 2002	
Hibiscus sp.	"Anjali," India, 1987; "Baekseol," "CHANGhae," "Daegoang," Korea, 1999, 2006, 2004	
Hoya carnosa (hoya)	"Compacta," "Compacta Regalis," USA, 1980, 1980	
Iris sp.	"Belyi Karlik," "Chistoe Pole," Russia, 1984, 1984	
Juncus effusus (mat rush)	"Chikugo-midori," Japan, 2001	
Lagerstroemia indica (crapemyrtle)	"Centennial Spirit," USA, ?	
Lilium sp. (lily)	"Coral Bouquet," Japan, 1999	
Limonium sp. (statice)	"Daifura Lavender," "Daifura Pink Super," "Dai-lady Rose," "Dai-lady White," Japan, 1999, 2000, 2000, 2001	
Pelargonium sp. (geranium)	"Capli Ice," "Capli Luluby," Japan, 1988, 1986	
Pelargonium grandiflorum (geranium)	"Dark Mozart," Germany, 1988	
Portulaca grandiflora (portulaca, moss rose)	"Chompoo Praparat," Thailand, 2000	
Rosa sp. (rose)	"Beijingzhichun," "Beiyumudan," "Binghua," "Caiyemingxin," "Chuanxiu 1," "Chuanxiu 2," "Chuanxiu 3," "Chuanxiu 4," "Chuanxiu 5," "Chuanxiu 6," "Chuanxiu 7," "Chunyanqifei," China, 1990, 1986, 1986, 1990, 1986, 1990, 1990, 1990, 1990, 1990, 1989; "Desi," Germany, 1965; "Abhisarika H.T., Angara," "Curio," India, 1975, 1975, 1986; "Banbina," "Bridal Sonia," Japan, 1995, 1985	
Streptocarpus sp.	"Aurora (=Neptun rot)," "Blue Windor," "Burgund," "Dark Windor," "Dolly," Germany, 1979, 1986, 1978, 1987, 1979; "Albatros," "Blue Nymph," "Cobalt Nymph," Netherlands, 1973, 1969, 1969	
Tibouchina organensis (glory bush)	"Caribbean Princess," Germany, 2000	
Tulipa sp. (tulip)	"Dominique," Netherlands, 1985; "Denj Pobedy," Russia, 1993	
Rhododendron sp. (azalea)	"Adinda," Belgium, 1972; "Cobalt," Japan, 1973; "Aleida," Netherlands, 1978	
Weigela sp. (weigela)	"Couleur d' Automne Courtatum," "Courtadur," France, 1979, 1980	

(Continued)

TABLE 3.4 (Continued)

A Small Selection of Crop Varieties of Different Genera and Species Bred by Induced Mutation Programs and Application of Different Mutagens

Fruits and fruit trees	*Citrus clementina* (clementine)	"Clemenverd," Spain, 2010
	Ficus carica (fig)	"Bol (Abundant)," Russia, 1979
	Fragaria × ananassa (strawberry)	"Akita Berry," "Anteher," Japan, 1992, 1994
	Malus pumila (apple)	"Donghenghongpingguo," China, 1987; "Belrene," "Blackjoin BA 2 520," "Courtagold," "Courtavel," France, 1970, 1970, 1972, 1972
	Musa sp.	"Al-Beely," Sudan 1992
	Olea europaea (olive)	"Briscola," Italy, 1981
	Prunus avium (cherry)	"Compact Lambert," "Compact Stella," Canada, 1964, 1974; "Burlat C1," Italy, 1983; "Aldamla," "Burak," Turkey, 2014, 2014
	Pyrus communis (pear)	"Chaofu 1," "Chaofu 10," "Chaofu 11," "Chaofu 2," China, 1989, 1989, 1989, 1989
	Ribes nigrum (black currant)	"Burga," France, 1979
	Rubus idaeus (raspberry)	"Colocolchnik," Russia, 1991
	Ziziphus mauritiana (Indian jujube)	"Dao tien," Vietnam, 1986
Forage crops	*Agropyron cristatum* (crested wheatgrass)	"CD-II," USA, 1996
	Agrostis sp. (creeping bent grass)	"Chiba Green B-2," Japan, 1997
	Alopecurus pratensis (meadow foxtail)	"Alko," Germany, 1983
	Eremochloa ophiuroides (centipedegrass)	"AU Centennial," USA, 1983
	Lupinus albus (white lupin)	"Dnepr," Russia, 1979
	Lupinus angustifolius (blue lupin)	"Chittick," Australia, 1982; "Bar," Poland, 1991
	Lupinus luteus (yellow lupin)	"Aga," Poland, 1981

(Continued)

TABLE 3.4 (Continued)
A Small Selection of Crop Varieties of Different Genera and Species Bred by Induced Mutation Programs and Application of Different Mutagens

Forage crops (*Continued*)	*Trifolium alexandrinum* (Egyptian clover)	"BL–22," India, 1984
	Trifolium incarnatum (crimson clover)	"Cardinal," Slovakia, ?
Industrial crops	*Corchorus olitorius* (tossa jute)	"Atompat-28," "Atompat-36," "Atompat-38," "Binadeshipat-2," "Binapatshak-1," Bangladesh, 1974, 1974, 1974, 1997, 2003
	Gossypium ssp. (cotton)	"013," "Chuanpei 1," China, 1985, 1982; "Badnawar-1," India, 1961; "Chandi 95," Pakistan, 1995; "Agdash 3," Russia, 1993; "AN-20," "AN-401," "C-3585," "C-7503," "C-7510," Uzbekistsn, 1986, 1971, 1979, 1980, 1991
	Humulus lupulus (hop)	"Crystal," USA, 1993
	Nicotiana tabacum (tobacco)	"Delhi 76," Canada, 1976; "Chlorina F1," Indonesia; "Baghdad-V77," Iraq, 1995; "American 307," "American Bahchysarajskii B," Russia, 1981, 1979
	Ppopulus trichocarpa (black cottonwood)	"Donetskii Zolotoi," Russia, 1977
	Saccharum officinarum (sugarcane)	"CCe 10582," "CCe 183," "CCe 283," "CCe 483," Cuba, 1990, 1993, 1993, 1993; "Co 6608," "Co 8153," "Co 85017," "Co 85035," "Co 997," India, 1966, 1981, 1985, 1985, 1967

Source: FAO/IAEA, Vienna, 2017.

of adequate proteins. It became an approach of development of antiallergic drugs. Pollen grains of grasses including rice are the major cause of hayfever and seasonal allergic asthma.

By the advances of molecular technology and bioinformatics, the precision of mutation breeding can be increased. A variety of techniques have been applied over the past 20 years for detecting induced mutations in plant populations, including denaturing HPLC, LI-COR DNA analysis, and high-resolution melting (HRM) analysis. As the cost of DNA sequencing dropped and its accuracy improved, the potential of this approach for mutation discovery has been explored. One of the first demonstrations of the use of next-generation sequencing technology to discover induced and natural mutations was conducted by the plant biotechnology company "Keygene" (Wageningen, the Netherlands). Using a Roche 454 platform, mutations in the *eIF4E* gene of tomato were detected by sequencing PCR-amplified DNA from 3000 EMS-mutagenized lines.

TSAI et al. (2011) developed TILLING by sequencing (TbyS) to identify induced mutations in crop species using an Illumina sequencing platform. In this application, a pool of PCR products is barcoded[8] with a unique DNA adapter sequence that facilitates mutant line identification in multidimensional pools. Bioinformatics is used to evaluate the frequency, sequencing quality, intersection pattern in pools, and statistical relevance of nucleotide changes. A comparison of mutation discovery by Illumina sequencing and traditional TILLING (DNA mismatch detection) was conducted. The level of sensitivity of the two approaches was found to be similar finding mutations in oilseed rape, but sequencing was more expensive.

Whole-genome sequencing is an alternative strategy being developed for the identification of induced mutations. Although it is currently cost prohibitive to carry out with large populations, whole-genome sequencing is already an option for mutation detection in the future. However, a real breakthrough in mutation research appeared during the recent years. Even a dream became true when plant scientists are inducing targeted point mutations (cf. Section 3.5.1.1). TbyS refers to the application of high-throughput sequencing technologies to mutagenized TILLING populations as a tool for functional genomics. TbyS can be used to identify and characterize induced variation in genes (controlling traits of interest) within large mutant populations, and is a powerful approach for the study and harnessing of genetic variation in crop breeding programs. The extension of existing TILLING platforms by TbyS will accelerate crop functional genomics studies, in concert with the rapid increase in genome editing capabilities and the number and quality of sequenced crop plant genomes.

Rice is probably the most amenable cereal crop when it comes to CRISPR/Cas9 (Section 4.2.2). It can generate mutations at target sites at nearly 100% efficiency whereas efficiency in wheat plants ranges from 1% to 7.5%. A CRISPR/Cas9 mutagenised rice line with enhanced blast resistance was recently released (Wang et al. 2016). Researchers based in Guangzhou (China) were able to target four genes known as regulators of grain number, panicle architecture, grain size, and plant architecture showing that different agronomic relevant traits can be quickly improved in a single cultivar. Scientists were also able to produce herbicide-resistant rice plants by CRISPR/Cas9-mediated homologous recombination. Finally, the use of an endonuclease alternative to Cas9 has been successfully explored: Cpf1 has been exploited in rice not only to mutagenize desired sites, but also as a transcriptional regulator (Sun et al. 2016). The first CRISPR/Cas9 mutagenised wheat plants were obtained in 2014 by researchers at the Chinese Academy of Sciences (Wang et al. 2014).

To date, 2337 mutation breeding varieties were officially released; almost 50% have been released during the last 30 years. For comprehensive information on mutant varieties, the database of the International Atomic Energy Agency (http://www-infocris.iaea.org), Vienna (Austria) is available.

A special experiment of utilizing mutants was started during the latter half of the 1920s. Considering the so-called "law of parallel mutations," the meritorious geneticist ▶ E. BAUR (1922) was convinced to select a sweet lupine among the wild types. An important prerequisite was a method for rapid

---

[8] DNA barcoding uses specific regions of DNA to identify species.

screening of bitter alkaloids in single seeds. In 1928, his collaborator ▶ R. von SENGBUSCH (1930) was able to screen 1,500,000 individuals; among them, he selected three showing low alkaloid content. In 1931, sweet lupine was grown on 2 ha; during the next seven years up to 100,000 ha.

Already C. DARWIN (1868) observed that in many tree species belonging to different genera frequently similar growth habits, such as pyramidal or pendulous, could be developed. He called this apparent phenomenon "analogous or parallel variation." He also tried to give quite a modern explanation:

> We can hardly account for the appearance of so many unusual characters by reversion to a single ancient form; but we must believe that all the members of the family have inherited a nearly similar constitution from an early progenitor. Our cereal and many other plants offer similar cases.

The Russian botanist ▶ N. I. VAVILOV assumed a general rule behind this. At the end of the 1920s, while he was working on his theory of centers of origin of crop plants (cf. Section 2.2), he also dealt with the "parallel variations" and summarized his ideas in the so-called "law of homologous series," which can be seen as a synonymous term to "parallel variation."

In the 1950s and 1960s, the U.S. government promoted the method as part of its "atoms for peace" program and had notable success. In 1960, disease heavily damaged the bean crop in Michigan— except for a promising new variety that had been made by radiation breeding. It and its offspring quickly replaced the old bean.

In the early 1970s, RUTGER et al. (1976) in Davis, California, fired gamma rays at rice. They found a semi-dwarf mutant that gave much higher yields, partly because it produced more grain. Its short size also meant it fell over less often, reducing spoilage. Known as "Calrose 76," it was released publicly in 1976. Recently, in Vietnam the IAEA, Vienna, has worked closely with local breeders to improve production of rice, a crop that accounts for nearly 70% of the public's food energy. One mutant had yields up to four times higher than its parent and grew well in acidic and saline soils, allowing farmers to use it in coastal regions, including the Mekong Delta. A team of 10 Vietnamese scientists reported in an agency journal, *Plant Mutation Reports*, that the nation had sown the new varieties across more than 1 million ha, or 3860 square miles. The new varieties, they added, have already produced remarkable economic and social impacts, contributing to poverty alleviation in some provinces.

A similar story unfolded in Texas. In 1929, farmers stumbled on the Ruby Red grapefruit, a natural mutant. Its flesh eventually faded to pink, however, and scientists fired radiation to produce mutants of deeper color—"Star Ruby," released in 1971, and "Rio Red," released in 1985. The mutant offspring now account for about 75% of all grapefruit grown in Texas.

Today, the process usually begins with cobalt-60, a highly radioactive material used in industrial radiography and medical radiotherapy. Its gamma rays, more energetic than X-rays, can travel many yards through the air and penetrate lead. Understandably, the exposure facilities for radiation breeding have layers of shielding. One runs small machines the size of water heaters that zap containers full of seeds, greenhouses that expose young plants, and special fields that radiate row upon row of mature plants.

In Japan, one circular field is more than 650 feet wide. A shielding dike some 20 meters high rises around its perimeter. In Japan, a rust fungus threatened the Japanese pear, a pear with the crisp texture characteristic of apples. But one irradiated tree had a branch that showed resistance. The Japanese cloned it, successfully started a new crop, and with the financial rewards paid for 30 years of research.

The payoff was even bigger in Europe, where breeders fired gamma rays at barley to produce "Golden Promise," a mutant variety with high yields and improved malting. After its debut in 1967, brewers in Ireland and Britain made it into premium beer and whiskey. It still finds wide use. The atomic agency had similar success in the Peruvian Andes, where some three million people live on subsistence farming. The region, nearly two miles high, has extremely harsh weather. But nine new varieties of barley improved harvests to the point that farmers had surplus crops to sell.

The atomic agency in Vienna has promoted the method since 1964 in outreach programs with the Food and Agriculture Organization of the United Nations, in Rome. Starting roughly a decade ago, for instance, the atomic agency helped plant pathologists fight a virus that was killing cocoa trees in Ghana, which produces about 15% of the world's chocolate. The virus was killing and crippling millions of trees. In the city of Accra on the Atlantic coast, at the laboratories of the Ghana Atomic Energy Commission, the scientists exposed cocoa plant buds to gamma rays. The mutants included one that endowed its offspring with better resistance to the killer virus. They planted the resistant variety on 25 farms across Ghana with no evidence of resurgence.

3.5.1.1 Point Mutation

The so-called gene cutter CRISPR/Cas9 is already longer hailed as a breakthrough in genetic research and gene therapy (cf. Section 4.2.2). During the years 2015–2016, U.S. researchers succeeded in modifying this tool so that it can even correct point mutations targeted. The gene cutter converts it to the faulty DNA base, without having to cut the DNA. In a first test, it already allowed the researchers trying to resolve two mutations in the APOE4 gene (KOMOR et al. 2016). They report the development of "base editing," a new approach to genome editing that enables the direct, irreversible conversion of one target DNA base into another in a programmable manner, without requiring dsDNA backbone cleavage or a donor template. Fusions of CRISPR/Cas9 and a cytidine deaminase enzyme were engineered that retain the ability to be programmed with a guide RNA, do not induce dsDNA breaks, and mediate the direct conversion of cytidine to uridine, thereby effecting a C→T (or G→A) substitution. The resulting "base editors" convert cytidines within a window of approximately five nucleotides, and can efficiently correct a variety of point mutations. In four transformed cell lines, second- and third-generation base editors that fuse uracil glycosylase inhibitor, and that use a Cas9 nickase targeting the nonedited strand, manipulate the cellular DNA repair response to favor desired base-editing outcomes, resulting in permanent correction of ~15%–75% of total cellular DNA with minimal indel[9] formation.

An example was published by a Swedish researcher: Professor Stefan JANSSON from Umeå University created a gene-edited cabbage that was fertile and even tasty (JANNSON 2016). Thus, base editing expands the scope and efficiency of genome editing of point mutations. Another example was reported by JUNG and AltPETER (2016). They deployed transcription activator-like (TAL) effector nuclease (TALEN) to induce mutations in a highly conserved region of the *caffeic acid O-methyltransferase* (*COMT*) of sugarcane. Capillary electrophoresis was validated by pyrosequencing as reliable and inexpensive high-throughput method for identification and quantitative characterization of TALEN-mediated mutations. Targeted *COMT* mutations were identified by capillary electrophoresis in up to 74% of the lines. In different events, 8%–99% of the wild-type *COMT* was converted to mutant *COMT* as revealed by pyrosequencing. Mutation frequencies among mutant lines were positively correlated to lignin reduction. Events with a mutation frequency of 99% displayed a 29%–32% reduction of the lignin content compared to nontransgenic controls along with significantly reduced S subunit content and elevated hemicellulose content. Soyk et al. (2016) induced by the same methodology variation in the flowering gene *SELF PRUNING 5G* (*SP5G*) that promotes day neutrality and early yield in tomato. Plants evolved so that their flowering is triggered by seasonal changes in day length.

However, day-length sensitivity in crops limits their geographical range of cultivation, and thus modification of the photoperiod response was critical for their domestication. They showed that loss of day-length-sensitive flowering in tomato is driven by the florigen paralog and flowering repressor *SP5G*. *SP5G* expression is brought by high levels during long days in wild species, but not in cultivated tomato because of *cis*-regulatory variation. CRISPR/Cas9-engineered mutations in *SP5G* cause rapid flowering and enhance the compact determinate growth habit of field tomatoes, resulting in a quick burst of flower production that translates to an early yield.

---

[9] Indel is a molecular biology term for the **in**sertion or the **del**etion of bases in the DNA of an organism.

3.5.2   SOMACLONAL VARIATION BY IN VITRO CULTURE

Another method for increasing the number of mutations in plants is tissue and/or cell culture (cf. Chapter 4). Tissue culture is a technique for growing cells, tissues, and whole plants on artificial nutrients under sterile conditions, often in small glass or plastic containers. Tissue culture was not developed with the intention of causing mutations, but the discovery that plant cells and tissues grown in tissue culture would mutate rapidly increased the range of methods available for mutation breeding.

Somaclonal variation was first defined as such by LARKIN and SCOWCROFT who reviewed the subject in 1981 and were among several authors at that time to draw attention to its potential use for crop improvement. For example, somaclonal variation was exploited in a novel way under a joint program of China and Australia to transfer from *Thinopyrum intermedium* resistance to barley yellow dwarf virus (BYDV) in wheat (*Triticum aestivum*). Single cell callus cultures from F_1 embryos rescued were initiated and induced to form plants showing somaclonal variation, which were then selected for BYDV resistance. Cytological analysis of the genotypes showing stable resistance revealed that chromosomal rearrangement of the chromosome carrying *Thinopyrum* introgressed segment had occurred during the tissue culture phase to confer the stability (BRETTELL et al. 1988).

Later, the Biotechnology Center at the Indian Agricultural Research Institute (IARI) has standardized the protocols of plant regeneration of *Brassica carinata* and is isolating somaclonal variants suitable for Indian conditions. Useful somaclonal variants for earliness, plant height, maturity, and so on have been induced in *B. juncea* and *B. napus*. In dicotyledonous species, EVANS and SHARP (1983) obtained several single gene mutations in tomato regenerated from tissue culture. Somaclonal variation in protoplast-derived plants has been reported in tobacco (LÖRZ and SCOWCROFT 1983), potato (SHEPARD et al. 1980), and tomato (SHAHIN and SPIVEY 1986). In monocotyledonous species, although it was generally difficult to regenerate plants from tissue and, later, protoplasts, plantlet regeneration from rice protoplasts was described early by FUJIMURA et al. (1985).

At the beginning of the 1960s, E. C. COCKING (University of Nottingham, UK) used an enzyme preparation for the degradation of cell walls and thus yielded single protoplasts from tomato root tips, though it took until 1968 before I. TAKEBE et al. (Institute for Plant Virus Research, Aobacho, Chiba/Japan) developed a technique for the production of large amounts of active protoplasts from mesophyll cells of *Nicotiana tabacum* that became standard. It spread fast due to its simple use, and within short time succeeded a number of laboratories in the yield of protoplasts from tissues of different plant species.

KYOZUKA et al. (1987) established a high-frequency plant regeneration system from rice protoplasts by novel nurse culture methods. "Hatsuyume" was the first rice variety developed by this procedure (SUKEKIYO et al. 1989). Among the progeny derived from "Koshihikari," one R_1 line was found to be superior to the original variety in several agronomic traits through R_1 and R_2 generation and selected. This line was registered as a new rice variety named in 1990.

An extensive number of reports soon followed in a wide range of species, indicating that somaclonal variation was widespread, and therefore accessible to all plant breeders (KARP 1991, 1995). The amount of references concerning somaclonal mutants and their utilization is big and permanently growing by new in vitro approaches.

The utilization of new genetic variability has become one of the major objectives of tissue culture. The assembly of genetic variability is vital for improvement of crop plants. Somaclones for the resistance of downy mildew, Fiji disease, and eye spot disease were identified. Application of somaclonal variation in crop improvement can produce increased yield and resistance to biotic as well as abiotic stresses. Some other crops where somaclones were produced are wheat, maize, tomato, geranium, sweet potato, sugarcane, celery, or brown mustard. To date, more than 1548 new, officially registered varieties of crop plants have been developed through somaclonal variation.

3.6 POLYPLOIDY AND BREEDING

E. STRASBURGER (1910) and H. WINKLER (1916) defined somatic cells and tissues as well as individuals as polyploid if they have three (triploid), four (tetraploid), five (pentaploid), or more complete chromosome sets instead of two as in diploids. WINKLER introduced the term "genome." The state of being polyploid is referred to as polyploidy and may arise spontaneously or is induced experimentally by mitotic poisons. Polyploidy is very common among the crop plants, for example, wheat, oats, flax, potato, sugarcane, cotton, coffee, banana, alfalfa, peanut, sweet potato, tobacco, plum, loganberry, or strawberry. A. LÖVE (1953) summarized that polyploids generally are more tolerant than diploids of climatic extremes at high elevation, even in Arctic.

Early in the twentieth century, related species with a common basic chromosome number were discovered in a number of plant species. This led to ▶ O. WINGE's hypothesis (1917) that polyploid species arose from hybrids between two diploid species where chromosome pairing either failed in the hybrid or was quite incomplete. T. H. GOODSPEED and J. CLAUSEN (1925) synthesized an artificial, yet new, fertile hexaploid species (*Nicotiana digluta*), which arose from a cross of diploid *N. glutinosa* with the tetraploid *N. tabacum*. When breeders learned in the 1920–1950s that several of their subjects are polyploid and often arose by either interspecific crossing or spontaneous doubling of chromosome sets, they recognized the chance of artificial resynthesis of polyploids as in tobacco, wheat, *Brassica*, or cotton and induction of new, larger, and vigorous polyploids than the naturally occurring diploids in the same groups. One of the earliest methods of obtaining polyploids was the twin method. In a low frequency of germinating seedlings, twin embryos are occasionally found, which give rise to heteroploid plants. ▶ A. MÜNTZING in Sweden (1937) was one of the first investigators to utilize this feature to obtain polyploids. L. F. RANDOLPH in the United States (1932) was able to induce chromosome doubling in maize by application temperature shocks to germinal tissue.

None of the forgoing procedures was very rewarding from the experimental point of view. It was not until the practical application of the colchicine technique that the way was paved for the production of polyploids in virtually unlimited numbers. It was proposed by A. F. BLAKESLEE and O. T. AVERY and by B. B. NEBEL (1937). Colchicine inhibits the spindle mechanism at mitosis and results in cells with double or more chromosome sets. It was recognized by A. P. DUSTIN in 1934. There are several techniques of application depending on the crop, tissue, cell, hybrid, and so on. Aneuploids (see Section 3.7.1) may occasionally arise after colchicine treatment because of anomalies in chromosome division and separation.

Natural autopolyploids are watermelon, strawberries, potato, and alfalfa, whereas from grape plants, rice, rape, einkorn, cotton, beets, buckwheat, clover, alfalfa, timothy gram, melons, apple, carnation, and many others artificial autopolyploids were produced. Doubling of crop plants offers more opportunity for effective recombination. However, selection is more complicate because of the tetra-allelic situation on one gene locus. Genetic stabilization of breeding populations can become extremely difficult. Moreover, the polysomic status leads to chromosomal irregularities during meiosis and mitosis (Figure 3.15). Aneuploids, seed shriveling, and partial sterility are the consequence. Therefore, vegetatively propagated crop plants are more suitable for autopolyploidization.

The two more or less universal effects of chromosomal doubling are increased cell size and decreased fertility. Consequently, crops that benefit most from increased cell size and suffer least from reduced fertility are inherently predisposed to benefit from polyploid breeding. Crops most amenable to improvement through chromosome doubling should

1. have a low chromosome number,
2. be harvested primarily for their vegetative parts, and
3. be cross-pollinating.

Two other conditions, the perennial habit and vegetative reproduction, have a bearing on the success of polyploid breeding by reducing a crop's dependence on seed production. In the 1950–1970s,

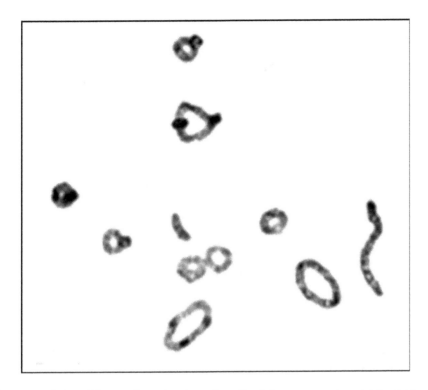

FIGURE 3.15 Meiotic pairing configuration (metaphase I) of chromosomes in autotetraploid rye (*Secale cereale*, $2n = 4x = 28$ chromosomes), showing one univalent, six ring bivalents, one chain trivalent, and three ring quadrivalents.

induced autopolyploidy in rye was considered to be an important breeding method. Rye showed good prerequisites for an autopolyploid crop. Russia, Poland, Germany, and Sweden paid considerable attention to the production of tetraploid rye. To overcome low fertility, seed shriveling, and aneuploid offspring—all three features were believed to be influenced by irregular meiotic chromosome pairing (see Figure 3.15)—two major approaches of chromosomal pairing regulation were investigated: The research group of H. REES at Aberystwyth (UK) (Hazarika and Rees 1967) favored an increased quadrivalent formation with convergent or parallel centromere co-orientation as a mean of reduced aneuploidy and consequently improved fertility. A "disjunction index" (number of pollen mother cells without univalents and trivalents divided by the total number of pollen mother cells) was used as a measure for meiotic pairing regularity.

They even demonstrated a positive correlation between the disjunction index and fertility.

Alternatively, J. SYBENGA at Wageningen (Netherlands, 1964) proposed a preferential bivalent pairing to reduce meiotic irregularities by induced allopolyploidization of an autotetraploid. Besides a genetic control of diploid-like chromosome pairing, a complex system of experimentally induced reciprocal translocations was intensively discussed. Despite strong efforts and tests over decades, no practical benefit could be achieved.

R. SCHLEGEL at Hohenthurm/Halle (Germany, 1976) introduced a method for mass production of autotetraploid rye by so-called valence crosses and utilization of gametes of first division restitution. Under microplot isolators, nonemasculated tetraploid genotypes (clonal plants) were crossed by spontaneous pollination with diploid genotypes (clonal plants). The tetraploids used as mother plants showed a recessive pale grain character, whereas the diploids used as male plants showed dominantly green seeds. In this way, green xenia could be selected among the pale grains after harvest of the mother plants. All the green xenia must be hybrids, either triploid or tetraploid.

Microscopic chromosome counting can differentiate them. The triploids result from fusion of a diploid female gamete and a haploid male gamete, whereas the tetraploids are derived from a diploid female and an unreduced male gamete (SCHLEGEL and METTIN 1975). In this way, diploid genotypes showing high chiasma frequencies per PMC (>17 chiasmata/pollen mother cell) and other useful agronomic characters could be transferred into the so-called half-meiotic tetraploids. The method proved to be efficient for broadening genetic variability of tetraploid breeding material without deleterious effects of colchicine treatment.

To date, only few crop varieties were released showing doubled genomes: tetraploid rye (*Secale cereale*, variety "Petkuser Tetraroggen," Germany, "Belta," Russia), tetraploid clover (*Trifolium pratense*, varieties "Marino," "Parenta"), tetraploid turnip (*Brassica rapa*, variety "Svalöf's Sirius"), triploid sugarbeet, triploid watermelons, triploid banana, and some ornamental plants.

Although most breeders cross-pollinate plants of a single species, some breeding methods rely on crosses that can be made between two species within the same genus. A cross between *Musa acuminata* and *M. balbisiana*, both members of the genus *Musa*, produced the bananas with which we are familiar, for example, the triploid banana, variety "Gros Michel." Less commonly, the cross is between members of two different genera.

Having only a single genome type and exhibiting multisomic inheritance, autotetraploids do not have intergenome heterozygosity. However, it may be possible to combine or pyramid blocks of genes containing diverse alleles into a single polyploid line, with the goal being to maximize allelic diversity. For example, CHASE (1964) proposed an "analytic breeding" method for autotetraploid potato, where improvements could be made at the diploid level and then transferred to the tetraploid level. Ironically, because autopolyploidy was thought to be maladaptive, the most dramatic example of increased heterozygosity is in autotetraploid alfalfa, where a single locus can potentially have up to four different alleles. BINGHAM et al. (1994) demonstrated that maximum heterozygosity was obtained after intermating double-cross progeny for one or more additional generations beyond the final F_1 line, resulting in a phenomenon they termed "progressive heterosis." Similar strategies to maximize allele diversity first at the diploid level have also been developed to improve banana and sweet potato.

The successful induction of allopolyploids as new crops either by recombining characteristics of separate species or by synthesizing new hybrids may often require more time and effort on the breeder's part. It is a truism that allelic copy number increases with ploidy level, potentially leading to novel phenotypes via dosage effects. Allelic diversity also increases during allopolyploidy, when two (or more) divergent genomes become joined in a common nucleus. This intergenomic heterozygosity will apply not only to single loci but to the entire genome, and hence to specific chromosome blocks of possible interest. For example, intergenomic heterozygosity has been shown to have positive effects on oilseed production in *B. napus*. OSBORN et al. (2003) found lower oilseed yields associated with a loss of intergenomic heterozygosity when recombinants of homoeologous recombination were evaluated alongside lines containing the parental chromosomal configurations. Even novel phenotypic variation is known to accompany polyploidization. In synthetic allotetraploid *Brassica*, for example, significant *de novo* variation was found for flowering time and for several life history traits (SCHRANZ and OSBORN, 2004).

Another example of an allopolyploid crop is triticale, an allopolyploid hybrid with the genome formula $2n = 8x = 56$ (AABBDDRR) derived from crosses of hexaploid wheat (*Triticum aestivum*, $2n = 6x = 42$, AABBDD) × diploid rye (*Secale cereale*, $2n = 2x = 14$, RR). Its history can be traced back to the end of the nineteenth century when a first (sterile) hybrid was reported (Chapter 3). About 100 years later, first (hexaploid) triticale varieties became successful in agriculture, although the optimum genome formula turned out as $2x = 6x = 42$ (AABBRR). It is claimed as the first man-made crop plant (SCHLEGEL 1996). Induced allopolyploids and autopolyploids were later used for successful transfers of useful characters between species and genera (see Section 3.7).

In conclusion, polyploidy provides genome buffering, increased allelic diversity and heterozygosity, and permits novel phenotypic variation to be generated. Polyploid formation is often accompanied with loss of duplicated chromatin, changes in gene expression, novel epistatic interactions

(Table 3.7), and endosperm effects. All of these factors need be considered in a genome-wide context for optimizing marker-assisted selection and further crop plant improvement (UDALL and WENDEL 2006). Polyploid-induced changes can generate individuals that are able to exploit new niches or to outcompete progenitor species. This process has been a major driving force behind the divergence of the angiosperms and their biodiversity, and thus for crop improvement as well.

3.7 CHROMOSOME MANIPULATIONS

Among many important advances that were made in the last century, it was cytogenetics. BOVERI (1902) and SUTTON (1903) both studied the cytology of meiosis and independently showed that chromosome behavior mimics MENDEL's law. They concluded that chromosomes must be the carriers of genetic information, which we now know are genes, and fully established the chromosomal basis of Mendelian inheritance. The classical time of cytogenetics (1900–1949) is characterized by the discovery and/or invention of DNA structure by WATSON and CRICK (1953); one began to know that DNA is organized into chromosomes. It became clear during the 1930–1940s that chromosome behavior and transmission is under genetic control as other morphological traits too. As ▶ M. M. RHOADES (1955) has stated:

> It is evident that genes not only determine the development of the plant as a whole but that the chromosomes themselves, which carry the genes, are also under genic control.

Even the stability of the karyotype is under genetic control in sexual reproduction of plants. This chromosome stability has a profound effect upon the regularity on inheritance of characteristics. Irregularities in the transmission of chromosomes introduce biases of concern to the breeder.

In hexaploid wheat (*Triticum aestivum*), it was demonstrated that a dominant gene locus *Ph1* on the long arm of chromosome 5B is responsible for the control of homoeologous, diploid-like chromosome pairing as happens in diploids. It restricts the pairing between homologous chromosomes in a polyploid (RILEY 1960). Similar mechanisms were later found in polyploid cotton, rapeseed, tobacco, peanut, cotton, oats, and polyploid grasses.

When a regular chromosome pairing is given, then whole chromosome additions can be established as allopolyploids. In the genus group of Triticeae, more than 716 allopolyploid combinations were produced involving genera such as *Aegilops*, *Triticum*, *Secale*, *Haynaldia*, *Hordeum*, or *Agropyron* (MAAN and GORDON 1988).

Autopolyploids and their backcrossing to the diploid parents produces a variety of aneuploid progeny from which trisomic, telotrisomic, or tetrasomic individuals can be selected. In 1920, A. F. BLAKESLEE discovered the first trisomics in the weed *Datura stramonium*. In 1921, he coined the term "monosomic" for diploid plants missing one chromosome. Because of the cytogenetic instability of trisomics, they do not have a direct breeding value. Their utilization for genetic analysis and chromosomal localization of important agricultural traits was for many years the only way of gene mapping, particularly in diploid crops. The pioneering work of ▶ B. McCLINTOCK (1902–1992)—Nobel Prize winner for Physiology in 1983 for discovery of mobile genetic elements, a discovery that heavily influenced molecular genetics—in utilizing trisomics in maize to associate linkage groups with specific chromosomes is well known (RHOADES 1955). In 1911, G. N. COLLINS and J. H. KEMPTON (1916) already demonstrated linkage of genes in maize and, in 1913, ▶ A. H. STURTEVANT (1891–1971) provided the experimental basis for the concept of linkage versus genetic map distance between loci. By analyzing mating results for fruit flies (*Drosophila melanogaster*) with six different mutant factors each known to be recessive and X-linked, he constructed the first gene map. He traced each mutation and its normal alternate in relation to each of the other mutants, and thus calculated the exact percentage of crossing-over between the genes. In 1929, B. McCLINTOCK was the first reported the basic chromosome number 10 ($2n = 2x = 20$) in maize. A tremendous impact of trisomic research was also given in tomato (RICK and BUTLER 1956), barley (TSUCHIYA and GUPTA 1991), or rye (MELZ et al. 1997).

The second period of cytogenetic advances (1950–present) became the time of molecular cytogenetics. Between 1950 and 1970, advances were made in cytogenetic mapping in crop plants, including wheat. Major conceptual advances in polyploidy cytogenetics also occurred, using wheat as an allopolyploid inheritance model and potato as an autopolyploid model (cf. Section 3.6). From the 1970s, there were many advances in chromosome banding techniques, *in situ* hybridization, microscopy, and DNA manipulation.

Different types of banding techniques allowed the identification of individual chromosomes and a revolution in plant cytogenetic identification began because chromosomes could be fingerprinted. Banding techniques provided insights into structure and function of chromosome landmarks, such as heterochromatin, rRNA genes, the centromere, and the telomere and/or subtelomere.

C-banding in wheat has been used to analyze the substructure of wheat chromosomes. It not only allows fast and reliable identification of all 21 chromosomes but also permits the identification of 38 of the 42 chromosome arms (GILL et al. 1988).

Fluorescence *in situ* hybridization (FISH) was originally derived from the *in situ* hybridization technique, which used isotopes to label probes to detect the DNA or RNA sequences in cytological preparations (GALL and PARDUE 1969). In 1982, a nonradioactive (immunological) FISH method using fluorochromes for signal detection was developed and, since then, has been used widely in plant research. Because of its specificity, clarity, and relative rapidity of detection, FISH remains the technique of choice for direct visualization of genomes, chromosomes, chromosome segments, genes, DNA sequences, and their order and orientation.

Genomic *in situ* hybridization (GISH), a special type of FISH that uses genomic DNA of a donor species as a probe in combination with an excess amount of unlabeled blocking DNA, provides a powerful technique to monitor chromatin introgression during interspecific hybridization. In addition, the GISH technique allows the study of genome affinity between polyploidy species and their progenitors. GISH is thus a valuable supplemental technique to traditional genome analysis such as conventional meiotic pairing analysis.

Latest example of whole-genome recombination is the Chinese wheat variety "Shumai 969" that was derived from a cross of *Triticum turgidum* (AABB) × *Aegilops tauschii* (DD). A chromosomal segment was transferred harboring the *Glu-D1* locus from *Ae. Tauschii* (Liu et al. 2016). In oat, LADIZINSKY (2000) generated a stable synthetic hexaploid variety, "Strimagdo" (A_sA_sCCDD), by crossing *Avena strigosa* × *A. magna*.

The turf industry was revolutionized with the development and release of "Tifgreen" bermudagrass, a triploid ($2n = 3x = 27$) interspecific hybrid resulting from a cross of *Cynodon transvaalensis* ($2x = 18$) and *C. dactylon* ($4x = 36$). The triploid combined the toughness, wear tolerance, cold tolerance, and disease resistance of *C. dactylon* with the fine texture of *C. transvaalensis* (HANNA et al. 2016). An early release in 1943 was the tetraploid forage variety "Costal."

Strawberry breeding strongly benefits from ploidy crosses. The commercial cultivars "Frel Pink Panda" and "Serenata" were derived from *Potentilla palustris* and *Fragaria* × *ananassa* (YANAGI and NOGUCHI 2016). An interspecific hybridization between the seed parent of *Fragaria* × *ananassa* var. "Toyonoka" (with good aroma) and a pollen parent of *F. nilgerrensis* yielded a vigorous pentaploid. After chromosome doubling, the allodecaploid strain "TB13-125" could be selected for the consumer showing several new characters including a peach-like aroma.

In ornamental plants, the tulip varieties "Rambo." "Zorro," and "Hunter" are outstanding. They are induced autotetraploids of diploid cultivars of the species *Tulipa gesneriana*, *T. fosteriana*, and *T. kaufmanniana*, respectively (CIOLAKOWSKA et al. 2016). Diploid × tetraploid crosses resulted in successful triploid varieties, such as "Lady Margot," "Benny Neymann," or "Sun Child," as well as in tetraploids "Riant," "Beauty of Canada," and "Peerless Yellow." A reddish/purple variety of ornamental grass "Princess Caroline" ($8x = 56$) was registered in 2010 in the United States (HANNA et al. 2016). It was derived from a cross of red tetraploid pearl millet (*Pennisetum glaucum*, $4x = 28$) and napiergrass (*P. squamulatum*, $4x = 28$).

The triploid hybrid variety "Astria" in poplar is 22% taller and 25% larger in diameter than the diploid parent. It was derived from a cross of *Populus tremula* (4*x*) × *P. tremuloides* (2*x*). Other varieties are "I-214" from Italy (*P.* × *canadensis*), "Sacrau 79" from Germany (*P.* × *canadensis*), "Zhonglin 46" from China (*P.* × *euramerica*), and "Wuhei 1" (*P.* × *euramerica*) from China (KANG 2016).

3.7.1 ANEUPLOIDS

A cell or plant whose nuclei possess a chromosome number that is greater or smaller by a certain number than the normal chromosome number of that species is called aneuploid. An aneuploid results from nonseparation of one or more pairs of homologous chromosomes during the first meiotic division. This event was described as nondisjunction by C. B. BRIDGES (1916). As a consequence, monosomic, nullisomic, double-monosomic, trisomic, tetrasomic, double-trisomic, or double-tetrasomic karyotypes may arise.

Before the first sets of aneuploids in allohexaploid wheat were developed in the beginning of 1940s by ► E. R. SEARS (1910–1991, USA), numerous studies were necessary to prepare the ground. In 1918, T. SAKAMURA made first efforts to classify the chromosome numbers in wheat species. In 1922, K. SAX suggested the "A" genome of bread wheat might be related to that in a diploid species, and in 1926, from Reading (UK) J. PERCIVAL (1863–1949) explained the origin of polyploid bread wheat. SEARS was appointed by USDA as a research geneticist and started his long association with the University of Missouri in 1936 till his retirement in 1980. By his both theoretical and practical contributions, he became the "Father of Wheat Cytogenetics" and stimulated related research in other crop plants.

The discovery of a low level of female fertility in a wheat haploid of the variety "Chinese Spring" and the recovery of aneuploid progeny led to the construction of a vast range of aneuploids that is unequalled in its versatility, practicality, and creativity in any other species known to humans.

"Chinese Spring" (wheat) is probably the most famous wheat variety worldwide. It is generally accepted as the standard variety for cytogenetic and molecular research. From this variety, more than 300 aneuploids were developed. Using traditional cytogenetic techniques, E. R. SEARS developed a series of unique and valuable cytogenetic stocks, which are still used widely and are very important for both cytogenetic and genetic studies in wheat. These stocks are a treasure for modern wheat cytogenetics. The great benefit of these aneuploids is that they provide cytogenetic markers for each of the 21 chromosomes and most of the 42 chromosome arms. Among these stocks, the most important and widely utilized stocks include nullisomic–tetrasomic, monosomic, ditelosomic, and double ditelosomic lines. Because of the nullisomic–tetrasomic lines, SEARS was able to place the 21 wheat chromosomes into three genomes and seven homoeologous groups. Monosomic and telosomic lines allowed researchers to locate genes and DNA markers to individual chromosomes and chromosome arms. At present, the ditelosomic stocks are being used for flow sorting and constructing chromosome arm-specific BAC libraries (DOLEZEL et al. 1998), which are crucial for sequencing the gene-rich regions of the individual chromosome arms.

Agronomically, however, "Chinese Spring" has some serious faults, such as shattering, susceptibility to almost all wheat diseases and insects, and a poor adaptation to the world's major wheat-growing regions. The answer to why the variety was chosen in which to produce monosomes, trisomes, tetrasomes, compensating nulli-tetrasomes (SEARS 1965), nullisomes, telosomes, ditelosomes, double ditelosomes, and almost all of the other possible aneuploids is simple (SEARS 1959). It came into cytogenetic use by accident. SEARS made hybrids between wheat and rye in 1936 in attempts to induce chromosome doubling by heat shocks. "Chinese Spring" was used since it was known to cross readily with rye. Among the wheat–rye hybrids, a few wheat haploids were obtained; one of these haploids was pollinated by euploid "Chinese Spring." Thirteen viable seeds were derived from this backcrossing, showing chromosome numbers of 2*n* = 41, 42, 43 and reciprocal translocations. Nullisomes were eventually obtained from the monosomes and one nullisome proved to be 3B, which is partially asynaptic and was therefore a good source of additional monosomes and trisomes.

It seems that "Chinese Spring" came to the western world from the Szechuan province of China. British representatives in foreign countries used to be encouraged to collect plants that would be of possible value in their homeland. It was ▶ R. BIFFEN, later Director of the Plant Breeding Institute at the University of Cambridge, who received from Szechuan at about the beginning of the nineteenth century a wheat, which he called "Chinese White." This type of wheat was at that time of interest because it was early maturing, set a high number of seeds per spikelet, and was tolerant to drought. In North America, "Chinese White" first appeared in North Dakota, where the pioneering wheat breeder L. R. WALDRON obtained it from R. BIFFEN in 1924. WALDRON shared it with other breeders, and one of them passed it on to J. B. HARRINGTON at Saskatoon (Canada). A sample of wheat called "Chinese Spring" came to University of Missouri in 1932 from Saskatoon. It was known at Missouri that this wheat was highly crossable with rye and therefore acquired by L. J. STADLER. STADLER was strongly interested in research on polyploidy and amphiploid hybrid production. It seems clear that the "Chinese Spring," which E. R. SEARS used in his crosses with rye, was the same as BIFFEN's "Chinese White." BIFFEN's "Chinese White" is still maintained at the John Innes Centre Collection, Norwich (UK), and is indistinguishable from "Chinese Spring." Recently, Y. ZOU (Oregon State University, USA) noticed a strong resemblance of "Chinese Spring" to certain Chinese varieties.

Utilizing "Chinese Spring," SEARS developed the concept that chromosomes from three species contributed their genomes to bread wheat, that is, chromosome 1 of species A codes for functions that are similar to chromosome 1 of species B, and chromosome 1 of species B codes for functions that are similar to chromosome 1 of species D. This concept of so-called homoeology is now fundamental to perception of all allopolyploid species. His joint work with M. OKAMOTO led to deeper understanding of how genes regulate chromosome pairing in polyploid wheat. The recognition of how homoeologous chromosomes are prohibited from pairing and thus immediately allowing a high degree of fertility in a polyploid plant became a cornerstone for modern chromosome manipulations (SEARS and OKAMOTO 1958) despite the influence on molecular genetics and biotechnology.

Chromosome deletion sets are the latest utilization of cytogenetic stocks. ENDO and GILL (1996) isolated more than 400 deletion stocks involving all 42 arms of wheat using the action of a gametocidal gene combined with chromosome banding. These deletion stocks, with various sized terminal deletions in individual chromosome arms, are useful for the targeted physical mapping of any gene or DNA sequence of interest to a defined chromosome bin and, therefore, provide a unique way of conducting physical mapping in wheat (FARIS et al. 2000). Cytogenetically based physical maps for all seven homoeologous groups of wheat have been constructed with the help of the deletion stocks. These deletion lines are currently a critical and powerful resource for wheat genome mapping projects. In addition, the deletion stocks were crucial in relating genetic maps to physical maps of chromosomes, map-based cloning of genes, studying the distribution of genes, and recombination frequency along the chromosomes.

3.7.2 Chromosome Additions

Interspecific chromosome addition lines in more or less stable disomic status are mostly known from wheat. K. SHEPHERD and A. ISLAM (1988) counted more than 200. Their cytological stability is insufficient for utilization in breeding but good enough for genetic studies. The wheat–barley addition line 5H or the wheat–Haynaldia villosa addition lines 1Ha and 3Ha are extremely difficult to propagate.

The alien chromosomes are subsequently eliminated on the male and female sides. A more rare situation is the preferential transmission of alien chromosomes first demonstrated in wheat–*Aegilops sharonensis* addition line 4S[l]. T. E. MILLER et al. (1982) called this particular chromosome "cuckoo" chromosome. Later, it was tried to recombine this alien chromosome with the standard wheat chromosome 4D to stabilize its sexual transmission. Chromosome 4D of wheat carries an important gene *Rht2* for straw length. In breeding populations, so-called off-types with longer culms can be identified because of the spontaneous loss of chromosome 4D.

FIGURE 3.16 Different spike morphology of disomic rye chromosome additions to hexaploid wheat var. "Holdfast" (left: control "Holdfast," followed by chromosome additions of diploid rye "King II" 1R, 2R, 3R, 4R, 5R, 6R, and 7R). (Courtesy of R. SCHLEGEL. With Permission.)

Thus, a 4D/4S^l recombination might be a way to contribute to more stable wheat varieties (KING et al. 1988). Other alien chromosomes of *Aegilops cylindrica, Ae. Triuncialis, Ae. Sharonensis*, and *Ae. Longissima* were recognized to carry so-called gametocidal genes causing several chromosome aberrations, such as translocations and deletions.

They heavily contributed to the establishment of series of wheat chromosome deletions. More than 436 deficient lines with defined chromosomal breakpoints became excellent tools for physical gene mapping for common traits as well as several molecular until recent time (ENDO and GILL 1996). Chromosomes added to polyploid wheat are meanwhile common. Complete sets of wheat–rye (Figure 3.16), wheat–barley, and wheat–*Agropyron* are intensively used in genetic studies. The most current addition series is the complete set of monosomic and disomic oat (*Avena sativa*, $2n = 6x = 42$)–maize (*Zea mays*, $2n = 2x = 20$) additions (KYNAST et al. 2001). However, alien chromosome additions in diploid crop plants are still seldom. Karyological modifications are less tolerated. The only example published to date is the monosomic and telosomic series of rye–wheat additions (SCHLEGEL 2005).

3.7.3 CHROMOSOME SUBSTITUTIONS AND TRANSLOCATIONS

A variation on the wide crossing procedure is to select plants that have single chromosomes or chromosome arms substituted from one species into another. Just for hexaploid wheat, more than 100 different substitution lines are described (SHEPHERD and ISLAM 1988). Donor species were *Secale cereale, S. montanum, Aegilops umbellulata, Ae. Variabilis, Ae. Caudata, Ae. Comosa, Ae. Longissima, Ae. Sharonensis, Ae. Bicornis, Agropyron elongatum, Ae. Intermedium, Haynaldia villosa, Hordeum vulgare, H. chilense, Triticum boeoticum, T. urartu, T. timopheevi*, and others.

After the first reports on spontaneous wheat–rye chromosome substitutions 5R(5A) by the KATTERMANN (1937), O'MARA (1947), and RILEY and CHAPMAN (1958), during the past four decades, particularly, 1R(1B) substitutions and 1RS.1BL translocations were described in more than 200 cultivars of bread wheat from all over the world (SCHLEGEL et al. 1994) demonstrating the enormous impact on wheat breeding (see Figure 3.17). Most of those interchromosomal rearrangements are so-called Robertsonian translocations; that is, the breakpoints are within the centromeric region. The centromere is a cytologically visible component of a chromosome appearing as a primary constriction at metaphase. It plays an essential role in the accurate segregation of chromosomes during mitosis and meiosis. In recent years, several centromere-associated repetitive sequences have been characterized and mapped to the centromeric regions of chromosomes of grass

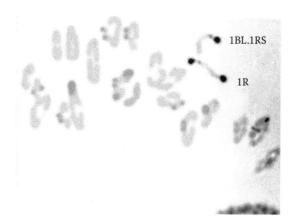

FIGURE 3.17 Metaphase I chromosome constitution with 2 univalents and 20 bivalents in a F_1 hybrid of hexaploid wheat from crossing of 1R(1B) substitution variety "Orlando" (Germany) × 1RS.1BL translocation variety "Kavkaz" (Russia) showing both rye chromosome 1R (lower heavy marked univalent) and the wheat–rye translocated chromosome 1RS.1BL (upper univalent with C-bands on the short arm only).

species by FISH. Among these sequences, only two are species-specific, that is, rye and sorghum. All the others are common to many grass species, including rice, maize, barley, and wheat. The presence of these centromere-specific repetitive sequences in different members of the Gramineae indicates that the cereal centromere may have evolved from a common progenitor before divergence about 60 million years ago. Although the function of these sequences remains unknown, they may be related to centromere function because of their location and high degree of repetition.

Using a common grass centromeric probe pRCS1 and a rye-specific centromeric probe pAWRC.1 in FISH experiments, ZHANG et al. (2001) demonstrated for the first time the compound structure of the centromere and the hybrid nature of centromeres in wheat–rye translocation chromosomes, indicating that centric breakage–fusion can occur at different positions within the primary constriction without influencing the centromere function and behavior.

The most important phenotypic deviation of translocated lines from common wheat cultivars has been the so-called wheat–rye resistance, that is, the presence of wide-range resistance to races of powdery mildew and rusts (BARTOS and BARES 1971, ZELLER 1973), which is linked with decreased breadmaking quality (Zeller et al. 1982), good ecological adaptability and yield performance (RAJARAM et al. 1983, SCHLEGEL and MEINEL 1994), and improved root growth (EHDAIE et al. 2003).

The origin of the alien rye chromosome was intensively discussed by genetic and historical reasons. It turned out that basically four sources exist—two in Germany, one in the United States, and one in Japan. The variety "Salmon" (1RS.1BL) is a representative of the latter and the variety "Amigo" (1RS.1AL) is a representative of the penultimate group, whereas almost all remaining cultivars can be traced back to one or to the other German origin.

There was no doubt so far that the Japanese and the American derivatives differ from another and from the German sources. Although on two places of Germany—Salzmünde near Halle/S. (breeder: RIEBESEL) and Weihenstephan near München (breeder: KATTERMANN)—wheat–rye crosses were already carried out since the 1920s and 1930s and independent pedigrees could be fragmentary reconstructed by the few reports left, some authors presumed only one German source (LEIN 1973).

By cytological and genetic means, it was not possible so far to verify whether or not there is a bi- or monophyletic origin of the 1RS.1BL wheat–rye translocation originating from Germany. However, latest molecular approaches offered a chance to study the identity of that particular wheat–rye translocation occurring in the Weihenstephan and Salzmünde wheat derivatives. A number of DNA probes were considered, (a) which were critical for the short arm of the rye chromosome 1R and (b) which should show specificity for the gene pool of Petkus rye. The DNA probe *CDO580* was revealed as a specific one. (1) It clearly differentiated the 1RS.1AL ("Amigo"), 1RS.1BL ("Salmon"), and 1RS.1DL ("Gabo") from the two German sources. (2) Both the deriving translocation wheats from the Weihenstephan and from the Salzmünde origin showed an identical DNA fragment that was typical for the gene pool of Petkus rye. Therefore, SCHLEGEL and KORZUN (1997) suggested that both German sources have one progenitor in common. So, a single rye chromosome spontaneously introduced into the wheat genome, probably by a single cross in Germany during the 1930s, spread through wheat breeding programs all over the world, and it still happens since 70 years!

Another example of a successful translocation is the 6VS.6AL wheat–*Haynaldia villosa* interchange. At least 10 commercial wheat cultivars of China carry it conferring powdery mildew resistance gene *Pm21*. It was derived from chromosome 6VS of *H. villosa* (LIU et al. 2016).

3.7.4 CHROMOSOME-MEDIATED GENE TRANSFER

The early studies on wheat–rye chromosome substitutions showed that almost exclusively dominant genes of the alien species are expressed in the recipient background (ZELLER and HSAM 1983). Moreover, only few of whole chromosome substitutions remained stable during the breeding cycles. Therefore, from the beginning there was a breeder's interest to transfer small segments of alien chromatin to increase their stable inheritance and to minimize the deleterious effects of donor chromosomes. It was the wheat–rye translocation line "Transec" (KNOTT 1971) that met these requirements. It was the first induced 4B/2R wheat–rye translocation involving a small piece of rye chromosome 2R carrying resistance genes against leaf rust (*Puccinia recondita*) and mildew (*Erysiphe graminis tritici*) in wheat (DRISCOLL and ▶ ANDERSON 1967). A second example was the induced translocation line "Transfer," carrying a dominant gene for leaf rust resistance of *Aegilops umbellulata* (SEARS 1961). A third 1AL.1RS translocation was produced in the United States for greenbug resistance (*Schizaphis graminum*) (SEBESTA and WOOD 1978).

By genetically induced homoeologous recombination, a fourth alien transfer line was established—"Compair" (RILEY et al. 1968). It showed yellow rust resistance of *Aegilops comosa*. Of particular breeding importance was a fifth 4BS.4BL–5RL wheat–rye translocation available in the wheat variety "Viking." The rye segment transferred carries genes for high micronutritional efficiency, such as copper, iron, and zinc efficiency, which gets pronounced attention in recent breeding programs of several of third world countries (SCHLEGEL et al. 1993). The latter "Viking" translocation and the "Transec" line also revealed that not only homoeologous recombination but also nonhomoeologous recombination could be mediated.

Those experiments of targeted alien gene transfer were only feasible by the high standard of cytogenetic research in wheat starting in 1920s and the specific homoeologous relationships within the *Triticeae*. They brought the first experiences for modern genetic engineering applying molecular techniques. The combination of both approaches resulted in a number of new transfers of resistance to leaf rust, barley yellow dwarf virus, powdery mildew, strip rust, eyespot, or wheat curl mite from *Aegilops*, *Thinopyrum*, and *Agropyron* species into hexaploid and tetraploid wheat.

Plant breeders in recent days will change their modus operandi with the development of objective marker-assisted introgression and selection methods. Backcross breeding will be shortened by eliminating undesired chromosome segments (also known as linkage drags) of the donor parent or selecting for more chromosome regions of the recurrent parent. Parents of elite crosses may be chosen based on a combination of DNA markers and phenotypic assessment in a selection index, such as best linear unbiased predictors (BERNARDO 1998). To achieve success in these endeavors, cheap, easy, decentralized, and rapid diagnostic marker procedures are required. The impact of introgression experiments is big. Various studies, mostly conducted on cereals, have estimated that more than 50% of the increase in crop production has been due to the improvement of crop varieties. It is brought about by transferring desirable genes and traits to crops from landraces and other more distant germplasm sources. Maize hybrids in the tropical and subtropical areas have been improved by 25%–40% during the past 20 years due to the introgression of a wider germplasm base!

3.7.4.1 Microprotoplast-Mediated Chromosome Transfer

Besides donor protoplast irradiation, micronuclei- and microprotoplast-mediated chromosome transfer has been considered as an alternative method for partial genome transfer. Mass induction of micronucleation and efficient isolation of microcells are key steps in any sort of this approach. Microtubules are involved in several processes such as chromosome migration, cell structure, cellulose microfibrils guidance and arrangement, cell wall formation, intracellular movement, and cell differentiation. Toxic substances such as antimitotic herbicides or colchicine prevent their normal polymerization. Application of these spindle toxins to synchronized cells generally blocks cells in metaphase and scatters chromosomes in the cytoplasm; subsequently, those decondense into micronuclei. Subsequently, these micronucleated cells are stripped of their cell wall, and the resulting microprotoplasts are ultracentrifuged to subdivide them into classes. These can be further filtered through sequential filters of smaller pore width. Some recent examples of suspension cell-derived microprotoplasts are *Citrus unshiu* and *Beta vulgaris*. In developing microspores of ornamental species such as *Lilium* and *Spathiphyllum*, micronuclei were induced through the action of mitosis arresting chemicals, without synchronization requirement. By using microspores instead of suspension cultures, the risk of mutation accumulation in suspension cells can be avoided. The maximal number of micronuclei observed was 12, whereas the haploid chromosome number amounts to 15. Genome fragmentation in *Beta* microprotoplasts can be additionally quantified through flow cytometry and confocal microscopy.

3.8   UTILIZATION OF HAPLOIDS IN BREEDING

3.8.1   DOUBLED HAPLOIDS

Haploidy, that is, the presence of the half of the common chromosome set, can be a normal as well as abnormal phenomenon in plants, at least in parts of the life cycle. Concerning ploidy mutation, haploids may arise spontaneously. It is necessary to differentiate between monohaploids (deriving from diploids) and polyhaploids (deriving from polyploids). Spontaneous haploids originate either from disturbed fertilization events or from parthenogenesis. They were recognized by their strikingly weaker appearance in wheat, maize, potato, tomato, cotton, *Datura*, and *Antirrhinum* crosses already with frequencies between 0.01% and 0.5% (LINDSTRÖM 1929, KOSTOFF 1929, KATAYAMA 1935, HARLAND 1936, SMITH 1943). The frequency of spontaneous haploids seemed to be genetically controlled. E. COE (1959) discovered an inbred line of maize showing 3% haploids. The production of first haploid plants of *Datura stramonium* was reported by A. F. BLAKESLEE in 1921, and from pollen grains in 1964 by the Indian scientists S. GUHA and S. C. MAHESHWARI.

Already during the 1960s some cytogeneticists pointed to the advantage of haploids for breeding. ▶ G. MELCHERS (1960) proposed their utilization in mutation breeding to reveal deleterious mutants or to select valuable genes in hemizygous allele constitutions. Doubling of haploid individuals would result in direct selection among fully homozygous segregants of cross populations. The significant reduction of the population size by use of homozygous doubled haploids instead of heterozygous segregants can be seen in Table 3.5.

Doubled haploids (DH) enable breeders to achieve homozygosity from early generation breeding material of (mainly) self-pollinating crops. The procedure eliminates several generations of selfing normally required before uniform lines. It reduces the selection period from 13 to 5 years or less. Besides, it provides more accurate and efficient selection of homozygous plants since both dominant and recessive genes are easily expressed in the first generation.

TABLE 3.5

Frequencies and Ratios of Genotypes in F_2 Progeny of Doubled Haploids, Diploids, and Tetraploids

Number of Alleles	Frequency of Recessive Plants as Determined by One of the Plants Given below (1/x)		
	Doubled Haploid	Diploid	Tetraploid
1	2	4	36
2	4	16	1,296
3	8	64	46,656
4	16	256	1,679,616
5	32	1,024	60,466,176
N	2^n	2^{2n}	6^{2n}

The first criterion for application of DH systems to breeding programs proposed by J. SNAPE et al. (1986) states that doubled haploid lines should be produced efficiently from all genotypes. However, the criterion is associated with two problems: the difficulty and efficiency of haploid production through a given DH system, and whether varietal differences in haploid production are negligible. Because of the species-specific and genotype-specific differences in production of haploids, basically three ways of haploid induction can be distinguished:

1. Selection of spontaneous haploids, using suitable genotypes and screening methods, for example, as in maize. Redoubling of haploid seedlings by colchicine.
2. In vitro microspore culture from unripe anthers (anther culture). Microspores are haploid, with n chromosomes. These microspores can be stimulated to chromosome redoubling by introducing specific biochemical treatments and colchicine, thus making it possible—after regeneration—to produce a genetically pure line. This regeneration of doubled haploid plants from microspores provides an opportunity for producing fertile, genetically pure (fully homozygous) progeny from one heterozygous parent in a single generation. Several techniques were developed for a great variety of crop plants, which are successfully utilized in research and breeding.
3. Interspecific and intergeneric crosses and subsequent chromosome elimination.
   a. The so-called "*Hordeum bulbosum* procedure" is a method for producing zygotic haploids in barley by crossing *Hordeum bulbosum* with *H. vulgare* genotypes. It was first described by ▶ KASHA and KAO (1970). After formation of zygotes, the *H. bulbosum* chromosomes are subsequently eliminated during embryogenesis, which results in haploid *H. vulgare* plants. After florets pollinated with *H. bulbosum*, they are cultured on modified N_6 medium containing 0.5 mg/l kinetin and 1.2 mg/l 2,4-D. Cultures were maintained at 25°C with a 16 h photoperiod for 9 days before embryo rescue. In a comparison of haploid production efficiency using five F_1 hybrids from winter × winter and winter × spring barley crosses, 41.6 haploid plants/100 florets pollinated were produced using floret culture. Using detached tiller culture, 13.5 haploid plants/100 florets pollinated were produced. Higher efficiencies achieved with floret culture are attributed to the formation of larger, differentiated embryos. Such embryos lead to higher frequencies of plant regeneration (CHEN and HAYES 1989). Resulting haploid seedlings are re-diploidized by colchicine treatment.
   b. Wide crosses. Haploids and doubled haploid wheat plants have been produced successfully through wide wheat × maize crosses (and embryo rescue system) since the first

TABLE 3.6
Number of Commercial DH Crops

Crop	Number of Varieties	Releasing Country
Barley	116	Canada: 36; USA: 3; Australia: 1
Rapeseed	47	Canada: 35
Wheat	21	UK: 5; Hungary: 5
Melon	9	Spain: 9
Pepper	8	Spain: 8
Rice	8	China: 6
Asparagus	7	Italy: 5
Tobacco	6	China: 6
Eggplant	5	Spain: 5

Source: AHLOOWALIA, B. S. et al., *Euphytica*, 135, 187–204, 2004.

report by D. LAURIE and M. BENNETT (1988). Although this system has been found to be less recipient-genotype dependent and more efficient in haploid production than the *Hordeum bulbosum* system, there is a considerable varietal difference in the efficiency among wheat varieties. It has been suggested that maize genotypes may affect the efficiency of haploid production too. The technology is not anymore restricted to wheat haploidization but also to other cereal species.

The production of doubled haploids has become a necessary tool in advanced plant breeding of many crop species. In 1980, "Mingo" was the first DH barley released. In 1985, "Florin" was the first released wheat, and "Dellmati" rice was released from Louisiana State (USA) in 2003. China has in 1990 already released doubled haploids in rice (grown on 800,000 ha) and wheat (grown on 650,000 ha). Until 2003 more than 230 varieties were registered that were derived from DH procedures (Table 3.6).

3.8.2 DIHAPLOIDS

Dihaploids of allo- or autotetraploids, that is, a haploid individual containing two haploid chromosome sets, are unique material for genetic research and breeding. They offer the advantage of disomic rather than tetrasomic inheritance patterns and a new approach to gene transfer from numerous wild and cultivated species. Particularly, in potato (*Solanum tuberosum*, $2n = 4x = 48$) they were utilized for several approaches as mentioned earlier. Potato breeding in the modern sense began in 1807 in England, when ▶ T. A. KNIGHT made deliberate hybridizations between different varieties by artificial pollination, and flourished during the second half of the nineteenth century when many cultivars were produced by farmers and hobby breeders. The Americans HOUGAS and PELOQUIN (1958) found them first in the progeny of interspecific crosses *S. tuberosum* × *S. phureja* ($2n = 2x = 24$). They appear spontaneously. Decapitation of the seed–parent has resulted in a 10–15-fold increase in fruits per pollination and a consequent increase of haploid production. By use of genetically different males, the haploid frequency again can result in about a 10-fold increase. CHASE (1963) proposed an analytical breeding scheme at the dihaploid level followed by resynthesis of the tetraploid.

During 40 years, those dihaploids heavily contributed to genetic analysis, gene mapping, taxonomic relationships, and, finally, potato breeding. Recent molecular studies in potato would not be feasible without the utilization of dihaploids.

3.9   GRAFTING METHODS

In the beginning, grafting was a horticultural practice of uniting parts of two plants so that they grow as one (Figure 2.9). The scion, the part grafted onto the stock or rooted part, may be a single bud or a cutting that has several buds. The stock may be a whole mature plant, such as an apple tree, or it may be a root (usually of a seedling). The most important reason for grafting is to propagate hybrid plants that do not bear seeds or plants that do not grow true from seed. It is also used in dwarfing and in tree surgery, to increase the productivity of fruit trees by adding to the number of buds, to adapt a plant to an unfamiliar soil or climate by using the roots of another plant that thrives in that environment, to combat diseases and pests (e.g., the phylloxera) by using a resistant stock, and to promote sexual growth of clones in forest trees. Grafting does not produce new varieties in terms of genetic recombination, since both stock and scion retain their characteristics. In the history of grafting, the transfer of traits of the rootstock to the scion was often a matter of controversial discussion. ▶ T. D. LYSENKO[10] (1898–1976), a Russian horticulturist, was one of the last advocates of Lamarckianism and the idea of gene transfer via grafting. Together with his political influence in Russia during the 1930–1950s, he misguided Russian genetics and breeding for several years. Recent molecular studies rehabilitate him and also ▶ I. V. MITSCHURIN (1855–1935), at least partially. STEGEMANN and BOCK (2009) stated that gene transfer occurs but is restricted to the contact zone between scion and stock indicating that the changes can become heritable only through a lateral shoot formation from the graft site. It confirms that the concept of graft hybridization has been tested and the existence of graft hybrids has been confirmed by several independent groups of scientists. By the molecular evidence, the former concept of graft hybridization is untenable. In the history of genetics, neglecting certain findings and phenomena is not uncommon. For example, both G. MENDEL's laws of heredity and B. McCLINTOCK's study on transposable elements were ignored for decades. Graft hybridization is potentially a simple and powerful means of plant breeding. It is to be hoped that the new results will once more direct the attention of geneticists to a challenging, yet unaccepted and unexplained, phenomenon. STEGEMANN and BOCK have confirmed the exchange of large pieces of plastid DNA at the immediate graft site by ordinary grafting in tobacco. Further experiments are needed to confirm whether both nuclear DNA and plastid DNA transport are possible over longer distances after mentor grafting.

Grafting is mainly used by nurserymen of ornamental and vegetable plants, or fruit and forest trees. In general, only closely related plants can be grafted successfully. As a rule, the process is begun when the scion is dormant and the stock is just resuming growth. There are many methods of grafting, all of which depend on the closest possible uniting of the cambium layers of both scion and stock.

The method of transplanting scions was already applied since pre-biblical and biblical times (cf. Genesis 2:15; 3:19, 23; or 4:2). As early as 1800 BC, a cuneiform ceramic fragment from Mesopotamia showed grape budwood. It suggests that the technique of grafting was known at that time. It was also known that the olive tree requires grafting. Ungrafted suckers produce a small worthless fruit. In 58 AD, this explains the power allegory of Paul in Romans:

> For if you have been cut from what is by nature a wild olive tree, and grafted, contrary to nature, into a cultivated olive tree, how much more will these natural branches be grafted back into their own olive tree (Romans 11:24).

Grafting was extensively employed in Roman times. In 323 BC, THEOPHRASTOS of Eresos (371–287) describes six varieties of apples and discusses why budding, grafting, and general tree

---

[10] LYSENKO became famous for the discovery of "vernalization," an agricultural technique that allowed winter crops to be obtained from summer planting by soaking and chilling the germinated seed for a determinate period of time. He was the first to use the term "vernalization" (in Russian, "jarovization" means inducing spring growth habit).

care are required for optimum production and says seeds almost always produce trees of inferior quality fruit. In his "Historica plantarum," he describes even more than 500 plant species. The Indian sugarcane is mentioned for the first time among them. Moreover, he discussed in detail which plants can be propagated by fruits and which do not grow when they multiplied by seeds, for example, grape wine, apple, olive, figs, quince, pears, or pomegranate. Almonds derived from seed propagation taste bad. The latter appraisal shows that grafting seemed to be a common method for clonal propagation in fruit trees during that time. C. PLINIUS (23–79 AD) describes in "Historia naturalis" the grafting of quince:

> When the common quince (*Cydonia oblonga*) is grafted on strutea (a greenish wild quince) one gets a separate species, the Mulvian quince (a variety), which is even ate raw as the only one of the genus.

He described many examples of grafting and emphasized the changes of features on scions, including the taste of the fruits after grafting.

> By common opinion the plane tree is most susceptible to scions, followed by winter oak; both of them spoil the taste of fruits however.

Grafting of herbaceous vegetable crops was described in the fifth century AD in China. Chinese horticulturists were also the first peony breeders and their work led to the introduction of many huge, double flowered tree and herbaceous peonies. It is believed that during the twelfth century, Chinese horticulturist began using grafting techniques to reproduce valued cultivars.

In the seventeenth century AD in Korea, a technique to obtain large gourds was described by approach grafting four root systems to a single shoot. The bottle gourd (*Lagenaria siceraria*) was used as a rootstock for watermelon in the 1920s to overcome yield decline from soil-borne diseases associated with successive cropping. The development of plastic films in the 1960s led to widespread production of grafted vegetables in Japan and Korea of solanums (tomato, eggplant, and pepper), cucurbits (melons and cucumber), and ornamental cactus. A number of unique advances were achieved, including seedless watermelon developed by ▶ H. KIHARA in Japan (1951).

The choice of appropriate rootstock permits resistance to soil-borne diseases, promotion of scion vigor and yield, and incorporation of stress tolerance. Graft technology of vegetables seedlings later has been adopted in Europe. When meristem tip culture fails, it is possible to graft small meristematic domes on young seedlings growing in vitro. It is called micrografting. In this way, NAVARRO et al. (1975) eradicated all the virus diseases from Spanish citrus orchards. This technique was also very successful in eliminating virus diseases from peach trees (MOSELLA et al. 1980). G. MOREL and C. MARTIN produced first successful micrografts in 1958.

3.10 QUANTITATIVE TERMS IN BREEDING AND GENETICS

The rediscovery of MENDEL's laws of inheritance in 1900 was the basis for determining the inheritance of quantitative traits and for developing plant breeding and selection methods. R. A. FISHER, S. WRIGHT, and J. B. S. HALDANE were the primary early contributors for developing the theory and methods for studying the inheritance of quantitative traits. Greater interest in the inheritance of quantitative traits in plants occurred after 1946, primarily because of the heterosis expressed in maize hybrids. During the past 60 years, extensive research has been conducted to determine the relative importance of different genetic effects in the inheritance of quantitative traits for most cultivated plant species (HALLAUER 2007). Quantitative genetic research has contributed extensive information to assist plant breeders in developing breeding and selection strategies. Its principles will have continued importance in the future, but at different levels. Information from latest molecular research is integrated with current knowledge at the phenotypic level to increase the effectiveness and efficiency of selection. Studies have been conducted to determine the usefulness of marker–QTL associations relative to the efficacy of early testing in relation to later generations of testing, relations of QTL for traits in testcrosses, and relations between lines *per se* and their testcrosses. Studies are conducted to determine the effectiveness of marker estimates on parental contributions to F_2 and backcross-derived inbreds in different plant species, and comparisons of pedigree information, biochemical, and morphological data to predict heterosis. Two topics have received major attention: the use of molecular markers (MAS) as aids in selection of QTLs and the determination of genetic diversity among inbred lines and cultivars, that is, classification of inbred lines in appropriate heterotic groups for breeding purposes.

3.10.1 PLOT DESIGN, FIELD EQUIPMENT, AND LABORATORY TESTING

By end of the nineteenth century, agronomy as a science was developed from the old-style variety trials, crop rotation fields, and soil culture experiments, when field culture was an empirical art. Research workers and other interested agronomists in the science of crops and soils formed their first own circles of discussion, such as the American Society of Agronomy in 1907. In this regard, CARLETON (1907) stated:

> As a science it investigates anything and everything concerned with the field crop, and this investigation is supposed to be made in a most thorough manner, just as would be done in any other science. Thus, agronomy is the laboratory and workshop of many sciences: agrobiology, chemistry, botany, ecology, genetics, pathology, physics, physiology, and others concerned with the problems of crops and soils.

In Europe, J. B. BOUSEINGAULT established the first experiment station in 1834, being the first to undertake field experiments on a practical scale (Table 3.1). He farmed land at Bechelbronne, Alsace (France). He investigated the source of nitrogen in plants, and systematically weighed the crops and the manure applied. He analyzed both and prepared a balance sheet. Furthermore, he studied the effects on plants when legumes were in the rotation. One of his conclusions was that plants obtain most of their nitrogen from the soil!

In 1915, J. A. HARRIS proposed a criterion for measuring soil variability that he called a "coefficient of soil heterogeneity." Five years later, HARRIS reported that soil heterogeneity is practically universal throughout the world:

> The demonstration that fields upon which the plot tests have been carried out in the past are practically without exception so heterogeneous as to influence profoundly the yields of the plots emphasizes the necessity for greater care in agronomic technique and more extensive use of the statistical method in the analysis of the data of plot trials if they are to be of value in the solution of agricultural problems.

J. B. LAWES established the Rothamsted Experimental Station on his farm in England in 1841. In 1845, the systematic field experiments begun. They continue to present day. For long these experiments have been models of carefully planned trials. J. H. GILBERT assisted LAWES over 57 years. In 1859, the "Kleinwanzlebener Saatzucht" (now "KWS") was founded near Magdeburg (Germany). M. RABBETHGE became the first person in Germany to select beets for seed extraction by ascertaining the specific weight according to sugar content. The first private breeding enterprises emerged in the second half of the nineteenth century. Many originated at farms and have specialized in specific plant species. Currently, there are about 50 companies in Germany with their own breeding programs. The predominantly medium-sized companies have amalgamated to form the Federal Association of German Plant Breeders (Germany). The latter derived from the first "Society to Promote Private German Plant Breeding" in 1908.

Plant breeding in Sweden was started by farmers. The principal force behind was B. WELINDER, the owner of the estate Heleneborg in Svalöf. Due to his initiative and under the chairmanship of Baron F. G. GYLLENKROOK, the South Swedish Association for Cultivation and Improvement of Seed was found on April 13, 1886. Only a few years later, the General Swedish Seed Company was established in 1891.

In the United States, California was the first state initiating an experiment station in 1875, influenced by Norrill Act signed by E. LINCOLN in 1862, and began field experiments on deep and shallow plowing for cereals. A station was started in North Carolina in 1877, after which many others followed. After the Hatch Act passed by Congress in 1887, the start of the present USDA stations was made. Already 12 were in existence at that time. The Adams Act provided increased funds in 1906. Among the contributions of field testing in maize have been the discoveries that the show-type ear is unrelated to the performance in the field, that ear-to-row breeding may not lead to improvement in yield, and that the combination of inbred lines in hybrids had resulted in higher yields. In 1907, the American J. P. NORTON substituted the rod-row technique for breeding small grains in place of small plot tests. In wheat, the discovery of rust and bunt resistance and of physiological races of the pathogen has enabled breeding for resistant varieties. "Marquis" wheat was one of the most widely known improved varieties in the United States.

Learning from several inaccessibilities of early experiments, new types of precise field testing were proposed. Among them, long-term experiments of different fertilization on yield, processing performance, or continuous crop rotation were introduced. ▶ J. KÜHN began such an experiment with rye in 1878 at the University of Halle/S. (Germany), which lasts until recent days ("Eternal Rye Cropping").

In 1911, MERCER and HALL discussed questions about number, plot shape and size, and replications in field trials. It remained as a permanent issue in breeding and research!

Later, ▶ R. A. FISHER in Rothamsted (UK) has brought modifications in the field common experiments to make them more amenable to statistical treatment. His basic idea was always to combine experimental planning, field design, and statistical treatment.

One of the simplest experimental designs for testing the yielding capacity of a group of varieties, particularly if the number is not unduly large, is that of randomized blocks. In such designs, the varieties are grown in random order in each of several complete replication series or blocks, the number of replications used depending on the degree of precision desired for the comparisons of the variety means. FISHER stated that randomization of the order of varieties in a block must be followed if an unbiased estimate of error is to be obtained.

Since the 1920s and influenced by statistical advances, the block method was stepwise introduced in experimental field design. Most popular became the so-called "Latin square" or "Latin rectangle." With heterosis breeding and polycross methods, the number of treatments increased. For that, so-called lattice designs were developed. Later, polyfactorial trials were developed to consider more than one varying factor. Meanwhile, a list of experimental design can be given:

- Orthogonal hierarchical designs (randomized blocks, split plots, or split–split plots)
- Factorial designs (with blocking)
- Fractional factorial designs (with blocking)
- Lattice squares
- Balanced Latin squares
- Semi-Latin squares
- Alpha designs
- Cyclic designs
- Balanced incomplete block designs
- Neighbor-balanced designs
- Central composite designs used to study multidimensional response surfaces

To cope with the many concerns of breeders, a special field-plot technique was developed over the years. Various types of seeding, harvesting, and threshing equipment were applied and tested (Figure 3.18). Already in the 1930s seeders were described in the United States showing an endless-belt mechanics. It permitted much more accurate seeding. The seed is distributed evenly over the belt distance corresponding to the length of the row to be seeded, and the ratio of the belt speed to ground travel determines the seeding rate. VOGEL (1933) designed a wheat nursery seeder adapted for one- or three-row seeding and space or drill planting. Space planting is accomplished almost as rapidly as drill planting, and the planter seeds to the last grain (Figure 3.19).

A four-row cereal-nursery seeder was developed at the South Dakota Experiment Station in 1949. It was a modification of the V-belt principle and the utilization of a small tractor for power. As for small seeds, a two-row press-wheel maize and cotton planter was adapted by QUINBY and STEPHENS in 1933. Plant hoppers were removed and a downsprout water pipe was used to seed into from the operator's lap. Each planter was equipped for two men to seed and carry the seed packets arranged in boxes.

For harvesting and threshing, a rotary-shear cutter for cereal rod rows was established by KEMP (1935) in the United States. In 1936, BROWN and THAYER designed a power machine for cutting rod rows and a bundle-tying device.

(a)

(b)

FIGURE 3.18   (a) Hand sowing of cereal trials 70 years ago (Courtesy of M. HÖLLER, Wintersteiger AG, Austria. With Permission.) and (b) modern pneumatic precision space planter for seeding experimental plots. (Courtesy of M. HÖLLER, Wintersteiger AG, Austria. With Permission.)

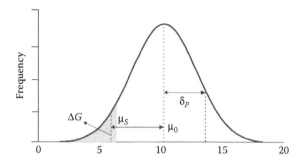

FIGURE 3.19 Schematic drawing of the genetic gain as a simple universal expression for expected genetic improvement. $\Delta G = h^2 \cdot \delta_p \cdot i/L$, where ΔG is the genetic gain, h^2 is the heritability, δ^2 is the additive variability of population ($= \delta^2_{genotype} - \delta^2_{dominance} - \delta^2_{epitasis}$), δ^2_p is the phenotypic variability of population ($= \delta^2_{genotype} + \delta^2_{environment} + \delta^2_{g+e} + \delta^2_{error}$), i is the selection intensity (proportion of population selected to produce the next generation), and L is the length of cycle interval, for example, one generation.

A mechanical harvester for small-grain nurseries was described by ATKINS in 1953. It was designed for cutting the two center rows of a four-row plot, although it could be used for cutting single rows as well. A first nursery thresher and an individual-plant thresher were described by KEMP (1935) in the United States. A cyclone-type thresher was designed by ARMSTRONG and COOPER in 1948. Its principal advantage was the ease of cleaning.

A furnace-type fan with rubber facing was mounted and powered through a four-step pulley. LILJEDAHL et al. (1951) constructed the first self-propelled combine–harvester for small-grain field plots.

3.10.2 STATISTICS IN BREEDING

The major contribution of modern plant breeding from a genetic point of view, as practiced over the last century, has been the maximization of the expression of traits of agronomic interest, principally the tolerance to biological and environmental stresses (DUVICK and CASSMAN 1999). This has been achieved primarily through the application of powerful tools in quantitative genetics and statistics (Table 1.1).

The progress of practical breeding, that is, selection among big numbers of individuals, handling of experimental plots, and description of numerous characters, required a more and more precise treatment of quantities. In 1812, C. F. GAUSS (1777–1855), the German mathematician, already suggested the theory of least squares. By 1820, he began to explain the idea of "probable error."

According to HACKING (1965), J. ARBUTHNOTT (1710) was the first to publish a test of a statistical hypothesis. HOGBEN (1957) attributes to J. GAVARRET (1840) the earliest use of the probable error as a form of significance test in biology. He also states that VENN (1888) was one of the earliest users of the terms "test" and "significant." The use of Chi-square for testing the goodness of fit for segregation ratios was already suggested by HARRIS in 1912.

F. GALTON (1822–1911), who is accepted as creator of empirical human genetics and who also introduced the term "eugenics," applied the twin studies for inheritance of human characters and biostatistic investigations for theoretical and practical genetics. The introduction of mathematical methods in breeding and genetics, particularly, is the merit of L. A. Quetelet (1796–1874). His mathematical treatment of variability founded the English school of biometricians. GALTON (1889) introduced the quantitative measurement in segregating populations. He treated populations as units and found several laws. The law of regression is one of them. K. PEARSON (1900), a student of GALTON and the founder of the journal *Biometrics*, further developed this law and made a contribution to mathematical aspects of testing. He introduced the term "standard deviation."

Based on a counterattack against biometricians of the late 1800s, who misunderstood MENDEL's law of segregation, W. E. CASTLE (1903) discovered an equilibrium law not only for the 1:2:1 ratio expected in F_2 generation but also for an infinity large class of ratios that could result from selection in any generation. Thus, he had unknowingly discovered the basic rule of population genetics, the HARDY–WEINBERG rule.

In 1908, the English mathematician G. H. HARDY (1877–1947, England) and, in 1909, the German physician W. R. WEINBERG (1862–1937, Germany) independently developed the law of equilibrium for populations. One locus with two alleles at the frequency of p and q will maintain those frequencies over an infinitely large random mating population with no mutation, selection, or gene flow. If p is the frequency of A alleles in the population and q is the frequency of A' alleles ($p + q = 1$), then the frequency of individuals in the next generation will be as follows: p^2 (AA individuals), q^2 ($A'A'$ individuals), and $2pq$ (AA' individuals), where $p^2 + 2pq + q^2 = 1$.

The importance of "standard deviation" was later proved by the Danish geneticist ▶ W. JOHANNSEN (1857–1927) in terms of dissociation of populations into pure lines. Experimental field design, sampling, analysis of variance, the theory of measuring observational error, and refining procedures are associated with longstanding director of the Agricultural Research Station Rothamsted (UK), ▶ R. A. FISHER (1890–1962). In 1918, he introduced the ideas of quantitative inheritance and correlation, and in 1922 the analysis of variance and other statistical methods:

> In general, the hypothesis of cumulative Mendelian factors, ... the influence of non-genetic causes such as environment, ... seems to fit the facts observed for quantitative traits very accurately.

The paper[11] became the fundament of quantitative genetics and had great impact on the development in 1950s and 1960s.

A promising tool to bridge the gap between Mendelian and quantitative genetics came with the development of methods for mapping of QTL during the 1990s. The basic idea of this approach traces back to Sax (1923). After S. WRIGHT's (1921) "Biometrical Relations Between Parent and Offspring," R. A. FISHER published his book "Statistical Methods" in 1936. It was standard for generations of students. According to E. S. BEAVEN (1935), ▶ T. B. WOOD and STRATTON were the first to determine probable errors in the context of replicated agricultural experiments. The foundations of modern hypothesis testing were laid by FISHER (1925), although the modifications were propounded by NEYMAN and PEARSON (1933).

One of the must applied test method, the so-called "t-test," was introduced under W. GOSSET's (1876–1937) pseudonym "Student" (therefore "Student test"). He established tables for estimating the probability that the mean of a unique sample of observations lies between "minus"/"infinity" and any given distance of the mean of the population from which the sample is drawn. If plots are too small, there is competition between them. In 1924, GOSSET recommended to discard the outer rows of all plots at harvest. Systematic layouts ABC...ABC... lead to bias in estimates of treatment differences. He recommends the half-drill strip. In 1938, he published a very important

---

[11] However, R. C. LEWONTIN (1977) mercilessly pointed to the misery of quantitative genetics during the First International Conference on Quantitative Genetics, held in 1976 at Iowa State University (USA). His major criticism was that it did not provide accurate answers to questions like: How many genes influence a trait? What is their contribution to the phenotypic and genotypic variance? What is their action and interaction? How are they organized within the genome? or How do they react in different environments?

paper. Until this point, he and R. A. FISHER had corresponded and agreed. He clearly showed that "balance" for a suspected trend leads to smaller bias in estimates of treatment effects but tends to overestimate error. This agrees with much of FISHER, but GOSSET concluded that one should balance, whereas FISHER said that one should block and randomize. GOSSET used uniformity data. He calculated the variance ratio for a particular layout and some particular values of the variance of treatment effects (assuming an additive model). So, he was assessing power, whereas WELCH and PITMAN had been doing randomization tests on uniformity data and thus were assessing significance.

After 1920, the biometrical genetics and statistics in breeding rapidly developed fast, so that a detailed discussion would burst the framework of the book.

E. S. PEARSON proposed the analysis of variance in cases of non-normal variation. In 1937, he assumed that the purpose of randomization is to do a randomization test. He points out that this test depends on the statistic used following some examples of R. A. FISHER. The so-called "z-test" was introduced for experiments with randomized blocks and Latin squares by WELCH (1937). He assumed arbitrary fixed plot effects additive with treatment effects. For randomized blocks, expectations of differences and of mean squares agree with those of normal theory. Because the plot effects are "fixed," the usual mean squares are not independent. The tail probabilities of the variance ratio may not be those given by normal theory, but several sets of uniformity data show good agreement. Actually, he used as statistic the treatment sum of squares, calculated formulas for its mean and variance, and applied a normal approximation to that. For Latin squares, he randomized by choosing at random from among all Latin squares of the same size. The variance of the treatment sum of squares depends on the number of intercalates in the squares.

Later, ▶ F. YATES (1939) discussed the point that the purposes of randomization are to avoid accidental bias, to make the results credible, and to obtain unbiased estimators both of treatment differences and of their variances. So, he is interested in estimation rather than testing. He supports GOSSET's criticisms of FISHER, but goes on to consider the possibility of randomizing designs, which have complicated ways of allowing for trend, such as the Latin square with balanced corners and the Latin square with split plots.

With the introduction of complex multifactorial trials, additional problems of estimation came up. R. L. PLACKETT and J. P. BURMAN (1946) optimized symmetrical factorial experiments with all treatment factors equireplicate. They introduced what is now called "orthogonal arrays of strength 2." For factors at two levels, they used what are now called "Hadamard matrices," constructed by PAYLEY's method. For p levels with p prime, affine geometries via sets of mutually orthogonal Latin squares have to be used. N. L. JOHNSON (1948) proposed alternative systems in the analysis of variance and discussed how the randomization procedure can justify the assumption of a simple model, in a simple case.

In 1951, D. R. COX claimed more systematic experimental designs. Plots should be in a line, with a low-order polynomial trend. He wanted treatment contrasts to be orthogonal to this trend as far as possible. If the design is symmetric about its midpoint, then treatment contrasts are orthogonal to odd-order orthogonal polynomials, so trial-and-error solutions orthogonal to the quadratic orthogonal polynomial are orthogonal to cubic trend, whereas H. D. PATTERSON (1952) favored the construction of balanced designs for experiments involving sequences of treatments, that is, change-over designs for first-order residual effects. Each treatment should occur equally often in each period. Subjects form a balanced (complete or incomplete) block design. Each treatment follows each other treatment equally often.

Statistical methods for design and analysis have made a major contribution to improving experimental efficiency since the basic principles of experimental method were established by ▶ R. A. FISHER and ▶ F. YATES in the 1920s and 1930s. These principles—to avoid bias, control haphazard error, take due account of treatment structure and experimental constraints, and ensure objectives are met—need to be interpreted for each situation. Experiments, which appear to be similar, may have very different objectives and thus require different approaches to design and analysis.

Despite the subject's long history, there are still many exciting opportunities for statistical research, as new fields of application open up and the objectives of experimental studies become increasingly complex. Recent advances in computer power have also led to a substantial increase in the range of feasible analytical techniques.

Much research has been stimulated by close links with plant breeding and variety testing organizations. This has ensured that the new techniques meet user needs and are rapidly taken up for routine use. The book "Statistical Methods for Plant Variety Evaluation" by R. KEMPTON and P. FOX (1997) has heavily contributed to the development and application of methods in sensory analysis. Computer algorithms partially developed by researchers of traditional Rothamsted Experimental Station (UK) for generating and analyzing designs have been incorporated into statistical packages, such as "GenStat," widely utilized in statistical analysis of field and genetic experiments.

Present plant breeding involves assessing the relative performance of large numbers of genotypes over several stages of selection. Trials may cover large areas of land, and comparisons between genotypes are then likely to be distorted by fertility trends or other sources of plot-to-plot variation. Much work has been done on modeling the variation in plot response and investigating control methods, which will lead to increased precision of genotype comparisons. The use of small blocks of plots, with genotypes allocated to blocks in an alpha design, was proposed in the 1990s. Latest developments include blocking in two dimensions using row and column designs, which we have found provide substantial additional gains in precision. Alpha designs are now widely used by national and international organizations throughout the world. Computer programs for generating alpha designs and analyzing their results are available from different providers.

Spatial modeling has been enthusiastically promoted since the early 1980s as an alternative to block designs for controlling plot variation. Current work is focused on the use of generalized additive models, which use the data directly to represent spatial variation across the trial, without assuming that these trends follow any fixed functional form.

Interference occurs when the response to a treatment applied to an experimental unit is affected by treatments on neighboring units. In variety testing, for example, interplot competition can lead to a reduction in the yield of shorter varieties due to shading from taller varieties in adjacent plots. This can produce substantial biases in the prediction of relative performance of varieties when in commercial production. Another application relates to the visual scoring of samples, for example, when screening plants for disease.

Since the 1960s, ▶ John W. DUDLEY has expanded the knowledge and application of the use of quantitative genetics in plant breeding. He has been a consistent contributor developing theoretical models, gathering, analyzing, and interpreting data relative to specific problems, analyzing long-term selection studies, and integrating molecular marker data with the expression of QTLs. DUDLEY started his career working on sugarbeets and alfalfa, but the majority of his activity was conducting basic research on the genetics and breeding of maize. He developed mating designs to study the inheritance of quantitative traits in autotetraploid crop species. The longest continuous selection study for oil and protein content of maize has been thoroughly evaluated by DUDLEY and his colleagues to measure rates of response, relative importance of types of genetic effects responding to selection, and estimates of the number of effective factors involved in the expression of oil and protein contents (Figure 3.10), and compare three methods for estimation of epistasis[12] (Table 3.7) among related and unrelated maize hybrids. Choices of germplasm to use in breeding programs to improve current germplasm sources have always been an important component in ultimate success to develop better cultivars. DUDLEY developed a unique method to identify alleles that are not present in current breeding materials (DUDLEY 1984).

---

[12] The term is recently used for all-allelic interactions between different genes that modify Mendelian ratios.

TABLE 3.7

Some Modified F₂ Segregations Deviating from Mendelian Ratio-Based Epistatic Effects

Genotype Interaction	Segregants			
	AABB	AAbb	aaBB	aabb
Dominant epistatic (A−)	12		3	1
Recessive epistatic (aa)	9	3	4	
Duplicated genes with cumulative effects	9	6		1
Duplicated dominant genes (A−, B−)	15			1
Duplicated recessive genes (aa, bb)	9	7		
Dominant and recessive interaction	13		3	1
Mendelian pattern	**9**	**3**	**3**	**1**

Quantitative genetic principles have been particularly powerful as the theoretical basis for both population improvement and methods of selecting and stabilizing desirable genotypes (HALLAUER 2007). An important concept in quantitative genetics and plant breeding is genetic gain (ΔG), which is the predicted change in the mean value of a trait within a population that occurs with selection. Regardless of species, the trait of interest, or the breeding methods employed, ΔG serves as a simple universal expression for expected genetic improvement (FALCONER and MACKAY 1996). Though clearly an oversimplification of the advanced quantitative genetic principles employed in plant breeding, the genetic gain equation effectively relates the four core factors that influence breeding progress: the degree of phenotypic variation present in the population (represented by its standard deviation, σ_p), the probability that a trait phenotype will be transmitted from parent to offspring (h^2 = heritability), the proportion of the population selected as parents for the next generation (i = selection intensity, expressed in units of standard deviation from the mean), and the length of time necessary to complete a cycle of selection (L). L is a function of not only how many generations are required to complete a selection cycle, but also how quickly the generations can be completed and how many generations can be completed per year (Figure 3.19).

A plenty of other proposals and tests were continuously introduced. They cannot be ruled out in detail. However, they were always associated with progress in breeding and complexity/specificity of field and laboratory testing.

3.10.3 Bioinformatics

Historically, the term "bioinformatics" did not mean what it means today. Paulien HOGEWEG and Ben HESPER coined it in 1970 to refer to the study of information processes in biotic systems. This definition placed bioinformatics as a field parallel to biophysics or biochemistry. Bioinformatics is now an interdisciplinary field that develops methods and software tools for understanding biological data. As an interdisciplinary field of science, bioinformatics combines computer science, statistics, mathematics, and engineering to analyze and interpret biological data. The actual process of analyzing and interpreting data is referred to as computational biology. Important subdisciplines within bioinformatics and computational biology include:

- Development and implementation of computer programs that enable efficient access to, use and management of, various types of information.
- Development of new algorithms and statistical measures that assess relationships among members of large data sets. For example, there are methods to locate a gene within a

sequence, to predict protein structure and/or function, and to cluster protein sequences into families of related sequences. Related fields are sequence analysis, DNA sequencing, sequence assembly, genome annotation, comparative genomics, analysis of gene expression, analysis of regulation, chromosome topology, high-throughput single cell data analysis, biodiversity informatics, and so on.

Common uses of bioinformatics include the identification of candidate genes and nucleotides (SNPs). Often, such identification is made with the aim of better understanding the genetic basis of disease, unique adaptations, desirable properties of crops, or differences between populations. In a less formal way, bioinformatics also tries to understand the organizational principles within nucleic acid and protein sequences, called proteomics.

A majority of economically important plant traits, such as grain or forage yield, can be classified as multigenic or quantitative. Even traits considered to be more simply inherited, such as disease resistance, may be "semiquantitative" for which trait expression is governed by several genes (e.g., a major gene plus several modifiers). The challenge to use strategically new technology (such as DNA-based markers) to increase the contribution of "science" to the "art plus science" equation for plant improvement therefore applies to most, if not all, traits of importance in plant breeding programs.

Historically, early researchers in quantitative genetics questioned whether the inheritance of these continuously distributed traits was Mendelian (COMSTOCK 1978). The answer to this question has major implications in the consideration of the use of markers for plant breeding programs. During the past century, plant geneticists have obtained convincing evidence that Mendelian principles apply to quantitative as well as to qualitative traits. This evidence has also shaped the general model that embraces the multiple-factor hypothesis for quantitative traits, that is, with genes located in chromosomes and hence sometimes linked, and incomplete heritability because of the contribution of environmental factors to total phenotypic variation.

The impact of the informatics revolution in crop improvement can be partially assessed by counting the number of publications indexed in Plant Breeding Abstracts (CAB International, Wallingford, UK). There was about 31-fold increase of publications in the 1930–2005 period. It was in the 1970s that indexed publications in plant breeding exceeded 10,000 per year. More publications and easy means for retrieving this information accounted for such growth of knowledge dissemination in plant genetics and breeding. Today, rapid information exchange has been facilitated with electronic mail and access to the Internet to read electronic publications. Information technology and DNA science are beginning to fuse into a single operation. Computers are deciphering and organizing the huge amount of genetic information that becomes "the raw resource of the emerging biotech economy" (RIFKIN 1998). Scientists working in the new field of "bioinformatics" are developing biological data banks to download the genetic information accumulated during millions of years of life evolution, and perhaps reconstruct some of the living organisms of the natural world (ORTIZ 1998). Bioinformatics may be simply described as the data repositories, data mining, and analysis tools designed to interpret the genome data that are currently available along with the data that will soon deluge the plant science community. Current public repositories include sequence databases, species-specific genome databases, and the germplasm databases. The sequence databases such as "GenBank" (nucleotides) and "SwissProt" (amino acids) are robust resources for sequence deposition and sequence analyses. They provide powerful online tools for sequence analysis and searching, where searches can be made for motifs and secondary structure as well as for amino acid or nucleotide similarities. Sequence databases can support record-to-record links to the species-specific databases. However, these links need to be specified by curators of the species-specific databases.

Species-specific databases exist for major crops, such as soybean, maize, wheat, rye, barley, oat, and rice, and for model organisms such as *Arabidopsis*. These genome databases integrate the map data for the species and provide documentation on the functionality of the genome. The data include the physical and genetic maps, clones and primers, QTLs, trait variances, references, images of pest and stress responses, and mutant phenotypes. They access the sequence information curated in the central sequence databases. The species-specific databases require scientific curation to ensure data quality, uniformity of gene and allele nomenclature, and accurate integration of data. The germplasm databases catalog information on available seed resources, along with certain agronomic and quality trait data. They presently do not contain any genome information, but linking to the genome databases is under investigation.

3.10.3.1 Molecular Markers

Genomic selection is a method for predicting genomic breeding values using molecular markers covering the whole genome. It is fast becoming popular in plant breeding, because of recent advances in high-throughput marker technologies and accompanying reduction in the costs of genotyping. The performance of genomic selection procedures is often assessed by k-fold cross-validation. Accurate evaluation of the performance of genomic selection is difficult in practice because true breeding values are typically unknown. As result, simulation modeling is often used to generate breeding values as a basis for assessing the accuracy of genomic prediction. Once the true breeding values are available, the accuracy of genomic prediction can be expressed as the correlation between the true and the predicted breeding values.

To evaluate estimates of predictive accuracy, at least, there are seven methods, four of which use an estimate of heritability to divide predictive ability computed by cross-validation. They cover balanced and unbalanced data sets as well as correlated and uncorrelated genotypes. The seven methods can be divided into four indirect methods and three direct methods (ESTAGHVIROU et al. 2013): (1) Standard measure—breeders compute heritability for a single trial using $H2m1 = \sigma^2_g \sigma^2_g + \sigma^2_e/r$, where σ^2_g is the genetic variance, r is the number of replicates, and σ^2_e is the variance of plot error. This estimator is valid for randomized complete block designs, but is an approximation for incomplete block designs. (2) A measure that uses the best linear unbiased estimations (BLUE) and is computed as $H2m2 = \sigma^2_g \sigma^2_g + v^-/2$, where v^- is the mean variance of a difference of two adjusted genotypic means and σ^2_g is the genetic variance estimated from assuming independent genotypic effects. (3) A measure that is based on BLUP assuming independent genotypic effects and is computed as $H2m3 = 1 - v^-_{BLUP}2\sigma^2_g$, where v^-_{BLUP} is the mean variance of a difference of the BLUP of two genotypic effects $g\hat{}_i$. (4) A measure for estimating heritability. The sample variance s^2_g of the true genetic breeding value g_1 can be written as $s^2_g = 1n - 1\Sigma I = 1n(gi - g^-)2 = gTPug$, where $Pu = 1n - 1(In - 1nJn)$, I_n is the n-dimensional identity matrix, and $J_n = n \times n$ is a square matrix of ones. (5) Estimating rg and $g\hat{}$. The method uses the ridge regression BLUP of g as the "phenotype" to compute an alternative estimator of predictive accuracy unlike that produced by the methods that require cross-validation and use the adjusted means as the phenotypes. Heritability can be computed as the square root of the estimated predictive accuracy. (6) Estimating rg and $g\hat{}$ by evaluating, that is, by computing $sg\hat{}$, p, and $s2g\hat{}$ directly from the data (p) and the predicted breeding values ($g\hat{}$). The approaches indirectly estimate predictive accuracy as the correlation between the predicted genetic and phenotypic values, called predictive ability, divided by the square root of heritability, separately for each of 15 three-fold cross-validation replicates.

The estimated heritability is closer to its true simulated value for methods 2, 3, and 5 than for methods 1 and 4 in terms of its minimum, maximum, mean, standard deviation, and mean squared deviation for all the four scenarios. All the five methods (methods 1 to 5) underestimate the minimum, maximum, and mean true heritability. Method 5 produced estimates closest to the true heritability for all the four scenarios.

The estimated predictive accuracy is more precise for methods 3 and 4, based on the large data set, than for methods 1 and 2. Methods 5 and 6 are the best overall and gave the least biased and most precise estimates of predictive accuracy; that is, they are the most computationally efficient to implement and gave consistently the most accurate, robust, and stable estimates of predictive accuracy of the seven methods. These properties argue for their routine use in assessing predictive accuracy in genomic selection studies. Among the five methods that use cross-validation, methods 4 and 6 performed better than methods 1, 2, and 3, but were clearly inferior to methods 5 and 6. Both the genetic variance and the number of genotypes exerted strong influences on predictive accuracy. Thus, predictive accuracy is higher for the larger data set. Furthermore, reducing the genetic variance degrades predictive accuracy much more for the smaller of the two data sets.

4 Biotechnology and Genetic Engineering

Advances in biology augur a third agricultural revolution involving "biotechnology," a catchall term that includes both cell and DNA manipulation. As a parable, once D. N. DUVICK designated the hybrid maize production as biotechnology of the 1930s. Nevertheless, a conventional baseline for the biotechnological revolution is 1953, the date of the WATSON and ▶ CRICK paper on the structure of DNA. However, the biotechnological revolution has no precise beginning because science is cumulative. One pathway developed from a series of investigations into gene function and structure and another from the culture and physiology of cells using microbial techniques. It was stressed several times in the previous chapters that recent advances in biotechnology often have their roots in ancient epochs.

The beginnings of inquiries into gene structure can be traced to the 1860s, when a Swiss, J. F. MIESCHER (1844–1895), described a substance he called "nuclein," which was derived from pus scraped from surgical bandages and later found in fish sperm. In 1889, R. ALTMANN (1852–1900), a student of MIESCHER, split nuclein into protein and named the obtained substance "nucleic acid." Two distinct kinds of nucleic acid were found in thymus and yeast. ASCOLI, in 1900, and LEVENE (1869–1940), in 1903, demonstrated the presence of adenine, cytosine, guanine, and thymine in thymus nucleic acid (now known as DNA) and, with uracil replacing thymine, in yeast nucleic acid (RNA). The original assumption that those bases were present in equal amounts and thus formed a "stupid" molecule proved to be faulty, but E. CHARGAFF (1905–2002), in 1950, demonstrated a key equality. In molar amount, cytosine equaled guanine and adenine equaled thymine. This suggestion was basic to the complementary replication of DNA and to the discovery of the structure of the double helix.

Genetic studies of the biochemistry of gene products date to 1902, when the British physician, A. GARROD (1857–1936), demonstrated that the human disease alkaptonuria was inherited and, moreover, was due to an alteration in nitrogen metabolism. In a paper titled "Inborn Errors of Metabolism," alkaptonuria was established as a gene-induced enzymatic block. This prescient study affected the course of biochemistry but remained unappreciated, if not unread, by early geneticists.

The genetic investigations of metabolism from 1900 to 1950 formed a subdiscipline: biochemical genetics. The one-gene–one-enzyme model predicted by GARROD was established as a dogma in the new catechism of biochemical genetics by ▶ G. W. BEADLE (1903–1989) and E. L. TATUM. The eventual move to bacteria and bacteriophage, with new and powerful techniques for recombinational analysis, changed the concept of the particulate gene. Long considered to be analogous to a bead on a string, it was finally shown to be more like a long molecule, which was first proposed by R. GOLDSCHMIDT (1878–1958).

The emergence of the power of microbial systems and the rise of bacteriophage, with its 20-min generation time, as a genetic subject, altered the experimental approaches. A clear distinction arose between the old and the new genetics (▶ M. DELBRÜCK (1906–1981), ▶ F. CRICK, S. BENZER, J. LEDERBERG, J. WATSON, and M. NIREMBERG). Analysis of the transformation principle in *Pneumococcus* by O. AVERY, C. MCLEOD, and M. MCCARTY and, in 1952, the subsequent phage manipulations by A. D. HERSHEY (1908–1997) and M. CHASE (1930–2003) proved that the genetic material was DNA; but details on its ability to replicate and to affect protein synthesis were unknown until the famous paper by WATSON and CRICK (1953). The resolution of the structure of DNA was followed by a race that unraveled the genetic code and, as if that were not sufficient, by the discovery of restriction endonucleases that snip gene sequences and plasmid vectors that transfer them across barriers considered unbridgeable by even the most credulous imaginations. In 1973, it became feasible to isolate genes and the application of DNA recombinant techniques stimulated a boom of research from 1980 onward.

By conventional breeding, developing a new variety can take up to 15 years for wheat, 18 years for potatoes, even longer for some crops. Because of the tremendous impact of biotechnology, the Nobel Peace Laureate, ▶ N. BORLAUG (1997), suggested the application of new biotechniques, in addition to conventional plant breeding, to boost yields of the crops that feed the world. The scope of conventional plant breeding has increased with improvements in technology. In the laboratory, chemical and mechanical techniques are used to speed up the selection process and remove natural barriers to cross-fertilization, for example, between different crop species.

4.1   IN VITRO TECHNIQUES

Tissue culture for breeding application was developed in the 1950s and became popular in the 1960s. Today, micropropagation and in vitro conservation are standard techniques in most important crops, especially those with vegetative propagation. The early history of cloning as a technical object began in two different experimental systems during the first two decades of the twentieth century. Like the concept of the gene, the concept of clone was first introduced to biology at the beginning of the twentieth century. The gene concept referred to an abstract or even an ideal unit (according to W. JOHANNSEN'S use of the term), whereas the concept of the clone from the beginning referred to a concrete material object. In 1903, H. J. WEBBER,[1] a botanist from the Plant Breeding Laboratory of the USDA, introduced the term "clone" to designate

> ... groups of plants that are propagated by the use of any form of vegetative parts... and which are simply parts of the same individual seedling.

In 1916, V. JOLLOS performed experiments on the influence of temperature and poison on "pure lines" of *Paramaecium caudatum*, which he also designated "clones." The context of this research was the debate about the inheritance of experimentally generated modifications. Thus, the clone concept was used synonymously with the concept of a "pure line," which was introduced originally by W. JOHANNSEN in 1903. The latter had defined "pure lines" as individuals (plants) that descended from a single self-fertilizing individual. These "pure lines" were the experimental material that led him to differentiate between phenotype and genotype, and hence to the introduction of the gene concept.

Although the concept was first developed in the context of horticultural breeding, the clone soon became something, which could be called a technical object of different experimental systems. In a broad sense, in vitro techniques include

- Seed culture,
- Embryo culture,
- Ovary or ovule culture,
- Anther or microspore culture,
- In vitro pollination,
- In vitro fertilization,

---

[1] H. J. WEBBER (1865–1946), a botanist from the Plant Breeding Laboratory of the USDA (USA), introduced the term "clone" to designate "groups of plants that are propagated by the use of any form of vegetative parts and which are simply parts of the same individual seedling." Though the concept was first developed in the context of horticultural breeding, the clone soon became something, which could be called a technical object of different experimental systems in the first decades of the twentieth century. Two scientists stimulated theoretical research on clones—V. JOLLOS in the 1910s and J. HAEMMERLING in the 1920s. Both were co-workers of M. HARTMANN at the Kaiser Wilhelm Institute for Biology in Berlin (Germany).

- Organ culture,
- Shoot apical meristem culture,
- Somatic embryogenesis, organogenesis, and enhanced axillary budding,
- Callus culture,
- In vitro production of secondary metabolites,
- Cell culture and in vitro selection at cellular level,
- Genetic and epigenetic somaclonal variations,
- In vitro mutagenesis,
- Protoplast isolation, culture, and fusion,
- Genetic transformation,
- In vitro flowering,
- Micrografting,
- Cryopreservation or storage at low temperature,
- Culture of protoplasts, cells, tissues, and organs, or
- Culture of hairy roots.

The history of plant tissue culture begins with the concept of the cell theory given independently by M. J. SCHLEIDEN (1838) and T. SCHWANN (1839), which implied that the cell is a functional unit. Pioneering studies of in vitro culture of monocot plant organs and tissues by G. HABERLANDT in 1902 predicted that the notion of producing plants from cultured cells would provide final confirmation of the cell theory, that is, the cellular totipotency. But only STEWARD in 1958 demonstrated it. His research team was the first to be able to transform a carrot cell line in some "artificial embryos," later called "somatic embryos." A very interesting possibility of the somatic embryogenesis technology has been developed during the past 15 years by REDENBAUGH and his team (1993). Since 1958, when the first plant embryos were obtained from somatic tissues of carrot cultured in vitro, ever-increasing numbers of species and tissues have been induced to form somatic embryos. REDENBAUGH et al. were able to encapsulate somatic embryos by hydrogel coatings (sodium alginate), producing single-embryo artificial seeds. To date, some improvements offer the possibility to directly plant artificial seeds in the greenhouse on special substrates. The technique of artificial seeds were successfully applied in alfalfa, asparagus, bamboo, caraway, carrot, celery, cinnamon, coffee, eggplant, geranium, ginger, horseradish, lettuce, lily, papaya, rice, sandalwood, tangerine, spruce, olive, or vanilla by coating somatic embryos with alginate, polyox, polyethyleneglycol, or cellulose. There was a slow but continuous progress. However, the methodology will provide a good technique in the future to reduce the cost of transplants.

In 1922, W. J. ROBBINS introduced procedures for the culture of roots and L. KNUDSON developed the aseptic germination of the embryo-like seed of orchids. In 1934, P. R. WHITE (1934) reported for the first time successful continuous cultures of tomato root tip in liquid culture and obtained indefinite growth. In 1939, he was able to regenerate the first shoot from tobacco callus. During the same year, NOBECOURT cultivated the plant tissues of *Nicotiana tabacum* and *Salpiglossis sinuata* for unlimited period.

The first plantlet formation in vitro was reported as early in 1940s. E. BELL (1946) reported it in *Tropaeolum* and *Lupinus*. CAPLIN and STEWARD (1948) used coconut milk for the first time in cultures of single cells and introduced various types of vessels for culture work, like rotating nipple flasks, and so on. R. J. GAUTHERET (1955) established habituated cultures of tumor cells and discussed the importance of light and temperature for root growth. The French botanists G. MOREL and R. A. WETMORE (1951) were the first to culture monocotyledonous tissue. He used meristem tip culture for the elimination of virus disease of orchids and discovered two unique opines of crown gall tissues. In orchids, plantlet formation was reported by MOREL during the 1960s in a program that became commercially viable.

The breakthrough in plant cell and tissue culture arose from a series of physiological investigations, principally by F. SKOOG and his coworkers, on the growth-regulating substances, including vitamins, hormones (particularly auxin and cytokinin), and organic complexes such as liquid coconut endosperm, and from the development of generalized tissue culture media by P. R. WHITE in the 1930s and 1940s and most successfully by T. MURASHIGE and F. SKOOG in 1962.

The demonstration of asexual embryos initiated in the cultures of carrot root cells in 1958 by J. REINERT, F. C. STEWARD, and K. MEARS was a confirmation of the concept of cell totipotency, that is, that each living cell contains all the genetic information. In 1959, REINERT showed a bipolar embryo formation in carrot and embryogenesis, the possibility of cryopreservation of cells and full regeneration of plantlets. VASIL and VASIL (1965) published a landmark work on single isolated cells of tobacco, which paved the way for cells cultures in genetics. It has contributed significantly to cell and protoplast cultures as well as somatic embryogenesis of cereals (Figure 4.1).

Extensive investigation continues to explore the potential of cell and tissue culture as an adjunct to crop improvement. Techniques include embryo rescue, freeing plants from virus and other pathogens, haploid induction, cryogenic storage of cells and meristem for germplasm preservation, the creation of new nuclear and cytoplasmic hybrids via protoplast fusion, and the exploitation of changes or dubbed somaclonal variation. It was recognized that cell and tissue culture technology would be required as an intermediary for the recombinant DNA technology.

FIGURE 4.1 In vitro culture of wheat anthers for callus formation, haploid plantlet regeneration, and doubled haploid production. (Courtesy of K. SOON-JONG, Suwon, South Korea. With Permission.)

4.1.1 Embryo Rescue

Failure to produce a hybrid may be due to pre- or postfertilization incompatibility. If fertilization is possible between two species or genera, the hybrid embryo aborts before maturation. When the cross is incompatible after fertilization, the embryo resulting from an interspecific or intergeneric cross can be rescued and cultured to produce a whole plant.

In 1873, P. von TIEGHEIM tried to culture immature embryos of *Mirabilis jalapa* on potato starch. In 1904, E. HANNIG using several crucifers made another attempt in embryo culture. 17 years later, M. MOLLIARD cultivated fragments of different plant embryos, whereas L. KNUDSON was able to demonstrate asymbiotic germination of orchid seeds in 1922. First crop species were embryo cultured in 1925 by F. LAIBACH (F1 embryos of interspecific crosses in *Linum* species). C. R. LARUE (1936) published a report on embryo culture of gymnosperms. The usefulness of coconut milk for growth and development of very young *Datura* embryos was discovered by J. OVERBEEK in 1941. Already in 1951, excised ovaries were cultured in vitro (NITSCH 1951), and K. KANTA demonstrated in 1960 the first test tube fertilization in *Papaver rhoeas*.

In 1970s, this technique has been used to produce new rice for Africa, an interspecific cross of Asian rice (*Oryza sativa*) and African rice (*O. glaberrima*). However, its application is mainly restricted to prebreeding programs, where exotic germplasm has to be introgressed to advanced breeding strains.

An exception was demonstrated by ▶ K. J. KASHA and K. N. KAO (1970). They applied the embryo rescue method in *Hordeum vulgare* × *H. bulbosum* crosses, although KONZAK et al. (1951) already used the technique for the same combination.

By chance, they discovered spontaneous chromosome elimination leading to haploid seedlings. This technique subsequently became routine in haploid production of barley breeding programs (Section 3.8.1).

4.1.2 Cell Fusion and Somatic Hybridization

As in sexual hybridization, cell fusion techniques also contribute to the combination of whole genomes. When sexual hybridization fails, then somatic hybridization can be an alternative. Different types of cells are dissociated from the tissues, walls are stripped from the cells and the membrane of the resulting protoplasts is in vitro cultured and regenerated into intact plants. This technique can produce novel combinations of nuclei, mitochondria, or chloroplasts.

Protoplasts are the smallest units able to regenerate a whole plant. Therefore, protoplasts cultures can serve to enlarge genetic variability by introducing somaclonal variation. However, the main interest of protoplasts is their capacity to fuse and to produce hybrids or cybrids, new organisms most often unknown in nature. Although, in nature, cybrids were made already million years ago by spontaneous cell fusion of algae and cyanobacteria resulting in plants able to photosynthesis, in 1909, first man-made fusion of plant protoplasts was attempted by E. KÜSTER, though the products failed to survive. However, P. S. CARLSON et al. (1972) succeeded.

Protoplasts of *Nicotiana glauca* and *N. langsdorffii* were isolated, fused, and induced to regenerate into plants. The somatic hybrids were recovered from a mixed population of parental and fused protoplasts by a selective screening method that relies on differential growth of the hybrid on defined culture media. The biochemical and morphological characteristics of the somatically produced hybrid were identical to those of the sexually produced amphiploid.

The first successful electrical fusion was demonstrated by the German U. ZIMMERMANN (1982) using mesophyll protoplasts of *Kalanchoe* or *Avena* species. In 1983, G. PELLETIER et al. produced an intergeneric cybrid using radish and rapeseed.

Indeed, protoplasts allow to transgress the barrier of the botanical species or genus. Naked protoplasts can accept without rejection external elements containing genetic information. The first protoplast fusion application for breeding was a cytoplasmic transfer from one genotype to another to induce male sterility (CMS) from mitochondrial origin. This type of male sterility is of particular interest for production of commercial hybrid seed, e.g. in *Brassiceae*. Another cytoplasmic transfer concerns the introduction of chloroplastic DNA to induce herbicide (atrazine) resistance, from *Solanum nigrum* to tomato (JAIN et al. 1988).

Already in 1989, in Sweden, ▶ SJÖDIN and GLIMELIUS successfully transferred resistance against *Phoma lingam* to *Brassica napus* by asymmetric somatic hybridization combined with toxin selection. Japan and China are particularly deeply involved in this work. During the past 20 years, successful protoplast cultures were reported in more than 100 plant species. Plants regenerated from protoplast have been obtained in about 60 species, including vegetables, medicinal plants, legumes, and other economic crops, as well as woody species such as poplar, elm, and rubber trees (Table 4.1).

For the first time, plants from the monocot *Polypogen fugax* were reported from China. Using cell fusion techniques, novel citrus hybrid varieties, such as "Oretachi" (orange plus trifoliate orange),

---

TABLE 4.1
Some Examples of Regenerated Plants Derived from Induced Somatic Protoplast Fusion

Fusion Species
Food crops

Hordeum vulgare + *Daucus carota*

Triticum aestivum + *Agropyron elongatum, Haynaldia villosa, Leymus chinensis*

Oryza sativa + *O. brachyantha, O. eichingeri, O. officinalis, O. perrieri, Echinochloa oryzicola, Hordeum vulgare, Porteresia coarctata*

Solanum tuberosum + *S. acaule, S. berthaultii, S. brevidens, S. bulbocastanum, S. cardiophyllum, S. phureja, S. commersonii, S. etuberosum, S. nigrum, S. pinnatisectum, S. papita, S. tarnii, S. verrucosum*

Solanum commersonii, S. phureja + *S. tuberosum*

Solanum melangena + *S. sanitwongsei, S. integrifolium*

Fodder plants

Medicago sativa + *M. arborea, M. coerulea, M. falcate, Lotus corniculatus*

Vegetables

Brassica campestris + *B. campestris, B. oleracea, B. nigra, Raphanus sativa, Moricandia arvensis, Sinapis turgida*

Daucus carota + *D. cupillifolius*

Diplotaxis catholica + *Brassica juncea*

Erucastrum gallicum + *Brassica campestris*

Lactuca sativa + *L. virosa*

Lycopersicon esculentum + *Solanum melongena, S. muricatum, S. ochranthum*

Moricandia arvensis + *Brassica juncea*

Phaseolus vulgaris + *P. coccineus, P. polyanthus*

Sinapis alba + *Brassica juncea*

Trachystoma balli + *Brassica juncea*

Monostroma oxyspermum + *Ulva reticulata*

Ornamental plants

Petunia hybrida + *P. parodii*

Fruits and forest trees

Actinidia deliciosa + *A. chinensis, A. kolomikta*

Citrus autrantium + *C. reticulata*

Citrus sinensis + *C. paradise, C. unshiu, C. sinensis* + *tangerine, C. sinensis* + *Poncirus trifoliata, P. trifoliata*

Source: NAKAJIMA, K., Biotechnology for crop improvement and production in Japan, Regional Expert Consultation on the Role of Biotechnology in Crop Production, FAO Regional Office for Asia and the Pacific, Bangkok, 1991; NAGATA, T. and BAJAJ, Y.P.S., *Somatic Hybridization in Crop Improvement*, Springer, New York, 2001.

FIGURE 4.2 Interspecific somatic hybrids in potato. From left to right: Solanum tuberosum, clone C-13, S. pinnatisectum, somatic hybrid between them (P7), Solanum tuberosum, clone C-13, S. etuberosum, somatic hybrid between them (E1-3). (Courtesy of Dr. A. K. SRIVASTAVA, Central Potato Research Institute, Shimla, India. With Permission.)

"Shuvel" (Satsuma mandarin plus navel orange), "Gravel" (grapefruit plus navel orange), "Murrel" (murcott plus navel orange), and "Yuvel" (Yuzu plus navel orange) were developed. However, in several cases, the regeneration frequency is low and has to be improved. TIWARI et al. (2015) reported several interspecific hybrids within the genus *Solanum* (Figure 4.2).

By asymmetric cell fusion techniques, the first commercial tobacco male sterile line was developed. Such asymmetric male sterile lines for commercial exploitation of F1 hybrids have been also bred in carrots, cabbages, and eggplants (NAKAJIMA 1991).

4.1.3 VIRUS FREEING

Twenty years after HABERLANDT'S fruitless in vitro studies, his successors developed the techniques useful in the investigations of plant development (Section 4.1). The character of studies dictated the use of small tissue fragments as experimental objects to have full control over their living processes. The small size of *explants* forced the introduction of culture sterility to protect them from invasions of microorganisms inhabiting plant surfaces and those present in air. The first real successes in keeping small fragments of plant tissues alive in vitro for a prolonged period came with the discovery of auxins in the late 1930s. The addition of indoleacetic acid (IAA) to the tissue culture media enabled cultivation of callus, roots, and, later, shoots for arbitrarily long time.

In the 1950s, new plant hormones, such as cytokinins and gibberellins were discovered. Now, researches were in almost full control of the tissue growth and development in sterile environment. Knowledge of plant development reached the level that permitted commercial applications of tissue culture. It started with freeing plants from viruses. This could be achieved by isolation of very small shoot apices (meristems) and growing them on sterile media, which provided for their transformation into small plants. These plants could then be taken out of sterile culture vessels and established in standard (unsterile) horticultural environment.

In 1952, MOREL and MARTIN were successful in regenerating a virus-free dahlia plant by the excision of some meristematic domes from virus-infected shoots. Three years later, the same authors were able to eliminate virus A and Y from a virus-infected potato. SEMAL and LEPOIVRE (1992) reported that a virus-free sweet potato was producing 40 t/ha in China by comparison of 20 t/ha produced before the meristem culture.

Virus eradication is dependent on several parameters. However, to take advantage of the nonuniform and imperfect virus distribution in the host plant body, the size of the excised meristem should be as small as possible. Only tips between 0.2 and 0.5 mm most frequently produce virus-free carnation plants. The explants smaller than 0.2 mm cannot survive and those larger than 0.7 produce plants that still contain mottle virus.

Today, no definitive explanation can be given to understand this virus eradication. Various explanations have been given, such as absence of plasmodesm in the meristematic domes, competition between synthesis of nucleoproteins for cellular division and viral replication, inhibitor substances, absence of enzymes present only in the cells of the meristematic zones, and suppression by excision of small meristematic domes. This last proposal could explain why some potato plant showing virus particles in the meristematic domes could regenerate a virus-free plant.

Besides dahlias, carnations, chrysanthemums, and cymbidiums were the first plants freed from viruses through the meristem culture. The development of cymbidium in tissue culture was more complicated. Small shoot tips, when placed on media, produced globular structures called protocorms. Protocorms could be easily stimulated to produce new protocorms by mechanical destruction of the growing point of shoots. In this way, the stock of protocorms grown in vitro could be easily increased from one to a few million during 1 year of culture.

Producers of planting material expected big money in the application of this propagation method and this stimulated fast development of commercial in vitro propagation methods in the late 1960s and early 1970s. To lower the costs of this labor-intensive production, specialized equipment and instruments were developed. Laminar airflow cabinets providing bacteria-free and fungi-free work environment were introduced in tissue culture production in the early 1970s. At the same time, the term "micropropagation" was coined, denoting vegetative propagation in vitro. Nowadays, hundreds of thousands of hectares are planted with virus-free plantlets.

In 2016, BETTONI et al. (2016) described an additional method of virus freeing. Cryotherapy is the promising tool. The main methods employed in cryotherapy are encapsulation–dehydration, vitrification, encapsulation–vitrification, and droplet vitrification, which are based on the immersion of preconditioned shoot tips in liquid nitrogen, followed by their recovery in vitro on to the culture media for regeneration of healthy plantlets. In this method, the biological material is exposed to liquid nitrogen at −196°C for a short period, normally 1 hour; under those conditions, infected cells in the vascular region and vacuolated and more differentiated cells of the apical dome are eliminated by the ultralow temperature due to ice crystallization, leaving only highly cytoplasmic cells in the meristematic region. Cryotherapy has been used successful in eradicating virus infections in several species of economic importance, such as *Prunus, Musa spp., Vitis vinífera, Fragaria ananassa, Solanum tuberosum, Rubus idaeus, Ipomea batatas, Dioscorea opposite,* and *Allium sativum.*

4.1.4 MICROPROPAGATION

It is the practice of rapidly multiplying stock plant material to produce a large number of progeny plants using modern plant tissue culture methods. Micropropagation is used to multiply noble plants such as those that have been genetically modified or bred through conventional plant breeding methods. It is also used to provide sufficient number of plantlets for planting from a stock plant, which does not produce seeds or does not respond well to vegetative reproduction.

Micropropagation is another technique of great significance to agriculture, which is based on in vitro culture of plant cell. Each and every cell of a plant has the potency to be regenerated into complete plant by tissue culture techniques and this property of the cell is known as totipotency. There are two basic strategies used to multiply plants in vitro:

1. proliferation of preformed buds and
2. bud induction termed adventitious regeneration.

The micropropagation process is conveniently divided into five stages: (a) stock plant selection, (b) establishment of aseptic cultures, (c) multiplication, (d) preparation for microplant establishment, and (e) microplant establishment.

The British botanist Frederick Campion STEWARD (1904–1993) discovered and pioneered micropropagation and plant tissue culture in the late 1950s and early 1960s. The 1970s were a decade of boom in the horticultural micropropagation. It was facilitated not only by technical improvements but also by the work of many researchers. The most important from a micropropagation point of view were investigations into the role of growth regulators in shoot apical dominance and shoot branching. Stimulation of shoot branching is the core of the most reliable micropropagation methods of today. From among hundreds of researches active in the field at that time, T. MURASHIGE and his students from the University of California (USA) distinguished themselves by developing practical and theoretical foundations for today's micropropagation. They outlined the method, dividing it into four stages. They also developed commercial micropropagation methods of many important ornamental plants. Media developed by them in early 1970s are still the most commonly used in many commercial laboratories.

The application of labor-intensive, high-tech, and, therefore, expensive micropropagation has been economically justified chiefly in the case of ornamental plants. This situation has not changed during the last 20 years. The need for sterility of cultures dictates the use of specialized and expensive equipment. Small dimensions of objects and the need for their careful handling during subcultures lower labor efficiency. Cost of labor constitutes 60%–70% of the total micropropagation costs.

Reduction in micropropagation costs can be achieved by its mechanization and automation. It is a difficult task because the structure of plant objects grown in vitro is very complicated. Only a few plants, like eucalypti or potato, produce simple elongated shoots with clearly visible internodes and leaves. Most ornamental plants branch intensively during culture, producing thickets of entangled shoots of different sizes. Efficient automatic division of such structures is impossible at the present levels of knowledge in three-dimensional computerized vision systems. Additionally, difficulties in sterilizing complicated robots and accessory equipment must be added. Using this technique, hundreds and thousands of seedlings can be generated in a very short time, starting from a limited number of explants. This method of multiplication has two major advantages. First, it accelerates multiplication of new elite cultivars. For example, after its release, a new sugarcane cultivar may take 5–10 years before sufficient planting material may be generated for a wide coverage in the farmer's field. But using micropropagation techniques, we may achieve a comparable level of multiplication within 2 years. The second advantage is that one can get a disease-free planting material because of the use of aseptic conditions during tissue culture.

For example, banana production is in decline in many regions because of pest and disease problems that cannot be addressed successfully through agrochemical control for reasons of cost and negative environmental effects. The problem is exacerbated because banana is reproduced clonally; the use of diseased mother plants therefore gives rise to diseased offspring. Micropropagation represents a means of regenerating disease-free banana plantlets from healthy tissue. In Kenya, banana shoot tips have been successfully tissue-cultured. An original shoot tip is heat-treated to destroy infective organisms and then used through many cycles of regeneration to produce daughter plants. A single section of tissue can be used to produce as many as 1500 new plants through 10 cycles of regeneration. Micropropagation of banana has had a tremendous impact in Kenya, among many other countries, contributing to improved food security and income generation. It has all the advantages of being a relatively cheap and easily applied technology and one that brings significant environmental benefits.

There are also reports of rejuvenation and increased vigor following tissue culture. This can also help in rescue of important materials, which are on the verge of extinction due to disease attack. In species like daffodils and gladioli, micropropagation techniques are being used to speed up the release of new varieties. In strawberry, millions of plants can be produced from a single mother plant in 1 year.

Another application was the in vitro selection. In maize, callus cultures were established resistant to T-toxin of *Helminthosporium maydis* (GENGENBACH and GREEN 1975). Similar experiments were carried out with potato against *Phytophthora infestans* (1979), with barley (1987) alfalfa against *Fusarium oxysporum* f. sp. *medicagnis* (1984), and in *Brassica napus* against *Phoma lingam* (1982), in tobacco against *Pseudomonas* and *Alternaria* toxins (1983).

Somatic embryogenesis is also being commercially applied. Synthetic seeds ("synseed") of rice and vegetables have been developed by the private sector in Japan. Triploid clones of rubber produced through somatic embryogenesis have out-yielded diploid standard clones by about 20% in China.

To facilitate reforestation and promote social forestry and agroforestry, India has established two tissue culture pilot plants, one at the National Chemical Laboratory, Pune, and the other at the Tata Energy Research Institute, New Delhi, with the production capacities of 5–10 million propagules/seedlings of elite/plus trees of several important species, such as *Eucalyptus tereticornis, E. camadulensis, Tectona grandis, Dendrocolamus strictus, Populus deltoides, Bambusa vulgaris,* and *B. tulda*. Micropropagation of teak, rattan, eucalyptus, bamboo, and other tree species has been adopted in several other countries.

Novel breeding strategies have also been tried to be applied in potato breeding by using microtubers for propagation and cropping, despite strategies based on genotypic selection. It concerns the so-called *true potato seed* (TPS). For the latter, the breeders are still skeptical about the place of TPS in highly developed markets of Europe and North America, although there are efforts to develop sexually propagated varieties. The basis for this scheme rests on the generation 2n gametes by the first meiotic division restitution (FDR).

Generally, one can distinguish several methods of micropropagation, for example, meristem culture, callus culture, suspension culture, embryo culture, or protoplast culture. Micropropagation has a number of advantages over traditional plant propagation techniques, but there are also disadvantages (Table 4.2).

TABLE 4.2
Advantages and Disadvantages of Micropropagation Techniques

Advantages	Disadvantages
Production of many plants that are clones of each other	Expensive, labor cost of more than 70%
To produce disease-free plants	A monoculture is produced after micropropagation, leading to a lack of overall disease resilience
High fecundity rate, producing thousands of propagules	Not all plants can be successfully tissue cultured
The only viable method of regenerating genetically modified cells or cells after protoplast fusion	Sometimes plants or cultivars do not come true to type after being tissue cultured
Multiplying plants which produce seeds in uneconomical amounts	Some plants are very difficult to disinfect of fungal organisms
When plants are sterile and do not produce viable seeds	
When seed cannot be stored	
Often produces more robust plants, leading to accelerated growth compared to similar plants produced by conventional methods	
Very small seeds (orchids) are most reliably grown from seed in sterile culture	
A greater number of plants can be produced per square meter	
Propagules can be stored longer and in a smaller area	

4.2 MOLECULAR TECHNIQUES IN PLANT BREEDING

Genetic engineering is at a similar stage of development today as were computers in the 1980s. Humans have been engineering life for thousands of years through selective breeding, but it was not until the discovery of DNA that we understood why and how these modifications occur. The past century saw a number of different experimental methods to change DNA in plants, ranging from exposing plants to high levels of radiation to modifying specific genes.

That era began in the early 1980s with the landmark reports of producing transgenic plants using *Agrobacterium* (MOOSE and MUMM 2008). An estimate by Syngenta Ltd. claims that genetic improvements have increased potential wheat yields by 20% since the early 1980s.

The first transgenic plant, a tobacco accession resistant to an antibiotic, was reported in 1983. Transgenic crops with herbicide, virus, or insect resistance, delayed fruit ripening, male sterility, and new chemical composition have been released to the market with growing intensity. For example, the *antisense* strategy was applied by Calgene Inc. in 1994, when the first commercial transgenic plant was created, a long shelf-life tomato, by the suppression of polygalacturonase activity due to an antisense gene (SMITH et al. 1988). However, this "Flavr Savr" tomato variety was removed from the trade 3 years later because of its disease susceptibility and lack of productivity! Later, other tomato varieties with long storage qualities were obtained by the utilization of an antisense RNA inhibition of ACC synthase or ACC oxidase, two ethylene precursors. By 1996, the commercialization of transgenic crops demonstrated the successful integration of biotechnology into plant-breeding and crop-improvement programs.

Breeding has become more perfect but soulless. During the past 30 years, the continued development and application of plant biotechnology, molecular markers, and genomics have established new tools for the creation, analysis, and manipulation of genetic variation and the development of improved cultivars. Molecular breeding is currently a standard practice in many crops.

4.2.1 MARKER-ASSISTED SELECTION

Marker-assisted selection (MAS) or marker-aided selection is an indirect selection process in which a trait of interest is selected based on a marker (morphological, biochemical, or DNA/RNA variation) linked to a trait of interest (e.g., productivity, disease resistance, abiotic stress tolerance, and quality), rather than on the trait itself. Sometimes it refers to SMART breeding (Selection with Markers and Advanced Reproductive Technologies) or "Precision breeding," that is, a genetic engineering method of reproducing a crop by retaining desirable traits and so produce a stronger hybrid. The technique was successfully used by ▶ Professor Nachum KEDAR (1920–2015), an Israeli scientist from the Hebrew University of Jerusalem, who applied the technique using beefsteak tomatoes to produce a fruit that would ripen on the vine and remain firm in transit. "Smart breeding" has been advanced as an alternative to transgenics as a way to produce plants that are resistant to various environmental problems. Molecular marker systems for crop plants were developed soon thereafter to create high-resolution genetic maps and exploit genetic linkage between markers and important crop traits (PATERSON et al. 1988).

For tracing quantitative traits, morphological markers were applied first. One of such approach was the development of mutations at the loci controlling plant morphology of maize (STADLER 1929). The variations represented through these mutations were observed as altered plant phenotypes that range from pigment differences and gross changes in the development to disease resistance response. Nevertheless, morphological markers have not been used extensively in practical breeding because of the limited availability of different mutants and that phenotype can be confirmed through inheritance analysis (Table 4.3).

Allozymes were available as the first biochemical genetic markers in the 1960s. Population geneticists took advantage of such marker system for their early research. Allozymes generally differ due to the substitution of single amino acid of different charge at a locus. Such changes in the amino acid composition often alter the charge or less often the conformation of the enzyme. This leads to a change in electrophoretic mobility of the enzymes (Table 4.3).

TABLE 4.3
Types and Characteristics of Several Markers in Breeding and Genetics

Characteristics	Morphological Traits	Isoenzymes	RFLPs	AFLPs and RAPDs	Microsatellites or Simple Sequence Repeats (SSRs, RAD, AAPs, SNPs, RRLs, CRoPS, ISSRs, and ISAPs)
Number of loci	Limited	Limited	Almost unlimited	Unlimited	High
Inheritance	Dominant	Codominant	Codominant	Codominant or dominant	Codominant
Positive features	Visible	Easy to detect	Utilized before the latest techniques were available; robust; Reliable; transferable across populations	Quick assay with many markers; multiple loci; high levels of polymorphism generated; quick and simple; inexpensive; multiple loci possible from a single primer possible; small amounts of DNA required	Well-distributed within the genome, many polymorphisms
Negative features	Possibly negative linkage to other characteristics	Possibly tissue-specific	Radioactivity required; time-consuming; laborious; rather expensive; large amounts of DNA required; limited polymorphism (especially in related lines)	High basic investment, patented; large amounts of DNA required; complicated methodology; problems with reproducibility; generally not transferable	Long development of the marker; expensive

Abbreviations: RFLPs, restriction fragment length polymorphisms; AFLPs, amplified fragment length polymorphisms; RAPDs, randomly amplified polymorphic DNAs; SSRs, simple sequence repeats; RAD, restriction-site associated DNA; AAPs, allele-specific associated primers; SNPs, single nucleotide polymorphisms; RRLs, reduced representation libraries; CRoPS, complexity reduction of polymorphic sequences; ISSRs, intersimple sequence repeats; ISAPs, inter-SINE amplified polymorphisms.

It was no later than 1974 that ▶ C. M. RICK and J. F. FOBES and 1980 that H. P. MEDINA-FINHO established a tight genetic linkage between a nematode resistance gene and an *Aps-1* isozyme allele in tomato, which opened the avenue of tagging gene(s) of agronomic importance. S. TANKSLEY and collaborators (1984) found peroxidases closely associated with male sterility or self-incompatibility in tomato. Soon after, similar markers were detected in rye, apple, maize, beans, and so on. The effect of isozymes and other proteins on plant's phenotype is usually neutral and both of them are often expressed co-dominantly, making the discrimination possible between homozygote and heterozygote. However, due to limited number of protein and isozyme markers and because of the requirement of a different protocol for each isozyme system, their utilization was also very limited in plant-breeding programs.

The discovery of enzymes, which cleave DNA at specific sequences and subsequently ligate to extrachromosomal DNAs of bacteria, permit the gene replication in a bacterial host, a process known as gene cloning. In the 1970s, *restriction fragment length polymorphisms* (RFLP) and Southern blotting were added to the toolbox of the geneticists. This marker system was thus developed in early 1980s (BOTSTEIN et al. 1980). RFLP markers are co-dominant and available in unlimited number because only a small (1000 nucleotide base pair) fragment is used for cloning from genomes that may contain a billion or more base pairs linearly arranged along the chromosomes (Table 4.3).

The *Taq* polymerase was found in the 1980s, and the *polymerase chain reaction* (PCR) developed shortly afterward. Since then, marker-aided analysis based on PCR has become routine in plant genetic research, and marker systems have shown their potential in plant breeding (PATERSON 1996). In order to start a PCR reaction, the following are needed:

1. A template: a DNA or RNA strand that we want to amplify.
2. Nucleotides (dCTP, dGTP, dATP, dGTP): the building blocks from which DNA is made.
3. Primers: short nucleotide strands that will anneal to the position at which we want to start and end the amplification.
4. DNA polymerase: the enzyme catalyzing the reaction.
5. Sometimes accessory additives like buffers and Mg^{2+}.

The amplification process:

1. Denaturation: the DNA molecule is split into two single strands of DNA by heating (carried out in a PCR thermocycler).
2. Hybridization: during a cool-down phase, the primers anneal to the target sequences of the single-stranded DNA.
3. Elongation: polymerase extends the DNA molecule starting at the primer sequences, so doubling the template numbers.
4. The process is repeated from Step 1 to achieve an exponential increase in the number of templates until the substrate runs out or the product accumulation halts the reaction.

PCR was originally developed to detect specific DNA sequences in a sample; though this is still the most used application, many other applications have been developed over the past decades. Nowadays, there exist different types of PCRs, for example, end-point PCR, qPCR (quantitative real-time PCR), RT-PCR (reverse transcription PCR), multiplex PCR, nested PCR, and digital PCR. Depending on the polymerases used, the PCR is featured with special properties like a hot start mechanism or proofreading activity to enable the best results.

With these technologies, new generations of DNA markers such as randomly amplified polymorphic DNA (RAPD), sequence-characterized amplified regions (SCARs), sequence-tagged sites (STS), single polymorphic amplification test (SPLAT), variable number of tandem repeats (VNTRs), amplified fragment length polymorphism (so-called AFLPs are fragments of DNA that have been amplified using directed primers from restriction digestion of genomic DNA; KARP et al. 1997) are obtained.

DNA amplification fingerprinting (DFA), single-strand conformational polymorphism (*SSCP*), single-nucleotide polymorphism (SNP), microsatellites or short tandem repeats (*STRs*), *DNA micro arrays*, and *rDNA ITS* (SCHENA et al. 1995) were introduced into the modern plant-breeding systems. Some of these markers are relatively simple, easy to use, automatable, often co-dominant, near infinite in number, and comparatively faster to assay (Table 4.3). Furthermore, new single-nucleotide polymorphic markers based on high-density DNA arrays, a technique known as "gene chips" (CHEE et al. 1996), have been developed. With gene chips, DNA belonging to thousands of genes can be arranged in small matrices (or chips) and probed with labeled cDNA from a tissue of choice. DNA chip technology uses microscopic arrays (or microarrays) of molecules immobilized on the solid surfaces for biochemical analysis (LEMIEUX et al. 1998). An electronic device connected to a computer may read this information, which will facilitate MAS in crop breeding. There are basically five eras in genetic marker evolution:

1. Morphology and cytology in early genetic until late 1950s,
2. Protein and allozyme electrophoresis in the "pre-recombinant DNA time" (1960–1970s),
3. RFLP and minisatellites in the "pre-PCR age" (1970–1985s),
4. Random amplified polymorphic DNA, microsatellites, expressed sequence tags, sequence tagged sites, and amplified fragment length polymorphism in the "Oligocene period" (1986–1995), and
5. Complete DNA sequences with known or unknown function as well as complete protein catalogs in the current computer robotic cyber genetics generation (1996 onward).

By 2005, the DNA of a number of organisms had been completely mapped, including bacteriophage, bacteria, yeast, nematode, or the plant *Arabidopsis thaliana*. Recent projects are concerned of major crop species, such as maize, rice, and wheat. Gene maps of individual chromosomes carrying important clusters of genes are available in tomato, maize, rice, and wheat.

Analysis of gene function indicates that all living organisms hold genes in common. Soon all our major crop plants will be mapped. The name of the next emerging field has already been coined: "proteonomics," which will unravel the protein changes involved with gene function and development. Although there have been numerous mapping and quantitative trait loci (QTLs) mapping studies for a wide range of traits in diverse crop species, relatively few markers have actually been implemented in practical breeding programs. The main reason for this lack of adoption is that the markers used have not been reliable in predicting the desired phenotype. In many cases, this would be attributable to a low accuracy of mapping studies or inadequate validation. However, despite the lack of examples of MAS being practiced, there is an optimism regarding the role in the future by leading researchers (COLLARD et al. 2005).

The fields of application of molecular markers are permanently growing: they are used in technical aspects of screenings, for characterization of germplasm, for cultivar improvement through DNA-content manipulation, for genetic dissection of traits, for linkage map construction, for detection of genetic linkages, for marker to gene (map-based) cloning, for MAS, or for commercial variety identification, protection, and purification.

The frequent application of such markers systems transforms conventional plant breeding into MAS and/or "molecular breeding" (Figures 4.3 and 4.4). However, before initiating large-scale utilization of markers in plant-breeding programs, it is still necessary to have clear concepts of gene and genetic marker and their characteristics and purposes, so that the technique can be used effectively. Although recent advances in molecular genetics have promised to revolutionize agricultural practices, there are several reasons why molecular genetics can never replace traditional methods of agricultural improvement. Despite important strides in marker technologies, the use of MAS has stagnated due to the improvement in quantitative traits. As common statistical methods, biparental

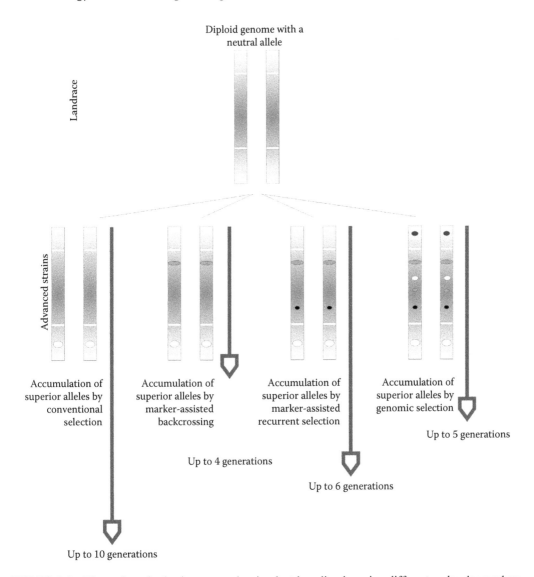

FIGURE 4.3 Illustration of selection approaches in plant breeding by using different molecular markers. (Redrawing after GUPTA, P. K. et al., *Plant Breed.*, 132, 431–432, 2013. With Permission.)

mating designs are less suitable for the detection of quantitative trait loci (QTL). Genomic selection (GS) has been proposed to address these deficiencies. GS predicts the breeding values of lines in a population by analyzing their phenotypes including their high-density marker scores. A key to the success of GS is that it incorporates all marker information in the prediction model, thereby avoiding biased marker effect estimates and capturing more of the variation due to small-effect QTLs. In simulations, the correlation between true breeding value and the genomic estimated breeding value (GEBV) has reached levels of 0.85 even for polygenic low heritability traits. This level of accuracy is sufficient to consider selecting for agronomic performance using marker information alone. Such selection would substantially accelerate the breeding cycle, enhancing gains per unit time. It would dramatically change the role of phenotyping, which would then serve to update prediction models and no longer to select lines (HEFFNER et al. 2009).

An impressive example of MAS was presented by ELLUR et al. (2016). MAS provided an unprecedented opportunity for precise transfer of genes responsible for biotic and abiotic stress-tolerant QTLs, that is, genes into various populations of Basmati rice varieties. The evolution of

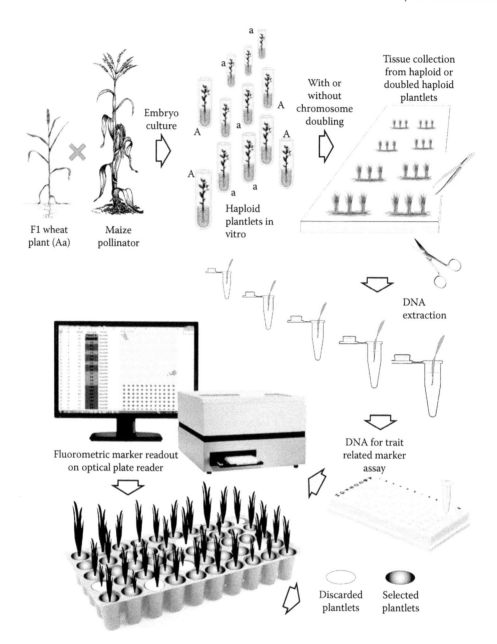

FIGURE 4.4 Scheme of MAS in wheat breeding including haploid and/or doubled haploid progeny. (Courtesy of R. SCHLEGEL. With Permission.)

marker technology revealed several Basmati quality traits. A recessive gene-controlling aroma was mapped on chromosome 8. The allele, *badh2*, has been cloned.

It was shown that the aroma is due to an 8bp deletion in aromatic varieties. Based on the *Badh* gene, a new marker, *nksdl*, was developed in order to differentiate between aromatic and non-aromatic genotypes. Another QTL for milled rice length was mapped on chromosome 3, which explained the phenotypic variance of as high as 75%. A major QTL conferring tolerance to phosphorus efficiency is identical with the *Pup1* locus, and so on. Moreover, the MAS enabled pyramiding of genes governing stress tolerance. By employing all that knowledge, the variety "Pusa Basmati 1" could be stepwise improved as "PB1121," "Pusa Basmati 6," and "Pusa Basmati 1," with resistance to blast and sheath blight, salinity tolerance, and others.

4.2.1.1 Plant Genomics

The neologism "omics" informally refers to a field of study in biology ending in -*omics*, such as genomics, proteomics, or metabolomics. The related suffix -*ome* is used to address the objects of study of such fields, such as the genome, proteome, or metabolome, respectively. Omics aims at the collective characterization and quantification of pools of biological molecules that translate into the structure, function, and dynamics of an organism. Functional genomics aims at identifying the functions of as many genes as possible of a given organism. It combines different -omics techniques such as transcriptomics and proteomics with saturated mutant collections.

Genomics[2] research generated new tools, such as functional molecular markers and informatics, as well as new knowledge about statistics and inheritance phenomena that increases the efficiency and precision of crop improvement. In particular, the elucidation of the fundamental mechanisms of heterosis and epigenetics, and their manipulation, shows great potential. Knowledge of the relative values of alleles at all loci segregating in a population could allow the breeder to design a genotype in silico and to practice whole genome selection. Costs currently limit the implementation of genomics-assisted crop improvement, particularly for inbreeding and/or minor crops. Nevertheless, marker-assisted breeding and selection will gradually evolve into genomics-assisted breeding.

4.2.1.1.1 Functional Markers

Functionally characterized genes, EST, and genome-sequencing projects have facilitated the development of molecular markers from the transcribed regions of the genome. Among the more important and popular molecular markers that can be developed from ESTs are SNPs, simple sequence repeats (SSRs), or conserved orthologous sets of markers. Putative functions can be deduced for the markers derived from ESTs or genes using homology searches (BLASTX) with protein databases (Figure 4.5).

4.2.1.1.2 Transcriptomics

The salient challenge of applied genetics and functional genomics is the identification of the genes underlying a trait of interest, so that they can be exploited in crop-improvement programs. Macro- and microarrays have been successfully used in many plant species to understand the basic physiology, developmental processes, and environmental stress responses and to identify and genotype mutations. However, use of these technologies for applied aspects in plant breeding has been limited because, except for near-isogenic lines, differential gene expression is caused not only by the trait of interest but also by the variation present in the genetic background. Therefore, background effects must be eliminated to establish a functional association between the level of gene expression and a given trait. In an early study, POTOKINA et al. (2004) used 10 barley genotypes that were characterized for six malting quality parameters and a cDNA array with 1400 unigenes to identify candidate genes for each of the six malting parameters. Such functional association analysis provides a useful link between functional genomics and plant breeding.

Microarray-based gene expression data between two genetically different lines can also be used to identify single feature polymorphisms (SFPs) for SNP detection. They can be exploited to

---

[2] The science of how the genes in organisms interact to express complex traits, that is, the field of research that aims to determine the function of newly discovered genes. It attempts to convert the molecular information represented by DNA into an understanding of gene functions and effects. It also entails research on the protein function (proteomics) or, even more broadly, the whole metabolism (metabolics) of an (plant) organism.

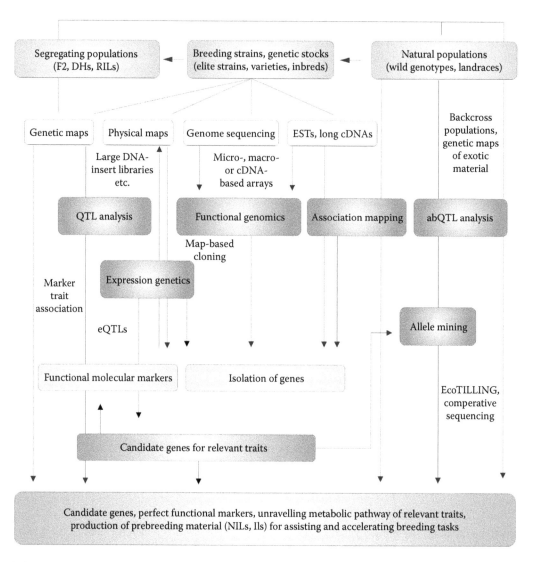

FIGURE 4.5 Schematic drawing of an integrated system of exploitation of genomics for different applications in research and breeding. (Redrawing after VARSHNEY, R. K. et al., *Trends Plant Sci.*, 10, 621–630, 2005. With Permission.)

develop functional markers. In a recent study using Affymetrix GeneChip[3] expression data, more than 10,000 SFPs have been identified between two genotypes of barley, a species with a large and complex genome. Meanwhile, the "Functional Genomics Group" at the University of Bristol, K. J. EDWARDS, UK, produced advanced genechips, particularly, for wheat. It provides information about the SNP markers; for example, the sequence upon which they are based, obtain primers used for their identification, and identify the haplotypes of common UK varieties. In 2014, both the 35K "Wheat Breeders Axiom Array" and the 820k "Wheat High Density Array" were offered from

---

[3] Affymetrix, Inc. is an American company that manufactures DNA microarrays. It was founded by Dr. Stephen FODOR in 1992. During the late 1980s, Affymetrix developed methods for fabricating DNA microarrays, called "GeneChip," using semiconductor manufacturing techniques.

Affymetrix. On January 1, 2015, the "Functional Genomics Group" developed the 820k Wheat HD array (equivalent to several hundred million data points).

4.2.1.1.3   Expression Genetics

Another approach was proposed by R. JANSEN and J.-P. NAP (2001). They used gene expression data in the QTL analysis. By analyzing the expression levels of genes or clusters of genes within a segregating population, it is possible to map the inheritance of that expression pattern (Figure 31). Expression QTLs (eQTLs) can be classified as *cis-* or *trans*-acting based on the location of the transcript compared with that of the eQTL-influencing expression of that transcript. Because of this feature, eQTL analysis makes it possible to identify factors influencing the level of mRNA expression. The regulatory factor (second-order effect) is of specific interest because more than one QTL can be putatively connected to a *trans*-acting factor. Thus, the mapping of eQTLs allows multifactorial dissection of the expression profile of a given mRNA, cDNA, protein, or metabolite into its underlying genetic components, as well as localization of these components on the genetic map. Subsequently, the eQTL analysis for each gene or gene product analyzed in the segregating population can identify the regions of the genome influencing its expression. Furthermore, for plant species in which the sequence of the whole genome is available, the annotation of those genomic regions is helpful for the identification of the genes and the regulatory sequences involved in their expression (VARSHNEY et al. 2005).

4.2.1.1.4   Backcross QTL

Many useful traits have been transferred from wild relatives into crop species, most of which are controlled by single genes or gene clusters conferring resistance to various diseases. For transferring the QTLs of agronomically important traits from a wild species into a crop variety, an approach named "advanced backcross QTL analysis" (abQTL) was proposed by Steven TANKSLEY and Clare NELSON (1996). In this approach, a wild species is backcrossed to a superior cultivar with the selection for domestication traits. Selection is imposed to retain individuals that exhibit domestication traits such as nonshattering. The segregating BC_2F_2 or BC_2F_3 population is then evaluated for traits of interest and genotyped with polymorphic molecular markers. These data are then used for QTL analysis, potentially resulting in the identification of QTLs while transferring these QTLs into adapted genetic backgrounds. The abQTL approach has been evaluated in many crop plant species to determine whether genomic regions (QTLs) derived from wild or nonadapted germplasm have the potential to improve the yield. However, the wild species' chromosome segments mask the magnitude of some favorable effects that were identified for certain introgressed alleles. Thus, the yield-promoting QTL did not make a substantial contribution to the phenotype and the best lines were inferior to commercial cultivars. However, in tomato, the pyramiding of independent yield-promoting chromosome segments resulted in new varieties with increased productivity under normal and stress conditions. One disadvantage is that the value of the wild accession for contributing useful QTL alleles is unknown before a major investment in mapping. Another major limitation to abQTL is the difficulty in maintaining an adequate population size in selected backcross populations so that useful alleles are not lost and the QTLs can be accurately mapped.

4.2.1.1.5   Exploitation of Germplasm

Most of the genetic variation present in wild species has a negative effect on the adaptation of plants to agricultural environments (cf. Chapter 7). The challenge is to identify and make use of the advantageous alleles in a breeding program. This is particularly the case for quantitative traits because the value of a wild or exotic accession for contributing useful alleles cannot be determined a priori with certainty. The concern that breeding is reducing genetic diversity is not always clear. Elena KHLESTKINA et al. (2004) examined the genetic diversity of cultivated wheat that was sampled over 50 years in Europe and Asia and found no significant change, whereas others reported a loss of 19% of SSR alleles over 100 years. A recent example provided by MIEDANER and colleagues (2016) demonstrated the correlated effects of exotic pollen-fertility restorer genes on the agronomic and quality traits of hybrid rye.

4.2.1.1.6   Association Mapping

The primary goal of association mapping is to detect correlations between genotypes and phenotypes in a sample of individuals based on the linkage disequilibrium (LD). Biparental populations such as doubled haploids (DHs), F2, or recombinant inbred lines (RILs) have been widely used to construct molecular marker maps and identify genes or QTLs for traits of interest. However, these mapping populations are the products of just one or a few cycles of meiotic recombination, limiting the resolution of genetic maps, and are often not representative of germplasm that is actively used in breeding programs (Figure 4.5). By contrast, the use of unrelated genotypes or natural populations in association mapping can provide a greater resolution for identifying genes responsible for variation in a quantitative trait.

4.2.1.1.7   Allele Mining

A breeder would ideally like to know the relative value of all alleles for genes of interest in the primary germplasm, an unlikely prospect. Information can be gathered for all alleles of a fully characterized gene in a germplasm collection and this process is known as "allele mining." A strategy based on targeting induced local lesions in genomes (TILLING), called EcoTILLING, was developed for detecting multiple types of polymorphisms (unknown SNPs) in germplasm collections (e.g., natural population, breeding, or genebank materials). EcoTILLING allows natural alleles at a locus to be characterized across many germplasms, enabling both SNP discovery and haplotyping. This can be done at a fraction of the cost of SNP genotyping or haplotyping methods, which require large-scale sequencing. Haplotypes generated after EcoTILLING across a range of germplasm can be binned (sorted into groups) and confirmatory sequencing done on only the unique haplotypes (COMAI et al. 2004).

EcoTILLING provides a series of alleles for those genes that are involved in important processes of the plant, even though the known variants for these genes have not been observed through genetic studies. Extensive information about the candidate genes in terms of structure and regulation or phenotypic expression is important for designing the primer pairs for EcoTILLING. The necessity of also screening regulatory regions, which are often distant from the effector genes, indicates that selecting the candidate sequences for EcoTILLING is not a trivial task. After identifying all alleles that are available, they must be evaluated for their relative value in adapted genotypes in the target environment. These analyses might help in designing synthetic alleles that are superior to those found in nature. This could be accomplished by recombining the coding regions of genes either randomly (e.g., by gene shuffling) or deliberately (e.g., by domain swapping).

This new term, defined by the development of biotechnology, refers to the investigations of whole genomes by integrating genetics with informatics and automated systems. Genomic research aims to elucidate the structure, function, and evolution of past and present genomes (LIU 1997). Some of the most dynamic fields concerning agriculture are the sequencing of plant genomes, comparative mapping across species with genetic markers, and objective-assisted breeding after identifying candidate genes or chromosome regions for further manipulations. As a result of genomics, the concept of gene pools has been enlarged to include transgenes and native exotic gene pools that are becoming available through comparative analysis of plant biological repertoires (LEE 1998). In twenty-first century, the improvement in sequencing technologies has allowed the complete deciphering of genomes in many species of agricultural interest, such as rice, maize, tomato, potato, pepper, eggplant, cucumber, melon, and others (MICHAEL and VANBUREN 2015).

Nowadays, the finding of new genes that add value to agricultural products seems to be very important in the agribusiness. Unique gene databases are being assembled by the industry with the massive amount of data generated by genomics research. A new term "biosource" was coined recently to refer to a fast and effective licensed technology of pinpointing genes.

With this method, a "benign" virus infects a plant with a specific gene that allows researchers to observe directly its phenotype. Biosource replaces the standard time-consuming approach of first mapping a gene to subsequently determine its exact function. Gene identification in DNA libraries coupled with biosource technology and an enhanced ability to put genes into plants will be routine for improving crops in the next decades.

Understanding the biological traits of one species may enhance the ability to achieve a high productivity or a better product quality in another organism. Today, DNA markers and gene sequencing provides quantitative means to determine the extent of genetic diversity and to establish objective phylogenetic relationships among organisms. "Gene chips" and transposon tagging will provide new dimensions for investigating gene expression. Molecular biologists study not only individual genes but also how circuits of interacting genes in different pathways control the spectrum of genetic diversity in any crop species. For example, more information will be available on why plant resistance genes are clustered together, or what candidate genes should be considered when manipulating QTLs for crop improvement.

Genomics may provide a means for the elucidation of important functions that are essential for crop adaptability. Regions of the world should be mapped by combining data of geographical information systems, crop performance, and genome characterization in each environment. In this way, plant breeders can develop new cultivars with the appropriate genes that improve fitness of the promising selections. Fine-tuning plant responses to distinct environments may enhance crop productivity. Development of cultivars with a wide range of adaptation will allow farming in marginal lands. Likewise, research advances in gene regulation, especially those processes concerning plant development patterns, will help breeders to fit genotypes in specific environments.

Photoperiod insensitivity, flowering initiation, vernalization, cold acclimation, heat tolerance, and host response to parasites and predators, are some of the characteristics in which advanced knowledge may be acquired by combining molecular biology, plant physiology and anatomy, crop protection, and genomics. Multidisciplinary cooperation among researchers will provide the required holistic approach to facilitate research progress in these subjects.

4.2.1.2   Genomic Selection

GS or genome-wide selection is a form of marker-based selection, referring to the simultaneous selection for many (tens or hundreds of thousands of) markers, which cover the entire genome in a dense manner, so that all genes are expected to be in linkage disequilibrium with at least some of the markers. In GS genotypic data, that is, genetic markers across the whole genome are used to predict complex traits with sufficient accuracy to allow selection on that prediction alone. Selection of desirable individuals is based on GEBV, which is a predicted breeding value calculated using an innovative method based on genome-wide dense DNA markers (NAKAYA and ISOBE 2012). GS does not need significant testing and identifying a subset of markers associated with the trait, that is, GS can remove the need to search for significant QTL–marker loci associations individually (DESTA and ORTIZ 2014). Thus, QTL mapping with populations derived from specific crosses can be avoided in GS. However, it does need to develop GS models, that is, the formulae for GEBV prediction. In this process, phenotypes and genome-wide genotypes are investigated in the training population, that is, a subset of a population, to predict significant relationships between phenotypes and genotypes using statistical approaches. Subsequently, GEBVs are used for the selection of desirable individuals in the breeding phase instead of the genotypes of markers used in traditional MAS. For accuracy of GEBV and GS, genome-wide genotype data are necessary and require high marker density in which all QTLs are in linkage disequilibrium with at least one marker.

The use of high-density markers is one of the fundamental features of GS. GS is possible only when high-throughput marker technologies, high-performance computing, and appropriate new statistical methods are available. This approach has become feasible due to the discovery and development of a large number of SNPs by genome sequencing and new methods to efficiently genotype large number of SNP markers. The ideal method to estimate the breeding value from genomic data is to calculate the conditional mean of the breeding value given the genotype at each QTL. This conditional mean can only be calculated by using a prior distribution of QTL effects, and thus this should be part of the research to implement GS. In practice, this method of estimating breeding values is approximated by using the marker genotypes instead of the QTL genotypes, but the ideal method is likely to be approached more closely as more sequence and SNP data are obtained (JIANG 2015).

4.2.2 Genetically Modified Crop Plants

The commercial application of genetically modified crops (GM crops) began in the mid-1990s. Since then, the technology has spread rapidly around the world, both in developed and developing countries (Figure 4.6). The mean impacts of GM crops are increase yield by about 20%, reduced pesticide quantity and or costs by about 40%, and increase in farmer profit by about 65%.

In 2016, North America was the largest market for genetically modified food and likely to remain as market leader in terms of revenue during the forecast period. China and India are the major countries in Asia-Pacific that are expected to emerge as major markets for genetically modified food market. Also, other developing countries such as Brazil are anticipated to grow at a robust CAGR over the forecast period. The US government is focused toward the safety of GM products. The industry is regulated by the Department of Agriculture, for farm biotechnology, and the Food and Drug Administration (FDA), which governs food and its ingredients. The developers of the genetically modified food products are intensely involved in certifying their safety.

For instance, FDA depends on a consultative process with developers who voluntarily present their plans to the agency before marketing the products. The market of GM food and food production, that is, high nutritional content in the food, high production quantity, and less requirement of pesticides, strongly drive the development of efficient methodology of breeding.

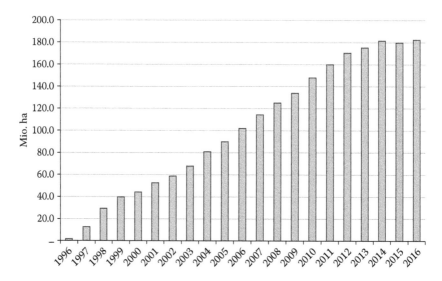

FIGURE 4.6 Increase of cultivated area with genetically modified crops wordwide between 1996 and 2016. (R. SCHLEGEL; compiled after data of Anonymous, GM Crops List, GM Approval Database-ISAAA.org; www.isaaa.org, Retrieved January 30, 2016, 2016a.)

4.2.2.1 Transgenesis

A transgene is a gene from a noncrossable species, or it is a synthetic gene. It represents a new gene pool for plant breeding. All crops have been genetically modified from their original wild state by domestication, selection, and controlled breeding over long periods of time. What is the major difference between "old" and "new" methods of crop improvement? Basically, it is the number of genes transferred to the offspring in each case. On average, plants contain approximately 80,000 genes, which can recombine during the process of sexual hybridization. The offspring may therefore inherit around 1000 new genes as a result of this recombination. This is equivalent to a 0.0125% change in the genome. By contrast, only one or two new genes are transferred during plant transformation. As this represents a 0.0025% change in the genetic information of the plant, it is argued that plant transformation provides a more precise approach to crop improvement than sexual hybridization.

When it became clear that DNA could be inserted into the DNA of higher plants by various techniques including the gene gun, a new era of research began. The most promising vector for dicotyledonous plants has been the tumor-inducing plasmid of *Agrobacterium tumefaciens*, a bacterium that normally incorporates its DNA in the host as part of the infection process. Even genetic engineering is not new!

The important prerequisites for successful integration of the *Ti* plasmid DNA from *Agrobacterium tumefaciens* in higher plants were published by CHILTON et al. (1977). In 1979, MARTON et al. developed the cocultivation procedure for the genetic transformation of plant protoplasts with *A. tumefaciens*, and the first naked DNA transformation of protoplasts was realized by KRENS et al. (1982).

Transgenic plants were first created in the early 1980s by four groups working independently at Washington University in St. Louis, Missouri, USA (1982), the Rijksuniversiteit in Ghent, Belgium, Monsanto Company in St. Louis, Missouri, and the University of Wisconsin, St. Paul, USA.

On the same day in January 1983, the first three groups announced at a conference in Miami (USA) that they had inserted bacterial genes into plants. The fourth group announced at a conference in Los Angeles (USA) in April 1983 that they had inserted a plant gene from one species into another species.

A group of Washington State University, headed by M.-D. CHILTON (BEVAN et al. 1983), had produced cells of *Nicotiana plumbaginifolia*, a close relative of ordinary tobacco, that were resistant to the antibiotic kanamycin (FRAMOND et al. 1983). ▶ J. SCHELL (1936–2003) and M. MONTAGU, working in Belgium (HERRERA-ESTRELLA et al. 1983), had produced tobacco plants that were resistant to kanamycin and to methotrexate, a drug used to treat cancer and rheumatoid arthritis (SCHELL et al. 1983). R. FRALEY, S. ROGERS, and R. HORSCH at Monsanto had produced *Petunia* plants that were resistant to kanamycin (FRALEY et al. 1983). The Wisconsin group, headed by J. KEMP and T. HALL, had inserted a bean gene into a sunflower plant (MURAI et al. 1983).

The major technical advance was the demonstration that transgenic plants could be regenerated from leaf disks following co-cultivation with *Agrobacterium*. Subsequently, transgenic plants have been produced from many families using this approach or modifications of it. Many other techniques have been tested for direct gene transfer in protoplasts. Electroporation was one of them, that is, the application of high-voltage electric pulses to cells to induce transient membrane pores, allowing entry of macromolecules including DNA. For the first time, FROMM et al. (1985) were successful in maize, rapeseed, soybean, rice, tobacco, and so on. The development of *Agrobacterium* as a vehicle for routine genetic transformation of plants was limited because of host-range limitations and difficulties with regeneration from protoplasts. POWELL-ABEL et al. (1986) developed the first transgenic plants using cDNA of coat protein gene of TMV virus-resistant tobacco and tomato. In the same year, Crossway et al. published transformation of tobacco protoplasts by direct DNA microinjection, and first field trials of transgenic plant were initiated.

In this context, bombardment of intact plant cells and tissues with high-velocity, DNA-coated microprojectiles were considered crude but effective (KLEIN et al. 1987). Already in 1966, plant virologists as MACKENZIE et al. have employed this technique to wound plant cells and facilitate entry of viral particles or nucleic acids. Recently, about 10 methods have been applied more or less efficiently depending on the recipient crop and the target genes available:

- Gene transfer by means of *Agrobacterium* and co-cultivation
- Gene transfer by means of viruses
- Gene transfer by microprotoplasts
- Gene transfer by means particle gun
- Gene transfer via microinjection
- Gene transfer by means of electroporation
- Gene transfer by utilization of micro LASER beams
- Gene transfer via chimeric DNA/RNA plasty
- Gene transfer by help of liposomes

As target of DNA introduction intact tissue, cell cultures, protoplasts, pollen cells, ovules, or chloroplasts are utilized. It is differentiated between input traits, such as resistance to herbicides, insects, viruses, fungi, bacteria, nematodes, or abiotic factors.

The early transgenic plants were laboratory specimens, but subsequent research has developed transgenic plants with commercially useful traits such as resistance to herbicides, insects, and viruses. With the advancement in DNA technology and understanding the structure and function of gene and its products, transgenics are coming out from various research program all over the world.

The ability to move new genes into old plants has led to imaginative flights of fancy: a new range of disease and stress-resistant plants, nitrogen fixation of nonlegumes, and amino acid-balanced plant protein.

The concept of improving agriculture in the traditional sense by recombinant DNA technology became a reality with two dramatic inventions. One was that soybeans could be induced to be resistant to the nonselective, environmentally benign herbicide glyphosate ("Roundup"). The other was that the gene *Bt*, from the bacteria *Bacillus thuringensis*, could be transferred to the crop plant. *B. thuringiensis* was used as an insecticide by the 1950s. The first gene encoding the *Bt* toxin was cloned by 1981. The regulation of *Bt* gene was known in 1986 and inserted into maize in 1990. *Bt* hybrids were first sold in 1997.

The creation of "Roundup-Ready" soybeans was to have an extremely rapid rate of adoption, unsurpassed in agriculture. *Bt* cotton was also rapidly adopted and *Bt* maize somewhat less because the cost–benefit ratio was not as high, as the maize rootworm incidence varied with location. By 1999, herbicide-resistant soybean accounted for 57% of the crop area, *Bt* cotton 55% and *Bt* maize 22% in the United States.

Many traits controlled by transgenes are gain-of-function traits. It often is these types of traits that provide the added value expected from improved cultivars. Strong promoters controlling constitutive expression, such as cauliflower mosaic virus 35S promoter, currently drive these transgenes.

The production of transgenic plants has to be seen in connection with traditional plant breeding, where humans since prehistoric time have selectively bred particular wild plants with good characteristics. Qualities such as strength, yield, resistance against noxious organisms, and the ability to withstand wind and weather were improved by crossing the best individuals with each other. It still takes 10–15 years to develop a new variety using traditional breeding methods. Gene-transfer techniques can reduce this time by a half and make it possible to selectively transfer genes so that it is possible to know exactly which characteristics have been introduced. It also gives the potential to introduce genes from nonrelated species or artificial genes.

By years, the boon of biotechnology has been confined to few developed countries, and developing countries are still in vogue to harvest the benefits of modern technology. Australia was the first country in the world to have engineered and released a microorganism for biological control of crown gall in plants. Australia's first field trial of transgenic plants took place in 1991, involving transgenic potato varieties resistant to leafroll virus, developed by the Commonwealth Scientific and Industrial Research Organization (CSIRO, Australia) Division of Plant Industry. By 1994, the first genetically engineered crop plant was approved for commercial marketing—the "Flavr-Savr" tomato—designed to slow fruit ripening and increase shop life (LEVETIN and McMAHON 1996). During 1988–1991, several laboratories in Japan, using electroporation or *Ti* or *Ri*, reported successful production of transgenics in *Oryza saliva, Citrus sinensis, Cucumis melo, Lactuca saliva, Solanum tuberosum, Nicotiana tabacum, Morus alba, Actinidia chinensis, Atropa belledona, Brassica oleracea, Lycopersicon esculentum, Vigna angularis,* and *Vitis vinifera* (NAKAJIMA 1991). For effective gene expression, promoter regions/genes have been identified for tissue-, age-, and pathogen-specific expressions of the genes under transfer. Stable and useful transgenics for protein quality and for viral resistance have been produced in rice, potato, tomato, melon, and tobacco.

US vegetables farmers are already benefiting from growing transgenic squash cultivars resistant to *zucchini yellow mosaic virus, watermelon mosaic virus,* and *cucumber mosaic virus,* which were deregulated and commercialized since 1996. *Bt*-sweet maize has also proven effective for control of some lepidopteran species and continues to be accepted in the fresh market in the USA, and *Bt*-fresh-market sweet maize hybrids are released almost every year. Likewise, transgenic *Bt*-eggplant bred to reduce pesticide use is now grown by farmers in Bangladesh. Transgenic papaya cultivars carrying the coat-protein gene provide effective protection against *Papaya ring spot virus* elsewhere. The transgenic "Honey Sweet" plum cultivar provides an interesting germplasm source for *Plum pox virus* control. Enhanced host plant resistance to *Xanthomonas campestris* pv. *musacearum,* which causes the devastating banana *Xanthomonas* wilt in the Great Lakes Region of Africa, was achieved by plant genetic engineering (Table 14). Similarly, resistance to the fungus *Puccinia triticina* in hexaploid wheat can be mediated by transgenesis expressing an antifungal defensin. The varieties "Bobwhite" and "Xin chin 9" were transformed with a chimeric gene encoding an apoplast-targeted antifungao plant defensin MtDEF4.2 from *Medicago truncatula* (KAUR et al. 2017).

Evidence from laboratory studies of plants and their fungal pathogens indicates that both parties can fling RNAs back and forth into the other's cells. The plant produces a small RNA precursor, either a long double-stranded RNA or a pre-microRNA, with sequence similarity to a fungal gene. Evidence points to the idea that the small RNA precursors can pass directly to the fungal cell or undergo processing into small RNAs prior to transfer. If the precursor leaves the plant intact, the fungus's processing machinery chops it up. In either case, the result is a plant small RNA inside the fungal cell, though the mechanism of transfer remains unknown. Upon additional processing in the fungal cell, a single strand of the small RNA becomes part of the RNA-induced silencing complex, which then destroys an mRNA with a matching sequence. If the transcript is essential to fungus growth, the pathogen dies and the plant staves off disease.

During the 20-year period from 1996 to 2016, according to the International Service for the Acquisition of Agri-Biotech Applications (ISAAA), global area of transgenic crops (GM) increased more than 106-fold from 1.7 million ha in 1996 to 179.7 million ha in 2015. In recent years, GM crops expanded rapidly in developing countries. In 2013, approximately 18 million farmers grew 54% of worldwide GM crops in developing countries. The largest increase in 2013

TABLE 4.4
World Acreage of Transgenic Crop Plants

Year	Area (million ha)	Countries[a]
1996	1.7	6
1997	12.7	7
1998	29.4	8
1999	39.8	12
2000	44.2	13
2001	52.6	13
2002	58.7	16
2003	67.7	18
2004	81.0	19
2015	179.7	28
2016	185.3	31

Source: Anonymous, GM Crops List, GM Approval Database-ISAAA.org;
www.isaaa.org, Retrieved January 30, 2016, 2016a.

[a] Argentina, Australia, Bangladesh, Bolivia, Brazil, Burkina Faso, Canada,
Chile, China, Columbia, Costa Rica, Cuba, Czech Republic, Honduras, India,
Japan, Malaysia, Mexico, Myanmar, Norway, Pakistan, Paraguay, Philippines,
Romania, Slovakia, South Africa, Spain, Sudan, Uruguay, USA.

was in Brazil (403,000 km^2 vs. 368,000 km^2 in 2012). GM cotton began growing in India in 2002, reaching 110,000 km^2 in 2013 (Tables 4.4 and 4.5). Meanwhile, more than 30 countries grow GM crops.

In 2004, four principal countries grew 99% of the global transgenic crop area: the USA grew 47.6 million ha, Argentina 16.2 million ha, Canada 5.4 million ha (7.8 in 2016), Brazil 5.0 million ha, and China 3.7 million ha. China had the highest year-on-year growth with a tripling of its *Bt* cotton area in 2004. By 2016, *Bt* cotton occupies 95% of the produce in India.

Other six countries that grew GM crops in 2005 were Mexico, Bulgaria, Uruguay, Romania, Spain, Indonesia, Kenya, and Germany. Indonesia reported commercializing a transgenic crop. Globally, the principal GM crops were soybean occupying 48.4 million ha in 2004 (60% of global area), followed by maize at 19.3 million ha (24%), cotton at 9.0 million ha (11%), and rapeseed at 4.3 million ha (5%). During the 9-year period 1996–2005, herbicide tolerance has consistently been the dominant trait with insect resistance second. There is a cautious optimism that global area and the number of farmers planting GM crops will continue to grow in 2017 (Table 4.4).

According to industry experts, whereas the USA will continue to account for the largest seed sales, China, India, and Brazil will present fast growth opportunities. Aventis Cropscience, Cargill, Delta, KWS, Pine Land, Dupont, Limagrain, Monsanto, Ciba-Geigy, Bayer, Unilever, Zeneca, or Syngenta are the leading global players in the GM crops segment. They cover a wide range of products, which include maize, soybean, cotton, tomato, potato alfalfa, petunia, rapeseed, rice, wheat, mustard, beet, barley, gram, cabbage, papaya, and tobacco (Table 4.5).

4.2.2.2   Gene Editing

Genome editing technologies comprise a set of molecular tools that lead to the targeted modification of a specific DNA sequence within the genome of interest. They are all based on the production of double-strand breaks at specific DNA sites that trigger the DNA-repairing system of the cells.

TABLE 4.5

Compilation of GM Crops, Their Release, Changed Characteristics, and Utilization

Crop	Trait Changed	Release	Note
Rapeseed	Herbicide tolerance	1995 (Canada, U.S.A.), 2003 (Australia)	Used as cooking oil, margarine
	High laurate	1994 (U.S.A.), Canada (1996)	Used as cooking oil, margarine
	Phytase production	1998 (U.S.A.)	Emulsifiers in packaged foods
Maize	Insect resistance	1995 (U.S.A.), 1996 (Mexico), 1997 (South Africa), 1998 (Argentina), 2002 (Philippines), 2003 (Columbia, Uruguay), 2005 (Brazil), 2007 (Paraguay)	
	Herbicide tolerance	1995 (U.S.A.), 1996 (Canada), 1998 (Argentina, Spain, Portugal, Czech Republic, Slovakia and Romania), 2001 (Honduras), 2002 (Philippines, South Africa), 2003 (Uruguay), 2007 (Brazil, Columbia), 2001 (Cuba), 2012 (Paraguay)	
	Increased lysine	2006 (Canada, U.S.A.)	Used for animal feed, high-fructose syrup, starch
	Drought tolerance	2010 (Canada), 2011 (U.S.A.)	
Soybean	Herbicide tolerance	1993 (U.S.A.), 1995 (Canada), 1996 (Argentina, Mexico, Uruguay), 1998 (Brazil), 2001 (Costa Rica, South Africa), 2005 (Bolivia), 2007 (Chile)	
	Increased oleic acid	1997 (U.S.A.), 2000 (Canada), 2015 (Argentina)	Animal feed, oil production
	Stearidonic acid	2011 (Canada, U.S.A.)	
Sugarbeet		1998, 2011 (U.S.A.), 2001 (Canada)	
Potato	Virus resistance	1997 (U.S.A.), 1999 (Canada)	Food production
	Modified starch	2014 (U.S.A.)	Industrial use
Alfalfa	Herbicide tolerance	2005, 2011 (U.S.A.)	Animal feed
Sugarcane	Drought tolerance	2013 (Indonesia)	Food production
Cotton	Herbicide tolerance	1994 (U.S.A.), 2000 (Mexico, South Africa), 2001 (Argentina), 2002 (Australia), 2004 (Columbia), 2008 (Brazil, Costa Rica), 2013 (Paraguay)	Fiber, oil
	Insect resistance	1995 (U.S.A.), 1996 (Mexico), 1997 (China, South Africa), 1998 (Argentina), 2002 (India), 2003 (Australia, Columbia), 2005 (Brazil), 2006 (Myanmar), 2007 (Paraguay), 2008 (Costa Rica), 2009 (Burkina Faso), 2010 (Pakistan)	
Squash	Virus resistance	1994 (U.S.A.)	Food production
Eggplant	Insect resistance	2913 (Bangladesh)	
Papaya	Virus resistance	1996 (U.S.A.), 2006 (China)	Food production
Tobacco	Nicotine reduction	2002 (U.S.A.)	Cigarettes
Carnation	Flower color	1995 (Australia), 1997 (Norway), 2000 (Columbia), 2004 (Japan), 2007 (European Union), 2012 (Malaysia)	
Petunia	Flower color	1997 (China)	
Rose	Flower color	2008 (Japan), Columbia (2010), 2011 (U.S.A.)	
Poplar	Fast growing lignin	1998 (China)	

Source: Anonymous 2016a, GM Crop Utilization, International Service for the Acquisition of Agri-biotech Applications (ISAAA, www.isaaa.org, Manila, Philippines).

In plants, this usually happens through nonhomologous end-joining, which is error-prone and thus exploited by breeders to inactivate target genes. Less frequently, plants repair double-strand breaks via homologous recombination using sister chromatids or DNA templates provided by the experimenter. Homologous recombination is error-free and could be used to precisely edit DNA sequences or to insert DNA fragments in a given genomic position.

4.2.2.2.1 Gene Silencing

Gene editing is the process of inserting, deleting and/or silencing, or replacing DNA sequences within a genome. Such techniques utilize methods that frequently can be observed in nature, for example, gene-silencing techniques make use of a mechanism found in the nematode worm *Caenorhabditis elegans*. Gene silencing or RNA interference technology (RNAi) refers to a series of molecular biology techniques that reduces the transcription or translation of a gene. In contrast with genetic mutations, gene silencing selectively targets the gene of interest with various silencing intensities, taking advantage of the native gene regulation.

RNA interference (RNAi) technology presents a new potential tool for plant breeding by introducing small noncoding RNA sequences with the ability to switch-off gene expression in a sequence-specific manner. The ability to suppress expression of a specific gene provides an opportunity to acquire a new trait by eliminating or accumulating certain plant traits, leading to biochemical or phenotypic changes that do not exist in nontransgenic plants. RNAi is an ancient evolutionary mechanism adopted by plants as a defense strategy against foreign invading genes, but it is used today as a tool for generating new quality traits in plants.

For example, soybean is a widely used crop in processed food and known to possess various quantities of "P34" proteins, recognized allergens to soybean-sensitive individuals. Gene silencing prevented the accumulation of such allergenic proteins, without affecting the germination or maturation of the transgenic plant. Potential allergens in rice, α-globulin and β-glyoxalase, were also silenced using RNAi constructs in a mutated α-amylase/trypsin inhibitor rice line, producing transgenic rice seeds that are free from the three potential allergens that are used in breeding programs. Other examples of reducing potential allergens in food crops include "Ara h2" protein in peanuts, "Mal d1" allergen in apple, and "Lyce 1.01" and "Lyce 1.02" allergens in tomatoes. Moreover, pollen grain allergens such as the Lol P5 protein, the major allergen in ryegrass pollen, were suppressed by using the antisense strand of the gene to reduce the allergen without affecting the fertility of the pollen grains.

Gene silencing by topical application of pathogen-specific double-stranded RNA (dsRNA) for virus resistance represents an attractive alternative to transgenic RNA interference (RNAi). However, the instability of naked dsRNA sprayed on plants has been a major challenge toward its practical application. Latest studies demonstrate that dsRNA can be loaded on designer, nontoxic, degradable, layered double hydroxide clay nanosheets. Once loaded on layered double hydroxide, the dsRNA does not wash off, shows sustained release, and can be detected on sprayed leaves even 30 days after application. In 2017, a single spray of dsRNA loaded on a layered double hydroxide afforded virus protection for at least 20 days when challenged on sprayed and newly emerged unsprayed leaves (MITTER et al. 2017).

By the silencing approach characters, such as high amylopectin, preventing enzymatic browning, limiting cold-induced degradation of starch, limiting acrylamide formation, or limiting acrylamide formation in potato, were already intragenically modified; by over expression of the polygalacturonase inhibiting protein in strawberry, the gray mold resistance was improved. A so-called cisgenic modification was achieved in potato for leaf blight resistance, in apple for scab resistance, or grapevine for fungal disease resistance.

The improvement of baking quality in bread wheat carrying the famous 1B-1R translocation could be a next project. CHAI et al. (2016) demonstrated that gliadin bands are expression products of four ω-secalin genes. Thus, by the RNA interference method, the expression of the ω-secalin genes might be silenced in favor of the expression of gliadins.

The first so-called CRISPR-edited crop was presented in 2016 to the US regulatory system. It was DuPont Pioneer high amylopectin maize (WALTZ 2016). It could be cultivated and sold without oversight by the US Department of Agriculture. This is the success story, which began about 10 years ago. In 2007, a US yogurt company identified an unexpected defense mechanism that its bacteria use to fight off viruses. A birth announcement came in 2012, followed by crucial first steps in 2013 and a massive growth spurt in 2014. It has matured into a molecular marvel, and geneticists are taking notice of the genome-editing method CRISPR, awarded by Science journal 2015 as Breakthrough of the Year. In addition, Florida citrus breeder reported in 2016 a new variety "Bingo," which is resistant to "fruit greening." It is a disease that constricts a tree's vascular system, shriveling fruit and eventually killing the tree. The bacterium is spread by a tiny insect called a psyllid.

4.2.2.2.2   CRISPR

CRISPR (clustered regularly interspaced short palindromic repeats) was originally discovered in *Escherichia coli* during the 1980s. It was only in 2007 that it was understood to be a bacterial defense mechanism against viruses and foreign DNA, as part of the CRISPR/Cas9 system.

The French microbiologist Emmanuelle M. CHARPENTIER (1968–) and the American biochemist Jennifer A. DOUDNA (1964–) published a few papers about the resistance mechanism of bacteria against phages. Their discoveries offer the new methods for DNA manipulation, now called genome editing. The method is similar to induced mutations by application of physical or chemical mutagens, but more precise. A new technology of genetic engineering entered crop breeding.

CHARPENTIER and DOUDNA discovered that a protein called Cas9 can easily cut desired gene or DNA sequence by combining with specific "guide RNA" (gRNA), which corresponds to the target DNA sequence.

In general, the CRISPR-Cas9 system mainly consists of two components, which are the guide RNA and a protein called Cas with endonuclease activity. The guide RNA is made up of a CRISPR-RNA (crRNA) sequence and trans-activating crRNA (trRNA). crRNA contains RNA sequence that is complementary to the target DNA and trRNA interacts with crRNA to form a complex structure (crRNA-trRNA) complex and promote engagement of the Cas9 protein. Once the Cas9–gRNA complex recognizes and binds to its target, the target DNA sequence (which matches up with the gRNA) will be cut by the endonuclease activity of Cas9. As a result, the target sequence was removed, and this allows scientists to study functions of gene easily.

They found out that Cas9 has the ability to interact with DNA and generate a double-stranded break in the DNA at sequences that match the sequence in a guide RNA (crRNA) (Figures 4.7 and 4.8). Moreover, this guides RNA base pairs with a second RNA named "tracer RNA" (tracrRNA) forming a structure that recruits the Cas9 protein. So those two single RNAs and a single protein are what are required in nature to recognize and destroy viral DNA to prevent infection.

Meanwhile, there are several newer techniques looking for options that could avoid the regulatory hurdles faced by transgenic crops. Among those methods are the site-directed nucleases (SDN) including zinc finger nuclease technology *CRISPR and TALENs), the oligonucleotide-directed mutagenesis, the cisgenesis, the intragenesis, agro-infiltration, RNA-dependent DNA methylation, or the reverse breeding.

4.2.2.2.3   Gene Transfer

Gene transfer, also known as horizontal gene transfer, is the process of inserting a specific coding sequence into a cell, either as a plasmid or as a genomic insert. The aim is to induce the synthesis of the target gene product in the cell for later analysis or, for example, for therapeutic purposes.

Traditional genetic engineering works via a relatively brute-force method of gene transfer. A harmless virus, or some other form of so-called vector, ferries a good copy of a gene into cells

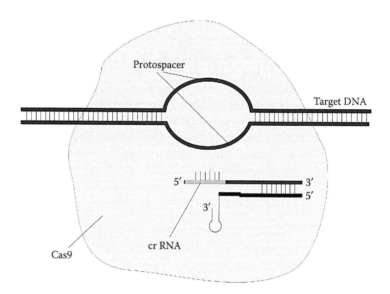

FIGURE 4.7 Schematic drawing of a bacterial CRISPR system with a single gene encoding a protein, Cas9, discovered by Charpentier and Doudna. (From R. SCHLEGEL 2017. With Permission.)

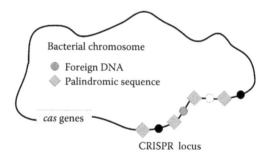

FIGURE 4.8 Schematic drawing of a bacterial and/or archaeal chromosome including palindromic sequences and foreign DNA segments. (From R. SCHLEGEL 2017. With Permission.)

that can compensate for a defective gene that is causing disease. But CRISPR technology can fix the flawed gene directly by snipping out bad DNA and replacing it with the correct sequence. In principle, that should work much better than adding a new gene because it eliminates the risk that a foreign gene will land in the wrong place of genome and turn on a deleterious function. Additionally, a CRISPR-repaired gene will be under the control of that gene's natural promoter, so the cell will not make too much or too little of its protein product.

4.2.2.2.4　Gene Modification

Gene modification refers to numerous molecular biological techniques that can introduce mutations into double-stranded DNA. Whereas the CRISPR/Cas system induces site-specific mutation in vivo, other genetic engineering techniques can introduce such changes during your subcloning steps or in vitro. Such gene modification techniques can also be used to change the specificity of RNA interference approaches.

4.2.2.2.5　Genome Editing

Genome editing with engineered nucleases represents a highly specific and efficient tool for crop improvement with the potential to rapidly generate useful novel phenotypes/traits.

Genome editing techniques initiate specifically targeted double-strand breaks facilitating DNA repair pathways that lead to base additions or deletions by nonhomologous end joining, as well as targeted gene replacements or transgene insertions involving homology-directed repair mechanisms.

Many of these techniques and the ancillary processes they employ generate phenotypic variation that is indistinguishable from that obtained through natural means or conventional mutagenesis; therefore, they do not readily fit current definitions of genetically engineered or genetically modified used within most regulatory regimes. Addressing ambiguities regarding the regulatory status of genome editing techniques is critical to their application for development of economically useful crop traits. Continued regulatory focus on the process used, rather than the nature of the novel phenotype developed, results in confusion on the part of regulators, product developers, and the public alike and creates uncertainty as of the use of genome engineering tools for crop improvement. For example, JIANG et al. (2017) used CRISPR/Cas9 to target the *FAD2* gene in *Arabidopsis thaliana* and in the closely related emerging oil seed plant, *Camelina sativa,* with the goal of improving seed oil composition. Guide RNAs were designed that simultaneously targeted all three homoeologous *FAD2* genes. The authors successfully obtained *C. sativa* seeds in which the oleic acid content was increased from 16% to over 50% of the fatty acid composition. These increases were associated with significant decreases in the less desirable polyunsaturated fatty acids, linoleic acid and linolenic acid. These changes resulted in oils that are superior on multiple levels: they are healthier, more oxidatively stable, and better suited for production of certain commercial chemicals, including biofuels.

4.2.2.2.6   Zinc Finger Nucleases

Zinc finger nucleases (ZFNs) is the oldest of the gene-editing technologies, developed in the 1990s and owned by Sangamo BioSciences Ltd. It has been primarily used in research for a variety of human diseases.

4.2.2.2.7   TALENs

In vitro *transcription* is a method that enables to synthesize RNA. These produced nucleotides can be used for a variety of methods such as nuclease protection assays and blot hybridizations. Larger transcription reactions may be used for expression studies such as in vitro translation and structural analyses. Transcription activator-like (TALs) effector nucleases (TALENs), developed in 2009, offered an easier method of gene editing. Its first reported success came in 2012, when disease-resistant rice was developed at Iowa State University. TALENs result in the fewest "off-target" effects, including through uncontrolled cellular growth, which occur when nucleotide sequences identical or similar to the target are cut unintentionally. Zinc fingers and TALENs made it possible to directly target a particular gene for the first time, but they are more time-consuming than CRISPR in their design and construction.

4.2.2.2.8   CRISP Mechanism

The so-called gene or genome editing is a broad category that includes several techniques allowing precise editing, including CRISPR. Clustered regulatory interspaced short palindromic repeats (CRISPR-Cas9), the newest of the gene-editing techniques, has been capturing the most attention (Figure 4.8).

4.2.2.2.9   Discovery

At the very beginning of the 1990s in Spain, FRANCISCO MOJICA and his colleagues were studying how could archaea adapt and survive in extremely saline environments. When analyzing a chromosomal region that they thought could be related to this adaptation, they noticed a very peculiar genomic structure that rapidly caught their attention: information that repeated several times, like a pedestrian zebra crossing. In 2003, they found that these sequences constituted an acquired immune system that allowed these microorganisms to fight against invaders such as viruses. The trick of this immune system, present in archaea as well as in bacteria, is that it allows viral DNA to be integrated into the chromosome and used as a bar code to recognize and fight against virus whose DNA was taken away in the first place.

When archaea and bacteria undergo a viral attack, they detect the foreign DNA that gets injected into the cell and integrate short pieces of it into the CRISPR locus within their chromosome. This integration does not occur randomly, but instead specifically as "spaces" between short palindromic repeats (Figure 4.9). Interesting thing to keep in mind about palindromic sequences is that when expressed on a single strand of RNA, they get to fold up on itself and can serve as kind of a handle for proteins to grab on to, as in the case of "Cas" endonucleases (Cas stands for CRISPR-associated proteins).

These foreign segments of DNA are further inherited through cell division and upon a second viral infection they are transcribed into pieces of RNA that are subsequently used together with Cas proteins to form an interference complex (Figure 4.9). This complex uses the information in these RNA molecules to base pair with the matching sequences in viral DNA and Cas proteins, which grab on to these RNAs, and then breaks down the viral DNA, thus preventing infection from going on.

Besides its mechanism, what is amazing of CRISPR is that elements of this system can be transferred to any cell. The purpose of this transference is that this system can be programmed to cut wherever it is wished to cut. One can edit the genome information. Once the cut is done, the DNA-repair system of the cell tries to repair it and by doing it, it introduces modifications. Moreover, these modifications or changes that can occur randomly can be directed by introducing a template.

CRISPR offers a relatively inexpensive, easy, and quick option to engineer changes. What makes the CRISPR system so special, in part, and so adaptable to the important task of gene editing? It is its relative simplicity. CRISPR is often described as being the equivalent of using a pair of molecular scissors to modify a specific portion of a gene. They can snip away weaknesses or insert strengths already found in the species being developed. The tool allows cutting out a section of DNA. Then, one of two things happen: the loose ends are essentially glued back together, eliminating the undesired trait or weakness, that is, a "repair" with a desired trait is inserted into the void.

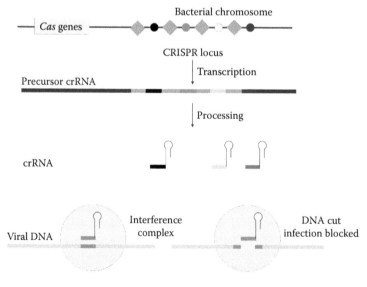

FIGURE 4.9 Schematic drawing of the bacterial RNA transcription and processing. (From R. SCHLEGEL 2017. With Permission.)

CRISPRs do not need to be paired with separate cleaving enzymes as other tools do. They can also easily be matched with tailor-made "guide" RNA (gRNA) sequences designed to lead them to their DNA targets. More than tens of thousands of such gRNA sequences have already been created and are available to the research community. CRISPR-Cas9 can also be used to target multiple genes simultaneously, which sets it apart from other gene-editing tools. CRISPR editing does not require the use of genetic material from a foreign species.

It can be concluded that gene-editing and other technologies can offer several advantages over traditional transgenic techniques. Among the most notable is that governments may regulate these products (when developed without foreign DNA) differently than they do traditional transgenics. There has been some movement this direction from the US Department of Agriculture, which has ruled in several instances that it does not need to approve new varieties made without transgenetics. Those products, however, still face scrutiny from the Food and Drug Administration if they are destined for the human food supply.

The accuracy of the new techniques also results in fewer unintended changes. There is still the chance of creating so-called off-target effects during the slicing and splicing. With that in mind, GMO critics argue that there is no reason to view gene-edited crops any differently than transgenic crops.

It is still not clear whether the US Department of Agriculture has jurisdiction, under current laws, over crops developed through gene editing that does not involve the insertion of foreign DNA. The agency has concluded that if an edit cannot be distinguished from a naturally occurring mutation, it is not a GMO.

The European Union, which heavily restricts many traditional GMOs, is assessing how to regulate plants developed through these techniques. The commission currently defines GMOs as organisms with alterations that cannot occur naturally, but it says it will clarify the issue, although Swedish authorities concluded that CRISPR-edited plants should not be defined as GMOs under EU legislation.

4.2.2.2.10 Cpf1 Protein

In September 2015, Zhang and colleagues (ZETSCHE et al. 2015) published a finding that changed the whole CRISPR discussion on patenting the CRISP system. It showed that a new protein other than Cas9 had been discovered to perform the gene editing, a protein called Cpf1. Because Cpf1 correspond to a new DNA-cutting protein, it is out of the scope of the CRISPR-Cas9 patents.

Zhang's team discovered Cpf1 by screening thousands of CRISPR systems in different bacteria in order to find enzymes that could be used in human cells. Cpf1 was finally found in the genus *Acidaminococcus* and *Lachnospiriceae*. The newly discovered CRISPR–Cpf1 system is a new class CRISPR system, even able to cut the target DNA substrate in human cells under the crRNA's guidance. Moreover, Cpf1 itself is a sequence-specific RNase, which is the only discovered nuclease that has both nuclease sequence specificity and the viability of DNase and RNase (Table 4.6).

In 2016, additional data revealed details about its structure and mechanism, and also provided the structural basis for its improvement, which could help the Cpf1 system to become the specific

TABLE 4.6
Comparison of Cas9 and Cbf1 Protein Characteristics of the CRISP System

Feature	Cas9	Cbf1
Cell type	Fast-growing cells	Nondividing cells
Size	Large	Smaller
Structure	Requires two RNA	Requires only one RNA
Cutting type	Creates "blunt ends"	Creates "sticky ends"
Cutting site	Proximal to recognition site	Distal from recognition site
Target site	G-rich PAM	T-rich PAM

and efficient new gene-editing system. Cpf1 and crRNA can be seen as a real couple, before and after association. Cpf1 is in a loose and flexible conformation, but after its combination with crRNA, Cpf1 has an obvious conformational change, into a compact triangular structure.

4.2.2.3 Nanotechnology

Nanotechnology is considered as one of the possible solutions to problems in food and agriculture. Just like biotechnology, issues of safety on health, biodiversity, and environment along with appropriate regulation are raised on nanotechnology. It is the ability to measure, manipulate, and manufacture things usually between 1 and 100 nanometers. A nanometer is one billionth of a meter; plant root hair cells vary between 15,000 and 17,000 nanometers in diameter.

The traditional plant-breeding methods take a long-term activity. Different nanodiagnostic methods such as nanofluidics, nanomaterials, or bioanalytical nanosensors have the potential for resistance breeding and can possibly be used in living plants in field-based assays for transgene expression. Nanofluidics such as "Open Array" or the "Fluidigm Dynamic Array" technologies supply automated PCR mixes for mega molecular breeding assays. Nanogenomics-based methods enable greater precision breeding. Nanoparticles, nanofibers, and nanocapsules can carry foreign DNA and chemicals into isolated plant cell and leaves that change genes, that is, nanoparticle mediated gene or DNA transfer (ABD-ELSALAM and ALGHUTHAYMI 2015). The honeycomb-like mesoporous silica nanoparticles (MSNs) system with 3 nm particle size provides this tool. MSN loaded with the gene keep the molecule from leaching out. Uncapped gold nanoparticles release the chemical and trigger the gene expression. Carbon-coated iron nanoparticles were inserted inside the internal hallow of the leaf petiole of pumpkin. Different types of microscopic methods were used to visualize and follow the transport and deposition of nanoparticles, as well as to verify the possibility of concentrating nanoparticles into targeted specific site of plant cell by small magnet. Starch nanoparticles effectively crossed the cell wall. The nanoparticle biomaterial is designed in such a way that it bonded the gene and transported it across the cell wall of plant cells by inducing the formation of transient membrane pores in cell wall, cell membrane, and nuclear membrane by using ultrasound method.

Single-walled carbon nanotubes/fluorescein isothiocyanate and SWNT/DNA conjugates also revealed the ability of nanotubes to act as nanotransporters in walled plant cells. Microinjection with carbon nanofibers containing foreign DNA has been used to genetically modify "Golden Rice" enriched with extra vitamin A (cf. Chapter 5). Method was successfully applied in *Arabidopsis*, maize, and tobacco too. Up to now, mainly calcium phosphate nanoparticles and poly dendrimer (DNA vector) entered into plants through simple co-culture methods and worked as carriers for gene delivery without any additional assistance. High transient delivery efficiency could be obtained.

Quick developments in next-generation sequencing technologies over the last decade have opened up many new chances to discover the relationship between genotype and phenotype. It consents the whole- or targeted genome resequencing of hundreds of genotypes and the mining of allele diversity in the populations of crop landraces and wild relatives. Knowledge of the structure and function of plant genomes in major agricultural and related plant genetic resources prompts the development of so-called second-generation biotechnologies and their application in breeding, that is, new plant-breeding techniques, aiming to the isolation of genes underlying the characters of interest and their precise modification or transfer into targeted varieties.

The third-generation systems quickly become common in general plant research and agronomy. Nanopore-based DNA sequencing protocols allow single-molecule, electrical detection of DNA sequence and have the potential of low sample preparation work, high speed, and low cost.

4.2.3 Transgenic Pyramiding Breeding

Three methods have been established to pyramid multiple transgenes; namely, multigene transformation, retransformation, and sexual hybridization of two or more transgenic events. The multigene transformation method can be further classified into two types: single-vector transformation and multi-vector co-transformation. For the single-vector transformation method, multiple genes are inserted within a single T-DNA, or multiple T-DNAs are included on a binary vector, the plant transformation of a single-vector is simple, but the assembly of multiple genes into a conventional vector is difficult owing to vector instability, the lack of unique restriction sites and the limited vector capacity. By this method, transgenes in the transgenic plants are tightly linked and co-integrated in the same loci in the plant genome, so they will not segregate in offspring. In multivector co-transformation methods, multiple genes are cloned on a different vector, respectively.

Multiple strains carrying a different vector are used for simultaneous infection of plant tissues, or multiple genes in different vectors are simultaneously transferred to plant cells. The transgenes from the co-transformation are usually co-integrated at the same position of the transgenic plant chromosomes, which may consequently be inherited together in the progeny.

Retransformation is a valid gene-pyramiding method with which multiple transgenes can be sequentially introduced into a plant. For retransformation, every transformation requires a new and different selectable marker from the original event, which is crucial in this process. The drawback of few available selective markers restricts the development of retransformation. The retransformant transgenes are randomly integrated and may segregate in the progeny.

Sexual hybridization, which is flexible for the combination of transgenes but time-consuming, is a conventional breeding technology that pyramids transgenes from different events. Considering that the transgenes from different events are not linked and will segregate in the progeny, a large breeding population is required to obtain the plant with stacked genes.

MAS is an efficient method for gene pyramiding in conventional breeding (Section 4.2.1). Generally, MAS-based pyramiding can be performed using two strategies: backcrossing and crossing.

The backcrossing is used for pyramiding genes into the genetic background of the recurrent parent to improve a few traits of the recurrent parent while the main traits of the recurrent parent remain unchanged. The crossing method is used for pyramiding genes into new genetic backgrounds through the genetic recombination of various parents to develop a new variety.

The Genuity®SmartStax™ maize developed by Monsanto Co. was derived from the stacking of four events (MON89034 × TC1507 × MON88017 × 59122) by conventional crossbreeding. Each event originated from the transformation of a vector with multiple genes. The event MON89034 was obtained from the transformation of vector PV-ZMIR254 containing *Lepidoptera* insect-resistance genes (*cry2Ab2* and *cry1A.105* from *Bacillus thuringiensis*). The event MON88017 was achieved from the transformation of vector PV-ZMIR39, which contains glyphosate-resistance gene "*cp4-epsps*" and *Coleoptera* insect-resistance gene "*cry3Bb1*." The event TC1507 was derived from the particle bombardment transformation of a linear fragment PHI8999A that comprises two expression cassettes of *Lepidoptera* insect-resistance gene "*cry1Fa*" and glufosinate-resistance gene "*pat*." The event 59122 contains *Coleoptera* insect resistance, particularly corn rootworm-resistance genes "*cry34Ab1*" and "*cry35Ab1*" and glufosinate-resistance gene "*pat*."

The Agrisure®Viptera™3111 maize developed by Syngenta International AG was also derived from four events (Bt11, GA21, MIR162, and MIR604) by conventional crossbreeding. The event Bt11 was obtained from the transformation of a vector with stacked genes "*cry1Ab*" and "*pat*," and has resistances to *Lepidoptera* insects and glufosinate herbicide. GA21 was acquired from the transformation of vector pDPG434 carrying glyphosate-resistance gene "*mepses*." The events, MIR162 and MIR604 were obtained from the transformation of vector pNOV1300 containing *Lepidoptera* insect-resistance gene "*vip3Aa20*" and "*pZM26*" containing *Coleoptera* insect-resistance gene "*mcry3A*."

A case of pyramiding of five genes for multiple resistances in rice was described by WAN et al. (2014). The elite restorer line "R022" of rice contains brown planthopper resistance genes "*Bph14*" and "*Bph15*." A backcrossing breeding scheme was designed to improve the resistance of "R022" against rice blast, leaf folder, stem borer, and herbicide. In this scheme, "R022" was used as the recurrent parent. The transgenic rice "T1C-19" was used as the donor of "*cry1C*" (encoding *Bacillus* *thuringiensis* toxin protein) and glufosinate resistance gene "*bar*" (encoding phosphinthricin acetyl-transferase). "T1C-19" was developed through the transformation of a construct containing "*cry1C*" and "*bar*" in the binary vector plasmid pC-1C*. The rice restorer line "R2047" was used as the donor of rice blast resistance gene "*Pi9*." Five genes, "*cry1C*," "*bar*," "*Pi9*," "*Bph14*," and "*Bph15*" were stepwise pyramided into "R022" through MAS-based crossing.

According to the examples provided, the transgenic pyramiding in commercialized GM crops with stacked traits is realized mainly through conventional crossbreeding of single events deriving from the vector transformation. Each event must be tested and approved before becoming commercially available, which is a complex, time-consuming, and arduous process. Considering that the single events used for crossbreeding have obtained approvals, the stacked products from the conventional crossbreeding do not involve new transformations, and no interaction effect among the stacked traits happens, the registration as a new GM product and new approvals are unnecessary for the stacked products from crossbreeding. Thus, this crossbreeding-based pyramiding can accelerate the commercialization of GM crops and reduce the research cost for GM crops.

Pyramiding by conventional crossbreeding also has disadvantages. Transgenes in different events are not linked and have a tendency to segregate in subsequent breeding progenies. In addition, transgene loci increase with increasing the crossing parents, resulting in the complex gene segregation. Therefore, a large and sufficient breeding population and more breeding generations are needed for the identification of plants with the stacked genes in offspring.

5 "Farmerceuticals," "Nutraceuticals," and Other Exotic Characters

If we bite into a lemon and it tastes sweet, then we have previously eaten a miracle berry (*Synsepalum dulcificum*). This effect is due to miraculin, a glycoprotein molecule, with some trailing carbohydrate chains. Sweet-tasting proteins and taste-modifying proteins have a great deal of potential in industry as substitutes for sugars and as artificial sweeteners. The taste-modifying protein, miraculin, functions to change the perception of a sour taste to a sweet one. This taste-modifying function can potentially be used not only as a low-calorie sweetener but also as a new seasoning that could be the basis of a new dietary lifestyle. However, miraculin is far from inexpensive, and its potential as a marketable product has not yet been fully developed. Japanese plant breeders produced a transgenic tomato, now delivering the miraculin more cheaply.

This sort of plant biofactories has progressed in recent years. Some recombinant plant-derived pharmaceutical products have already reached the marketplace. However, with the exception of drugs and vaccines, a strong effort has not yet been made to bring recombinant products to market, as cost-effectiveness is critically important for commercialization.

Breeding for chemical modifications in crops has been practiced for a long time. Better tasting fruits is one approach of that. However, the modern era brought the right breakthrough, namely by the development of chemical screening techniques. It began with the selection of low-bitterness, that is, alkaloid-poor lupines by the German ▶ R. VON SENGBUSCH (1898–1985) during the 1930s (SENGBUSCH 1930). The first low-alkaloid forms of *Lupinus albus* were selected from von SENGBUSCH in 1931–1935, as in *Lupinus luteus* and *Lupinus angustifolius*, but these were not further developed into varieties because of inconsistency. "Sweet" varieties of the White Lupine succeeded a little later in Landsberg/Warte (Germany) waiting with early, short, and medium-high types: "Pflugs Gela," "Pflugs Ultra," and "Plugs Hansa." Later came "Hadmerslebener Kraftquelle" and "Hadmerslebener Nährquelle," also Germany.

The origin of the White Lupine is in the Mediterranean region. Here the domestication took place already in Egyptian, Greek, and Roman antiquity. Wild forms of *Lupinus albus* are no longer present, only primitive forms (*L. graecus and L. termis*) are still found. In ancient Greece, it was an important crop, as are the written testimonies on the cultivation and use of the ancient writers: HIPPOCRATES (400–356 BC) and THEOPHRAST (372–288 BC). The significance of the white lupines for poor soils and as precursors for crops is also emphasized in the later Roman literature (CATO, 234–149 BC: "Lupine is one of the crops that fertilize the seed"). As early as 218 AD, FLORENTINUS describes the debilitating of the seeds for the nutrition of humans and animals.

A more recent example of chemical modification is rapeseed (*Brassica napus*). Original rapeseed was quite unsuitable for human consumption. It contained ingredients that affect the function of the thyroid gland. With their reduction through breeding in the 1970s, it became possible to make food-quality oil. In cereals, the selection for high protein content and specific protein composition has also become routine (Figure 3.9).

The production of plant-made pharmaceuticals and technical proteins is known as molecular farming. The objective is to harness the power of agriculture to cultivate and to harvest plants or plant cells producing recombinant therapeutics, diagnostics, industrial enzymes, and "green chemicals."

Modern crops are increasingly used for various industrial applications. They contain specific raw materials like fibers, oils, and starches for industrial products. Tailor-made plant compounds for specific industrial processes are a big challenge and may hopefully contribute to a more sustainable production. In the near future, plants might be used as factories for vaccines and pharmaceuticals (Table 5.1). The main focus is on diseases more or less eradicated in industrialized nations but hard to combat in developing countries because of the lack of money and infra structure.

TABLE 5.1

Renewable Crops for Better Nutritional Value and Future Production of Pharmaceuticals or Other Drugs

Crop	Trait	Expected Benefit
Soybean and rapeseed	Improved fatty acid; amino acid composition; reduced phytate content	Heart disease prevention; increased nutritional value for animals
Tomato	Delayed ripening	Increased shelf life
	Vitamin and mineral content	Sources of vitamin C, potassium, folic acid and carotenoids (β-carotene)
	Lycopene	Reduced cancer cells
Strawberries	Freeze-thaw tolerance	Better taste
Cassava	Improved protein content, elimination of cyanide	Increased nutritional value, reduced toxicity
Potato	Accelerated starch synthesis; multicomponent vaccine; spider silk production	Better digestibility; human immunization; artificial fibers with novel characteristics
	Hepatitis B surface antigen	Vaccine against hepatitis B
	Norwalk virus capsid protein	Norovirus vaccine
Soybean	Increased sweetness	Sweeter taste
Rice, wheat	Provitamin A and iron content; elimination of allergens	Increased nutritional value; allergy prevention
	Apo-A1^{Milano} protein	Therapeutic protein in cardiovascular disease
	Lactoferrin, lysozyme	Dietary against GI infections in infants
Cereals (wheat, rye, oat)	Gluten removal	Allergy prevention
Maize	Avidin, bovine trypsin	
	Collagen, structural protein	Reconstructive surgery; therapeutic enzyme in cystic fibrosis, pancreatitis
Linseed, flax	Human serum albumin	Therapeuticum in maintenance of blood plasma pressure
Banana	Cholera vaccine	Drug production
Tobacco	Antibodies	Clinical diagnostics
	2G12 IgG protein	Antibody for HIV prophylaxis
	rhinoRx protein	Antibody in rhinovirus prophylaxis
	caroRx protein	Antibody in dental caries
	Anti-idiotype IgG antibodies	Vaccine against Non-Hodgkin's lymphomas
Carrot	Glucocerebrosidase	Therapeutic enzyme for Gaucher disease
	Alpha-galactosidase	Therapeutic enzyme for Fabry disease
	Acetylcholesterase	Therapeutic enzyme for biodefence
	Antitumor necrosis	Antibody for arthritis
Spinach	Rabies glycoprotein	Vaccine against rabies

Source: Compiled from Economic Research Service, US Department of Agriculture and International Food Information Council, Washington, DC. https://www.ers.usda.gov/

Field experiments have already been carried out on potatoes, bananas, and pineapples containing cholera, *Escherichia coli* and hepatitis B vaccines.

Researchers already succeeded in immunizing mice with transgenic potatoes. As cereal grains are deficient in certain essential micronutrients, including iron or zinc, several approaches have been used to increase accumulation and, for example, alter iron metabolism. Because ferritin is a general iron storage protein in all living organisms, ferritin genes have been introduced into rice plants. Transgenic rice plants can express soybean and common bean ferritin under the control of the seed-specific rice Glu-B1 promoter. They accumulate up to three times more iron as wild-type seeds. A similar approach was successful in wheat: SINGH et al. (2017) developed wheat lines expressing rice "nicotianamine synthase 2" (*OsNAS2*) and bean ferritin (*PvFERRITIN*) as single genes as well as in combination. Nicotianamine synthase catalyzes the biosynthesis of nicotianamine (NA), which is a precursor of the iron chelator deoxymugeneic acid (DMA) required for long distance iron translocation. Ferritin is important for iron storage in plants because it can store up to 4500 iron ions. The authors increased the iron and zinc content in wheat grains of plants expressing either *OsNAS2* or *PvFERRITIN*, or both genes. In particular, wheat lines expressing *OsNAS2* greatly surpass the "HarvestPlus" recommended target level of 30% dietary estimated average requirement for iron and 40% for zinc, with lines containing 93.1 µg/g of iron and 140.6 µg/g of zinc in the grains, respectively.

In order to not only increase iron accumulation but also improve its absorption in the human intestine, two approaches have been adopted. Either the level of the main inhibitor of iron absorption, phytic acid, is decreased by the introduction of a heat-tolerant phytase from *Aspergillus fumigatus*, or a cystein-rich metallothionine-like protein gene (*rgMt*) is overexpressed in the recipient plant. It can be expected that high-phytase rice, with increased iron content and rich in cysteine-peptide, greatly improves iron supply in rice-consuming countries. In wheat, transgenic plants expressing *Aspergillus japonicas* phytase gene (*phyA*) significantly increased the bioavailability of iron and zinc (ABID et al. 2017).

In addition, the amino acid composition of crop plants can be modified. For example, potatoes contain limited amounts of the essential amino acids lysine, tryptophan, methionine, and cysteine. In order to improve the nutritional value of potatoes, a nonallergenic seed albumin gene *AmA1* from *Amaranthus hypochondriacus* was transferred to potato plants. The seed-specific albumin was under the control of a tuber-specific and a constitutive gene promoter, respectively. In transgenic lines, a 35%–45% increase in total protein content was possible, which corresponds to an increase in most of the essential amino acids.

The reduction of antinutritive factors is as feasible. For instance, cassava is one of the few plants that contains toxic cyanogenic glycosides in the leaves and shoots. Therefore, the genes were isolated that are responsible for the production of the glycosides. Subsequently, cassava plants were transformed with antisense constructs of the respective genes *Cyp79D1* and *Cyp79D2*. Another strategy for reducing the cyanide toxicity is to introduce a gene that codes for the enzyme hydroxynitrile lyase. This enzyme breaks down the major cyanogen acetone cyanohydrin.

In 2012, the US Food and Drug Administration approved the first drug for humans produced by a genetically modified plant. Made by Israeli biotechnological company "Protalix Biotherapeutics" and licensed in the USA by "Pfizer," "Elelyso" is an enzyme replacement therapy for Gaucher disease, a rare genetic disorder in which individuals do not produce enough of an enzyme called glucocerebrosidase, resulting in the buildup of fatty materials in the spleen, liver, and other organs (Table 5.1).

The development started in the 1980s. In 1983, first industrial production of secondary metabolites by suspension cultures of *Lithospermum* spp. was carried out by MITSUI Petrochemicals (Japan). The beneficial use of elicitors in cell suspension cultures was early recognized by B. WOLTERS and U. EILERT. The hairy root production of metabolites in *Hyoscyamus muticus* was demonstrated by H. E. FLORES and P. FILNER (1985). These roots produced more hyoscyamine than whole plants. The hairy root method provides several new approaches for breeding and biotechnology, such as better understanding and manipulating root-specific metabolism, co-culture with VAM fungi to increase secondary metabolite production, co-culture with insects to study pathogenesis, commercial exploitation of bioactive root exudates, or co-culture with shooty teratomas to exploit both root- and shoot-based metabolism for biotransformation.

In 1986, it was shown by BARTA et al. that tobacco plants and sunflower calluses could express recombinant human growth hormone as a fusion protein. In 1989, first plant-derived recombinant antibody, the full-size IgG was produced in tobacco (HIATT et al. 1989), and, in 1990, the first native human protein (serum albumin) in tobacco and potato (SIJMONS et al. 1990). A first plant-derived protein polymer (artificial elastin) could be harvested from recombinant tobacco (ZHANG et al. 1996). Even more exiting is the production of spider silk proteins in transgenic tobacco and potato (SCHELLER et al. 2001, 2004). The expression level of best plants reached up to 4% of the total soluble protein. It was stable in tobacco leaves and in potato leaves and tubers.

"Green biotechnology" has been used to engineer plants that contain a gene derived from a human pathogen (TACKER et al. 1998). An antigenic protein encoded by this foreign DNA can accumulate in the resultant plant tissues. Results from preclinical trials showed that antigenic proteins harvested from transgenic plants were able to keep the immunogenic properties if purified. ARAKAWA et al. (1998) demonstrated the ability of transgenic food crops to induce protective immunity in mice against a bacterial enterotoxin, such as cholera toxin B subunit pentamer with affinity for ganglioside. Also, potato tubers have been used successfully as a biofactory for high-level output of a recombinant single-chain antibody (ARTSAENKO et al. 1998).

The other example of high provitamin A "Golden Rice" came to mind of YE et al. (2000). Deficiencies in dietary vitamin A are one of the leading causes of child blindness in the developing world. "Golden Rice" was engineered to contain provitamin A based in the introduction of two genes from daffodil and one from the bacterium *Erwinia uredovora*. Common rice contains no provitamin A. Thus, the ability to introduce these genes to create a new biochemical pathway in rice is an exciting development. However, 5 years after initiating the project, the international "Greenpeace" organization claims that this project is a technical failure, not suited to overcome malnutrition and worse, and it is drawing funding and attention away from the real solutions to combat vitamin A deficiency.

Another example is transgenic tomato that received the bacterial carotenoid gene *crt1* encoding for the phytoene desaturase showed an increase of β-carotene content up top 45% of the total carotenoid content. D'AOST et al. (2004) developed a production system for recombinant protein C5-1, a diagnostic anti-human IgC for phenotyping and cross-matching red blood cells from donors and recipients in blood banks, based on leaves of alfalfa (*Medicago sativa*). The perennial alfalfa can be easily propagated by stem cutting to create large populations. In greenhouses, these populations can be harvested 10 times per year, and the recombinant plants can be maintained for more than 5 years for protein extraction.

KOIVU (2004) proposed a novel sprouting technology for recombinant protein production. Sprouted seeds or sprouts from oilseed rape (*Brassica napus*) are renowned for their excellent nutritional properties. They are often utilized as a protein-rich food, and some beneficial phytochemicals, such as vitamins A, E, and C, which are antioxidants. In a contained system, transgenic seeds carrying the gene encoding a specific protein of interest are first produced and harvested in a greenhouse and then sprouted in an airlift tank. The recombinant protein is produced during sprouting and extracted from the sprouts. Alternatively, protein can be removed directly from the growth medium.

For industrial countries, a low-calorie sugar could be of interest. New sugarbeet plants were engineered that produce fructan, a low-calorie sweetener, by inserting a single gene from Jerusalem artichoke (*Helianthus tuberosus*, Compositae). The gene encodes an enzyme for converting sucrose to fructan. Short-chain fructans have the sweetness of sucrose but provide no calories as humans lack the fructan-degrading enzymes necessary to digest them.

Vaccine production in plants is tried for anti-allergic medicals. Current allergy immunotherapy involving injections of a crude pollen (allergen) extract has the disadvantage that it can lead to severe and life-threatening anaphylactic side effects. Recently, nonallergenic forms of grass pollen allergens were produced via site-directed mutagenesis. Low toxicity, solubility, and production of high amounts of correctly folded recombinant proteins at low cost are prerequisites for developing new vaccination strategies. Therefore, plant-based expression systems are developed for modified (engineered) allergen potentially suitable for immunotherapy (BHALLA and SINGH 2004).

In 2005, an American research group was even able to produce a vaccine in transformed tobacco and tomato plants against the SARS virus that, in 2002, was identified in China for the first time. Another American group was able to transfer a microbal gene, the superoxide reductase, to plants for future (Mars) space missions. W. BOSS and A. GRUNDEN form North Carolina State University, Raleigh, have combined beneficial characteristics from an extremophile sea-dwelling, single-celled organism, *Pyrococcus furiosus*, into model plants like tobacco and *Arabidopsis thaliana*. They hope to improve extreme-temperature protection of plants.

5.1 NEUTRACEUTICALS

A functional food is a food given an additional function (often one related to health promotion or disease prevention) by adding new ingredients or more of existing ingredients. The term may also apply to traits purposely bred into existing edible plants, such as purple or gold potatoes having enriched anthocyanin or carotenoid contents, respectively. Functional foods may be "designed to have physiological benefits and/or reduce the risk of chronic disease beyond basic nutritional functions, and may be similar in appearance to conventional food and consumed as part of a regular diet." The term was first used in Japan in the 1980s, where there is a government approval process for functional foods called Foods for Specified Health Use.

The development started in 1950s when oranges were selected for juice production with high content of vitamin C. In sweet pepper, mutant genes for high carotene content were used for new varieties of hybrids with high content on carotene and vitamin C, plus tolerance to verticillium diseases. The orange appearance of the fruit was due either to the accumulation of β-carotene, or in two cases, due to only the accumulation of red and yellow carotenoids. Four carotenoid biosynthetic genes, *Psy*, *Lcyb*, *CrtZ-2*, and *Ccs* were cloned and sequenced from cultivars (GUZMAN et al. 2010), later used for targeted selection.

Recent studies show that the new vegetable plants should not contain just one specific nutrient to produce a kind of drug, but the complexity of the nutrients of a particular vegetable is important. Thus, one should add or reduce one or a few components to the existing mixture of nutrients of a given vegetable. This breeding approach should be more promising (VATTEM and MAITIN 2015), for example, Pak-choi salad with enriched content of antioxitants.

6 Intellectual Property Rights, Plant Variety Protection, and Patenting

Intellectual property rights (IPRs) are legal and institutional devices to protect creations of the mind such as inventions, patents, and plant breeders' rights. They have never been more economically and politically important than they are in today's industrial societies and neither have they ever been so controversial. Patents provide inventors with legal rights to prevent others from using, selling, or importing their inventions for a fixed period, nowadays normally 20 years. In a modern sense, patenting of plants, seeds, and plant parts does not appear to go much further back than the 1930s.

Nevertheless, a first plant patent seems to be granted to red currant (*Ribes rubrum*) in 1621 when in England Sir Francis BACON (scientist, philosopher, and Lord Chancellor of Her Majesty) confirmed a patent privilege to the English Crown to trade with currant, salt, starch, herrings, vinegar, and so on. In 1716, a further patent was granted in England for squeezing oil from sunflower seed, and, in 1724, T. GREENING was granted a patent for "grafting or budding the English elm upon the stock of the Dutch elm." In 1785, P. LE BROCQ from Jersey (Channel Islands) acquired a patent for "rearing, cultivating, training, and bringing to perfection, all kinds of fruit trees, shrubs, and plants; protecting their leaves, blossoms, flowers, and fruits."

However, the history of IPRs for plants is mainly connected with the development of breeding institutions during the past two centuries (cf. Section 3.2). In 1895, the first "Law over the forgery of agricultural products" (Seed law) became effective in the Austrian–Hungarian monarchy. In Germany, registered breeding stations and elite breeds were kept since 1888. In 1896, the German Agricultural Society (DLG = Deutsche Landwirtschaftsgesellschaft) established the first "Rules of Seed Registering." It led to the elite register of 1905, which also contained regulations for variety testing and purity.

Before, in 1869, ► F. NOBBE, Head of the Plant Physiology Experiment Tharandt (Saxonia, Germany), established the first official and independent seed control system worldwide. Seed morphology, purity, and germability were investigated and certified. Regulations were described in the "Statut betreffend der Controle landwirtschaftlicher Saatwaren" (Statute concerning control of agricultural seedware). Soon after, similar testing institutes followed in Denmark, Austria, Hungary, Belgium, Italy, and the USA. NOBBE's "Handbook of Seed Science" (1876) was a standard textbook at the time.

In the USA, land-grant institutions were already established in 1860s. This situation discouraged private investments in plant breeding because it was difficult to maintain control over sales and markets and recoup investments. With introduction of hybrid maize in the USA during the 1930s, saved seed was no longer an option. The hybrid varieties could be protected through trade secrecy. Proprietary maize hybrids were initially based on public inbreds. The pressures from investors to develop and support commercial plant breeding increased. However, all released varieties and hybrids could be used as breeding materials yet. During this period, trade secrets and contracts are often used as low-cost alternative to more formal means of protection. Prior to 1930, farmers still had direct access to seed and germplasm because most crops were "true breeding" and seed was easily saved. Moreover, most breeding was publicly financed.

However, the use and access to breeding strains, germplasm, and varieties may be limited or restricted, even during cooperative testing. In 1930, the "Plant Patent Act" was the first legislation in the USA to protect horticultural plants and nursery stocks of asexually propagated crops. Potato was still excluded. For differentiation, the variety must be "distinct" and "new." The control was administered through the US Patent Office. In Europe, patents were occasionally granted at around this time. In Germany, W. LAUBE (1892–1963), the breeder of the first tetraploid rye variety "Petkuser Tetraroggen" (1951), initiated a similar law ("Sortenschutzgesetz") protecting breeders from plagiarism. It was passed in 1932. Later, the German Appeal Board accepted an application for a patent that claimed seed material of certain varieties of lupine (cf. Section 3.5). Notwithstanding, the US "Plant Patent Act" stimulated one of the European initiatives leading on November 17, 1938, to the foundation of Association Internationale des Sélectionneurs pour la Protection des Obtentions Végétales (ASSINSEL) at Amsterdam (the Netherlands) with Denmark, France, Holland, and Germany as first member countries. During the 1940s and early 1950s, countries adopted a range of approaches, from denying all intellectual property protection (e.g., UK and Denmark), to allowing patents (Italy from 1948, France from 1949, Belgium from 1950), and creating specific intellectual property systems for plant varieties (e.g., the Netherlands 1941, Austria 1946, Germany 1953). In 1952, South Africa introduced a similar but modified patent system for plants as the USA.

In 1970, the American "Plant Variety Protection Act (PVPA)" was passed. Sexually propagated varieties could be protected. The primary goal was to promote commercial investments in plant breeding and to provide a "patent-like" protection for plants reproduced by seed. The right of owners was still limited. The protection included a limitation to entire plant and harvested material for 20 years. USDA issued it. The seed sale was only allowed through authorized dealers.

In 1961, in Europe a similar system was put into law by "Union pour la Protection des Obtentions Vegetales (UPOV)." Member countries were Belgium, Denmark, Germany, France, Great Britain, Italy, the Netherlands, and Switzerland. In France, "distinction," "homogeneity," and "stability" constituted the criteria for the first system for the protection of breeder's rights in 1920. This French system was adopted by the European Union in connection with the UPOV treaty. Ironically, the negotiators at the UPOV treaty failed to define "a variety," the object which in fact they sought to protect! Easy to understand, "Distinction," "homogeneity," and "stability" define a clone, the contrary to a variety.

Nevertheless, after this, breeders have the right to exploit products of their profession. It included authorization for plant production, sale, and marketing. The variety proposed must be "distinct, uniform, stable," and novel in at least one trait. However, problems remained, such as widespread "brown-bagging," that is, illegal sales and use, erratic and inadequate enforcement, enforcement responsibility of patent holder, only minor penalties for violation or infringement, and concern over impact on restriction of germplasm exchange and crop diversity.

This law led to an increase in private breeding, but only for a few crops. Market size and profit margins were primary determinants of commercial success. Crops, such as autogamous wheat and barley, faced with low profit margins for seed and extensive pirating, received only limited private investments. By the time, many government-owned plant breeding institutes had been privatized, and large chemical corporations mostly bought them. The breeding policy of the new owners was to stress neither kind of resistance, nor the grounds that crop protection chemicals can replace resistance entirely. Many of these corporations had also been buying seed production and distribution organizations. This process of re-organization, concentration, and monopolization of breeding is still ongoing.

In 1980, the BAYH-DOLE Act, and in 1986, the Technology Transfer Act, had additional impact on public plant breeding and research in the USA. It established that universities have the right to obtain patents and commercialize inventions created under government grants, that licenses and royalties can be sought by universities as means to generate revenue as government support declines, and that as an inventor, a university employee has the right to receive a portion of the royalties on the invention.

The Canadian Plant Breeders' Rights Act (PBR Act) came into force on August 1, 1990. The PBR Act is administered by the Plant Breeders' Rights Office, which is part of the CFIA. The PBR Act allows the developers of new varieties to recover their investment in research and development by giving them control over the multiplication and sale of the reproductive material of a new variety. The rights also include the ability to charge a royalty. In order to receive a grant of rights, varieties must be new, distinct, uniform, and stable. Two notable exceptions to a holders' rights are that protected varieties may be used for breeding and developing new plant varieties, and that farmers may save and use their own seed of protected varieties without infringing on the holders' rights. This second exception is referred to as Farmers' Privilege.

In 1994, an amendment of PVPA eliminated "saved-seed" provision of PVPA; that is, the farmer can save only for own on-farm replanting. This amendment brought the PVPA into accordance with UPOV in Europe. In the same year, the EC in Brussels (Belgium) passed a similar law (2100/94) that unified the variety protection European-wide.

In 2001, plant patenting reaffirmed that utility patents, plant patents, and plant variety protection are different, but to be treated as complementary (BARTON and BERGER 2001).

At present, in the USA, intellectual property protection for plants is provided through plant patents, plant variety protection, and utility patents. Plant patents provide protection for asexually reproduced varieties excluding tubers. Plant variety protection provides protection for sexually reproduced varieties including tubers, F1 hybrids, and essentially derived varieties. Utility patents currently offer protection for any plant type or plant parts. A plant variety can also receive double protection under a utility patent and plant variety protection.

On the other hand, European governments defied the EU Commission in 2005 by voting for the right to keep bans on patented GM crops and food. Five member states—Austria, Luxembourg, Germany, France, and Greece—were under pressure to give up their current bans because of a trade dispute in which the USA claims they are illegal. The bans were imposed in the five countries between 1997 and 2000 due to safety reasons.

In 1980, the first utility patent of "living organisms" was issued in the USA. It was a landmark decision of the US Supreme Court in a case "Diamond vs. Chakrabarty" and the US Patent and Trademark Office. It broadens the patent law to encompass living organisms and established that anything made by man is patentable, that is, ownership of plant varieties, traits, parts, and processes. The claims can be broad based, including entire species, plant parts, seeds, cell cultures, plant tissues, transformed cells, expressed proteins, threshold traits, and genes themselves. The standards for issuance of a utility patent are the novel character in relation to "prior art," the usefulness, and the obviously innovative step. Thus, it utility patent provides more IPRs than a plant variety patent, but at a higher cost and standard for issuance. It allows prohibition of farm-saved seed and prohibition of use in breeding.

In this context, the so-called *terminator technology* should be mentioned. Basically, two methods are available: (1) "Traitor"—officially known as Trait-specific Genetic Use Restriction Technology (T-GURT). It incorporates a control mechanism that requires yearly applications of a proprietary chemical to activate desirable traits in the crop. The farmer can save and replant seeds, but cannot gain the benefits of the controlled traits unless he pays for the activating chemical each year. (2) "Terminator"—officially named the Technology Protection System (TPS), incorporates a trait that kills developing plant embryos, so seeds cannot be saved and replanted in subsequent years.

Both methods avoid the difficulties associated with enforcing "no replanting" agreements, and ensure the seed companies the investments in their new varieties.

A first Utility Patent for genetically engineered maize was granted in 1985, also in the USA. The chief interest in Utility Patent came from inventors of biotechnology products and processes. However, seed companies have looked toward Utility Patent for additional protection beyond that afforded by Plant Variety Protection. By 1988, over 40 patents on crop plants had been issued. To date, there are more than 2000 US patents with claims to plants, seeds, or plant parts.

Main concerns remained about plant patents in spite of restrictions placed on patented varieties for subsequent use in breeding, of potential negative impact on crop diversity, and of increased domination of breeding by larger companies. In course of the last three generations, about 75% of crop and horticultural plants have vanished from fields and gardens! One study estimates that 34,000 species of plants (~12.5% of the world's flora) are facing extinction. Another FAO report (ANONYMOUS 1998) stated for instance that from about 10,000 wheat varieties in use in China in 1049, only 1000 remained in the 1970s; in the USA 95% of the cabbage, 91% of the maize, 94% of the pea, and 81% of the tomato varieties cultivated in the nineteenth century have been lost.

Modern plant breeding, whether classical or through genetic engineering, comes with issues of concern, particularly with regard to food crops. The question of whether breeding can have a negative effect on nutritional value is a part of the declining genetic crop and genotype diversity. By some reports, the majority of the indigenous landraces that have developed since man began to cultivate the environment have been lost due to disuse or neglect. During the past century, therefore, various national governments and international organizations developed a range of programs and institutes to identify and preserve plant genetic resources either in situ or ex situ. Although these collections are now quite large, with more than 7 million discrete accessions globally (in ~1300 genebanks of ~80 countries), it is unclear whether the agronomic vitality or traditional knowledge surrounding plant genetic resources is being preserved. Moreover, there is significant debate about who owns the rights to determine access and use of the materials and who should benefit from any discoveries or inventions using plant material, particularly, the development of new biotechnological tools has lowered the costs and widened the possibilities of finding economically valuable traits in indigenous plants.

Although relatively little direct research in this area has been done, there are scientific indications that, by favoring certain aspects of a plant's development and marketing, other aspects may be retarded. Nutritional analysis of vegetables of the USA done in 1950 and in 1999 found substantial decreases in 6 of 13 nutrients measured, including 6% of protein and 38% of riboflavin. Reductions in calcium, phosphorus, iron, and ascorbic acid were also found. It is explained by reduction and changes in cultivated varieties between 1950 and 1999.

6.1 PROTECTION OF NEW PLANT VARIETIES

The International Union for the Protection of New Plant Varieties (UPOV) and the Trade-Related Aspects of Intellectual Property Agreement (TRIPS) provide protection of new plant varieties (IPR). Like patent, copyright, trademark, and industrial design protection, it is a special protection for new plant varieties and available since 2007. The Act of UPOV Convention in 1991 gave this important protection to the plant breeders. The UPOV office is situated in Geneva and coordinates the Protection of New Plant species. There is a debate in many countries that the plants developed do not satisfy the nonobvious requirement in a patent application system, as existing techniques are used and the new breed is obvious. Therefore, a unique sui generis system is utilized for the Protection of New Plant Varieties. The TRIPS provides a protection for 25 years in case of trees and vines, whereas the protection for other plants is 20 years.

Meanwhile, several methods are in use that can be described as follows (Table 6.1):

TABLE 6.1

Types of Certification of Germplasm and Varieties

Type of Breed Recognition	Notes
Public domain	Germplasm belongs to the public and is freely available for utilization. Derivatives can potentially be appropriated and protected with IPR from further utilization.
Open source	Germplasm freely available for use with the restriction that there can be no further restriction. It is a new form of publicly accessible germplasm.
Plant variety protection	Since 1970, it is available through the US Plant Variety Protection Office. It requires that varieties be distinct, uniform, and stable. Certification for new varieties of sexual-reproducing plants lasts 17 years. It allows the breeder to determine, who is able to sell the seed and to charge the royalties. The varieties can be used in breeding programs and seed saving.
Plant patent	It is available through the US Patent and Trademark Office that exists since 1930; certification lasts 20 years and cover both sexually reproducing and asexually propagated plants. It allows the breeder to charge the royalties on propagation and sales of the variety.
Utility patent	Since 1990, it is available through the US Patent and Trademark Office. It requires that the inventions are novel, nonobvious, and useful. Although the utility patent before 1990 was not common for plants, several varieties and traits have been granted. The utility patent excludes others from using protected varieties for further breeding without permission or license from the holder of the patent. Certification lasts 20 years.
Trade secret	It allows the breeder to maintain the control of an F1 hybrid variety by keeping the inbred parent line secret to the public. It is not a formal legal protection. Because F1 hybrid varieties cannot breed true, the focus of the protection lies on the parental lines.
License	Beside the patent law, licenses, and contracts can be an additional form of private order. Restrictions are specific to the details of the agreement, for example, seed saving, breeding, transfer of seed to a third party, "bag tag" declaration, and research utilization.

6.1.1 OPEN SOURCE SEED INITIATIVE

For millennia, seeds have been freely available to use for farming and plant breeding without restriction. Within the past century, however, IPRs have threatened this tradition due to the commercialization of breeding by large companies. In response, a movement has emerged to counter the trend toward increasing consolidation of control and ownership of plant germplasm. The Open Source Seed Initiative (OSSI, www.osseeds.org) aims to ensure access to crop genetic resources by embracing an open-source mechanism that fosters exchange and innovation among farmers, plant breeders, and seed companies. Plant breeders across many sectors have taken the OSSI Pledge to create a protected commons of plant germplasm for future generations (LUBY and GOLDMAN 2016). Critical to supporting these goals is the use of an open-source mechanism for creation of a protected commons for crop plant germplasm that fosters exchange and innovation among farmers, plant breeders, seed companies, and consumers in a viral fashion without restrictions on further breeding.

The idea of a protected commons was coined by Richard JEFFERSON of CAMBIA[1] (Canberra, Australia) in an effort to integrate biotechnology and biological research with "copyleft" licensing arrangements that "support both freedom to operate, and freedom to cooperate" in a "protected

---

[1] CAMBIA is an independent nonprofit institute creating new technologies, tools, and paradigms to promote change and enable innovation.

commons." A protected commons allows for public access and improvement of a resource without imposing restrictions on downstream use. The OSSI created a pledge that embraces this approach. Users that accept the OSSI Pledge agree that no further restrictions can be made. OSSI promotes farmers' and gardeners' rights to save and replant seed of crops grown on their land and open access to plant germplasm for breeding. Both of these values are also shared by the UPOV and the US Plant Variety Protection Act (PVPA) (Table 6.1). What distinguishes the OSSI from either the UPOV or PVPA is that no restrictions can be placed on the derivative use of seeds distributed under the OSSI Pledge. Thus, OSSI Pledged seeds can be breed, sold, shared, and reproduced as long as any subsequent derivative or reproduction of that seed also carries the same freedoms.

7 Germplasm Maintenance

7.1 GENERAL REMARKS

The taste of today's tomatoes possibly is a very obvious example to the readers. Modern commercial tomato varieties are substantially less flavorful than heirloom varieties. When flavor-associated chemicals in 398 modern, heirloom, and wild accessions of tomato were quantified, it was found that modern commercial varieties contain significantly lower amounts of many of these important flavor chemicals than older varieties. Whole-genome sequencing and a genome-wide association study permitted identification of genetic loci that affect most of the target flavor chemicals, including sugars, acids, and volatiles (TIEMAN et al. 2017). Some of alleles got lost by breeding. But consumers require the recovery of good flavor—a challenge for molecular breeding.

The reduction of genetic diversity by modern agriculture is a well-known phenomenon. The following example is a good proof of this: successful crop genetic improvement is demonstrated by maize breeding in the USA. Maize hybrids released around 2000 yielded 10 tons per ha, more than twice the yield of hybrids released in 1930s and 1940s, when tested in historical variety trials. Modern varieties in the USA are based on six founder genotypes, which contributed on average 90% of the pedigrees of maize hybrids developed for the west-central US Corn Belt around 2000. Allelic diversity, measured by the average number of alleles per locus at 298 simple sequence repeat (SSR) loci, was less than 1.5 in the male and 2.0 in the female parents of hybrids released around 2000, but averaged 4.0 at the same loci in male and female parents of F_1 hybrids released in the 1930s and 1940s (DUVICK et al. 2004). Genetic improvement in maize in the USA over 70 years has been accompanied by a significant loss of genetic diversity at SSR loci that are most likely selectively neutral. This is a result of random genetic drift due to low effective population size, which in turn is determined by the number of founder parents and low rates of migration or mutation during 70 years of maize breeding.

Another example is related to the founding of the International Rice Research Institute (IRRI) in the Philippines in 1959. Similar to wheat, rice was also bred for high yield, semi-dwarf growth and susceptibility to high fertilizer application. In addition, a reduced growth period as well as a lower sensitivity to daily length was achieved. The latter allowed the shifting of sowing time and thus to an increase of harvests per year. The prototype of such a rice cultivar of was "IR8" (cf. Section 3.2.2). Its successor "IR36" showed a ripening period of 110 days, which resulted in two harvests within the monsoon season. The associated expansion of the cultivation area in southeast Asia with concomitant genetic depletion led to an extreme pest infestation, in particular with the brown-winged rice cicada (*Nilaparvata lugens*). Pesticides had to be used to achieve a minimum of yields. It began an intensive search for resistant genotypes in the available rice collections. One was found that resisted the rice cycad. In 1975, intensive breeding resulted in the variety "PB26." However, the resistance lasted only 2 years; the resistance of another bred "PB36" only 6 years (HERDT and CAPULE 1983).

Monoculture is also the reason of a repeated threat to banana plantations caused by the Panama disease—a plant disease of the roots of banana plants. It is a type of Fusarium wilt, caused by the fungal pathogen *Fusarium oxysporum*. The pathogen is resistant to fungicide and cannot be controlled chemically. During the 1950s, the tropical race 1 (TR1) of Panama disease wiped out most commercial "Gros Michel" banana production. The "Gros Michel" banana was the dominant cultivar of bananas, and the blight inflicted enormous costs and forced producers to switch to other, disease-resistant cultivars. New strains of Panama disease currently threaten the production of today's most popular cultivar, "Cavendish," which replaced "Gros Michel" on many places. The "Cavendish" was then thought to be immune, but it was immune only to the strain of the fungus that destroyed the "Gros Michel." The version that annihilated the "Gros Michel" was found in only the Western Hemisphere, but the version found in Malaysian soil was different, and the Cavendish is susceptible to it. TR4 was identified in 1990 in samples from Taiwan. It killed and spread faster, inspiring more panic than its earlier counterpart in Panama. The newly discovered strain was named TR4. Again, a race began in order to find resistance among banana collections (cf. Sections 3.6 and 7.1.4).

These examples demonstrate why the utilization of genetic resources is mandatory for future development of crops.

Fortunately, genetic resources are generally considered a public good and shared internationally. Wild relatives of crop species and their derivatives represent the reservoir of genetic diversity that will help to meet the food demands of 9 billion people by 2050. The dramatic increase in the number of GM crops reveals the value of new genetic resources. Genetic resources will provide a gateway to a new era of global food security.

The major activities for ex situ genebanks include assembling, conserving, characterizing, and providing easy access to germplasm for scientists and breeders. Although 7.4 million plant accessions are stored in 1750 germplasm banks around the world, only a small portion of the accessions has been used so far to produce commercial varieties. The reason is the long-lasting introgression process. COWLING et al. (2016) simulated prebreeding in evolving genebanks—populations of exotic and crop types undergoing optimal contribution selection for long-term genetic gain and management of population genetic diversity. The founder population was based on crosses between elite crop varieties and exotic lines of field pea from the primary genepool, and was subjected to 30 cycles of recurrent selection for an economic index composed of four traits with low heritability: black spot resistance, flowering time and stem strength, and grain yield. They compared a small population with low selection pressure, a large population with high selection pressure, and a large population with moderate selection pressure. Single-seed descent was compared with S_0-derived recurrent selection. Optimal contribution selection achieved higher index and lower population co-ancestry than truncation selection,[1] which reached a plateau in index improvement after 40 years in the large population with high selection pressure. With optimal contribution selection, index doubled in 38 years in the small population with low selection pressure and 27–28 years in the large population with moderate selection pressure. Single-seed descent increased the rate of improvement in index per cycle but also increased cycle time. This long breeding period, of course, frightens the breeders.

7.1.1 CORE COLLECTION

A core set of crop genetic resources are considered "first-look-sources" of trait-specific accessions for use in crop-breeding programs. FRANKEL and BROWN (1984) described the concept of the "core collections." Given the need for economic size, it was argued that a collection could be pruned to what was termed a "core collection," which would represent with a minimum of repetitiveness, the genetic diversity of a crop species and its relatives. The accessions not included in the core would be retained as the "reserve collections." The main purpose of the core fraction is to provide efficient access to the whole collection, which should be representative of the diversity at hand. The accessions in a core collection should be selected toward a manageable collection scaled down to the needs of the breeder and an inclusion of the widest possible range of variability.

To encourage greater use of a germplasm collection by breeders, it is suggested that a core collection needs to be designated by the curator. This core would represent the genetic diversity in the collection and its selection does require quality passport and characterization data.

---

[1] A breeding method in which individuals in whom quantitative expression of a phenotype is above or below a certain desired value (i.e., truncation point) are selected as parents for the next generation.

7.1.2 Prebreeding

Many improvement programs concerned with the utilization of plant germplasm include the process of prebreeding, also called developmental breeding or germplasm enhancement. So prebreeding is the early phase of any breeding program cultivar development utilizing germplasm. Though the end products of prebreeding are usually deficient in certain desirable characters, they are attractive to breeders due to their greater potential for direct utilization in a project than the original unadapted exotic sources. One of the main objectives of the IBPGR is to evaluate and characterize the available germplasm and to coordinate such activities with other crop-based institutes, coordinated projects, agricultural universities, and international institutes and to help in preparing inventories and catalogs on available genetic resources. The work on germplasm evaluation and characterization is being carried out at the Bureau's Headquarters and their Regional Stations (located in different eco/agroclimatic zones) for more than 75 major and minor crops. Crop curators for all major crops have been identified within International Board for Plant Genetic Resources (IBPGR)[2] and also in the National Boards for Plant Genetic Resources (NBPGR).

7.1.3 Evaluation

Preliminary evaluation and seed increase is done by the crop curators by growing one or two rows of the germplasm in an augmented design using two to three locally adapted checks for a period of minimum 2 years. The germplasm is being evaluated based on IBPGR descriptors list. Based on the evaluated data, over the years, several crop catalogs and/or inventories on wheat, barley, maize, amaranth, tomato, cluster bean, French bean, winged bean, cowpea, field bean, moth bean, soybean, lentil, *Sesbania*, *Trigonella*, opium poppy, safflower, sunflower, sesame, oat, okra, and so on, have been prepared. These crop catalogs are distributed to concerned plant breeders for identifying the useful germplasm for utilization in their breeding programs. In recent years, much emphasis has been placed on multilocational evaluation of germplasm.

7.1.4 Examples

In rice, during 1911–1956, about 400 cultivars were released in India through pure line selection of the traditional cultivars. These improved local types were virtually the pool of the traditional rice germplasm of India and made 10%–20% increase in yield over the traditional types under local agronomic practices and ecological conditions. They have continued to play a significant role in the varietal improvement of rice even to the present day by providing a well-adapted genetic background for incorporating other desirable characters. Many of these cultivars were adapted to and selected for upland and/or drought conditions, deep water and/or flood conditions, and saline soils.

In banana (*Musa* sp.), improved tropical plantain tetraploid hybrids, such as tetraploid cultivars "PITA 9" and "PITA 14," the tetraploid starchy banana "PITA 3," and diploid stocks strongly contributed to banana breeding. The tropical plantain tetraploid hybrids combine black sigatoka resistance, regulated suckering, and high bunch weight (ORTIZ 2016).

Successful evaluation of germplasm in kiwifruit (*Actinidia chinensis*) and introgression resulted in modern varieties, such as "Hayward," "Zespri Gold Kiwifruit," "Zespri Sungold Kiwifruit," or "Jingold" (FERGUSON and WU 2016). Interspecific crosses between *A. arguta* and *A. melanandra* led to red-skin and red-flesh varieties "Hortem Rua" in New Zealand and crosses between *A. argute* and *A. purpurea* to the varieties "Figurnaja," "Kiewskaja Gibridnaja," and "Kiewskaja Krupnoploidnaja" in Russia.

---

[2] In 2014, Bioversity International marked 40 years of operations. Bioversity International was originally established by the CGIAR as the International Board for Plant Genetic Resources (IBPGR) in 1974. In October 1991, IBPGR became the International Plant Genetic Resources Institute (IPGRI) and in 1994 IPGRI began independent operation as one of the centers of the CGIAR.

BITOCCHI et al. (2016) report varieties of bean (*Phaseolus vulgaris*), such as "NUA 35" and "NUA 56" that were selected from crosses with Andean germplasm.

The most widely grown varieties of Bahiagrass (*Paspalum* sp.), "Pensacola," "Paraguay 22," and "Wilmington" from Paraguay as well as "Argentine" and "Boyero" from Argentina are apomictic tetraploids deriving from crosses with South American germplasm populations (HANNA et al. 2016). In the 1950s, similar approaches led to the variety "Meyer" a US-released turfgrass-type of *Zoysia japonica*. From *Z. japonica* × *Z. tenuifolia*, the variety "Emerald" was established in 1955.

7.2   NEXT-GENERATION GENEBANKING

In general, the global activities are focused on the status of diversity in the food system. Where are the greatest vulnerabilities? Where the genetic diversity can be found in order to increase the crop productivity? What steps are needed to improve the ability of breeders to access genetic resources?

Advances in sequencing technologies have made it possible to analyze a large amount of germplasm accessions against low production costs. The new developments have an impact on the user-oriented activities than the housekeeping operations of genebanks. To better facilitate the user community, genebanks may have to strengthen their core collections, in particular, by improving quality management procedures and by providing access to a wider diversity of a crop's gene pool. In addition, genebanks may have to provide novel services, such as the introduction of specific user-oriented collection types, including research populations and genetically purified lines, and the establishment of novel information services, including plant genetic resources portals that can guide users to the information and materials of interest. To improve their user-oriented services, genebanks increase their communication and collaboration with the user community and develop strategic alliances with this sector (VAN TREUREN and VAN HINTUM 2014).

7.2.1   Screening Candidate Genes

Application of TILLING by sequencing methodology can determine levels of diversity within species and identify those populations that will represent priorities for conservation efforts, including germplasm for forestry and tree biomass production. For example, the re-sequencing of the genomes of wild crop relatives became a rapid method to determine the likely utility of the germplasm in crop improvement. The conservation of genetic resources both *in situ* and *ex situ* are guided by information on the novelty of specific populations at the whole-genome and specific allele levels. Thus, the analysis of wild relatives of crops supports crop improvement. Screening techniques for finding induced mutations is used to discover natural mutations in candidate genes of germplasm collections. The application of this reverse-genetics strategy to natural mutations has been termed ecotype TILLING (ecoTILLING) or breeding with rare defective alleles (BRDA). These approaches have been taken to search germplasm collections of a number of plant species for novel alleles in candidate genes. Among those species are rice, melon, *Brassica* sp., maize, barley, wheat, pepper, sugarbeet, or poplar (TILL 2014, VANHOLME et al. 2013). Next-generation sequencing, high-resolution melting, and LI-COR® analysis systems have been used to identify natural mutations, but procedures are modified for screening germplasm collections to account for the increased genetic heterogeneity (cf. Section 3.5.1).

7.3   OLDEST PLANT SELECTIONS AND CULTIVARS

About 10,000 years ago, wild progenitors gave rise to the first primitive varieties or primitive forms, such as landraces. Landrace is a seed-propagated crop available as population, including agricultural, horticultural, and ornamental plants. Medicinal and aromatic plants, fruit trees, and grapes are included. A landrace is identifiable and usually has a local name. It lacks formal crop improvement and is characterized by a specific adaptation to the environmental conditions of the area of cultivation.

It is closely associated with the uses, knowledge, habits, dialects, and celebrations of the people, who developed and continue to grow it. In a similar sense an heirloom variety is defined. It is an old cultivar that is maintained by gardeners and farmers. Both landraces and heirloom varieties have been commonly grown during earlier periods of human history. These may have been commonly grown during earlier periods in human history, but are not used in modern large-scale agriculture.

7.3.1 CEREALS AND SMALL GRAIN CROPS

7.3.1.1 Wheat (*Triticum* ssp.)

Einkorn (*Triticum monococcum*, domesticated species) is the most original form of cultivated wheat; *T. boeoticum* is the wild form. One still finds wild forms of the Einkorn today, so the domestication by means of human selection appears clear. In prehistoric times, the tetraploid emmer (*T. dicoccum*) developed by combination of diploid einkorn with another diploid game-grass (*Aegilops speltoides*). From wild tetraploid emmer, the durum wheat (*T. durum*) and kamut[3] are derived by breeding. Khorasan wheat or oriental wheat (*Triticum turgidum*) is related to kamut wheat. Khorasan refers to a historical region in modern-day Iran in the northeast. The wheat of Rome was emmer, called "far." "Farro" is an ethnobotanical term derived from Italian Latin for a group of three wheat species: spelt, emmer, and einkorn. The Italian word farro is derived from the presumed Latin word farrum, from Standard Latin *far* or *farris* "a kind of wheat." Far in turn is derived from the Indo-European root *bar-es*, that is, spelt, which also gave rise to the English word barley or Albanian bar.

The hexaploid wheat (*T. aestivum*) was produced by the inclusion of the entire gene set of the wild grass *Aegilops squarrosa* into the emmer. The hexaploid spelt (*T. spelta*), also known as dinkel wheat, or hulled wheat, is a species of bread wheat cultivated since 5000 BC. Some of the earliest recordings of spelt appear in the Bible (Exodus 9:30, Isaiah 28:25, and Ezekiel 4:9). In Greek mythology, spelt ("zeiá" in Greek) was a gift to the Greeks from the goddess Demeter and was first used by the Greeks. The ancient Greek word "zeidoros" meaning "life giving" comes from this particular cereal.

Naming of wheats came much later. In June 30, 1794, the "American Mercury" published an article titled, "Forward, an Account of a New Species of Wheat." It describes a new, hard winter wheat "Forward" that matured some 15–20 days earlier than other and showed a large plump grain.

Its hardness reduced the risk of premature germination before harvest. It originated 7 years earlier. A Mr. ISBILL of Caroline County, Virginia (USA) observed a single spike in a field of mixed wheat. Over the following years, ISBILL bulked it up, shared it with neighbors, so that in 1794 it could be distributed among the farmers in the county. "Chidham" wheat was one of the most widely grown landrace over the whole of Britain between 1800 and 1880. Mr. WOODS, a farmer in Sussex in southern England, found a single plant in a hedge. It contained 30 fair spikes, in which were found 1400 seeds. After several multiplications, it was sold to other farmers. The cultivar was recognized for its large-size spikes. It followed "Mungowells wheat," based on one plant from 1819, stable after the fourth generation, "Hopetoun wheat," based on one plant from 1832, "Shirreff's bearded red wheat" (by 1860 distributed for sale), "Shirreff's bearded white wheat" (by 1860 distributed for sale), and "Shirreff" Squarehead wheat (1882). The varieties "April Bearded" of Wales and "Rivet" are also centuries old.

Another example concerns a Russian wheat introduced from Ukraine in 1842. The original strain was called "Halychanka" and later known as "Red Fife" in Canada. Records on the "Halychanka" variety were found in the Galician Chronicle as far back as the time of King Yaroslav OSMOMYSL (1171–1187). The fact that the genetic characteristics of the "Halychanka" variety are based on a selection process reaching back to the twelfth century shows why this variety is so unique. Its

---

[3] Kamut derives from the ancient Egyptian word for wheat—said to have been derived from seed found in the Egyptian pyramids. It appeared on the market in 1980 and marketed as a new cereal; however, it is an ancient relative of modern durum wheat.

stability (genetic homozygosity) deserved the attention of every plant breeder. The variety was cultivated over centuries in the fields of Ukrainian peasants and known there as a local variety. The strain was used widely after 1848, becoming the leader in Ontario by 1851 and virtually replacing all others there by 1860.

7.3.1.2   Barley (*Hordeum* ssp.)

In 2016, an international team of researchers has succeeded for the first time in sequencing the genome of Chalcolithic barley grains. This is the oldest plant genome to be reconstructed to date. The 6000 years old seeds were retrieved from Yoram Cave in the southern cliff of Masada fortress in the Judean Desert in Israel, close to the Dead Sea. Genetically, the prehistoric barley is very similar to present-day barley grown in the Southern Levant, supporting the existing hypothesis of barley domestication having occurred in the Upper Jordan Valley (MASCHER et al. 2016).

As less old appears a landrace from Scotland, called "Bere," six-row barley currently cultivated in Orkney. It is also grown on Shetland, Caithness, and on a very small scale by a few crofters on some of the Western Isles. It is probably Britain's oldest cereal in continuous commercial cultivation. "Bere" is a landrace adapted to growing on soils of a low pH and to a short growing season with long hours of daylight. It is sown in the spring and harvested in the summer. Because of its very rapid growth rate, it is sown late but is often the first crop to be harvested. It is known locally as "the 90-day barley." Originally *bere* or *beir* or *bear* was a generic Lowland Scots word for barley of any kind and was used throughout the country. Now it is used mainly in the north of Scotland. Bere is a very old grain that may have been brought to Britain by Vikings in the eighth century or even from an earlier wave of settlement. In its early days, it was also called "bygge" or "big," probably originating from *bygg*, the Old Norse term for barley. It became well-adapted to the far north of Britain as successive generations of farmers grew it, selecting each year's seeds from the best plants of the previous year. Bere has a long history of use in breadmaking (Figure 7.1) and in making alcoholic beverages. Historical accounts from the fifteenth century onward show that Orkney produced large amounts of malt and beer, most of it probably from "Bere."

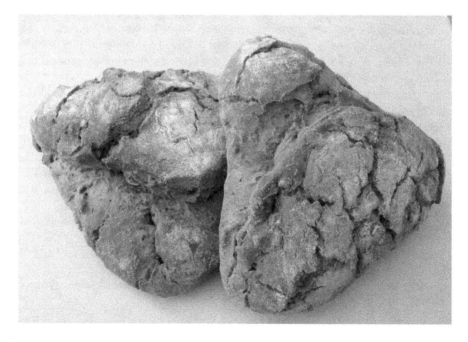

FIGURE 7.1   Traditional beremeal bannock, as made from bere barley on Orkney, Scotland.

7.3.1.3 Oats (*Avena* ssp.)

Little history of oat is known prior to the time of Christ. Oats did not become important to man as early as wheat or barley. Oats probably persisted as a weed-like plant in other cereals for centuries prior to being cultivated by it. Modern oats (white oats, *Avena sativa*) probably originated from the Asian wild red oat (*A. byzantina*). The oldest known oat grains were found in Egypt among remains of the 12th Dynasty, about 2000 BC. These probably were weeds and not actually cultivated by the Egyptians. The oldest known cultivated oats were found in caves in Switzerland that are believed to belong to the Bronze Age (around 1500 BC). C. PLINIUS, the Elder (23–79 AD), published in his "Historia naturalis" the idea that oats are a diseased form of wheat. Oats were first brought to North America by Scottish settlers with other grains in 1602 and planted on the Elizabeth Islands off the coast of Massachusetts. As early as 1786, George WASHINGTON sowed 580 acres to oats.

"Shetland oat" and "Small oat" (or "Black small oat," "Little oat," *Avena strigosa*) are described before seventeenth century in Western Islands of Great Britain. Later the varieties "Fine fellow oats" (by 1865 distributed for sale), "Long fellow oats" (by 1865 distributed for sale), "Early Angus oats" (by 1865 distributed for sale), or "Early fellow oats" (by 1865 distributed for sale) followed. In Europe, "Probsteier' oat" is assumed as an old local variety of Germany (Figure 7.2), similar as "Schatilow ovjos" from Mochowje/Tula in Russia.

Oat also has been cultivated in China for more than 2000 years. According to descriptions in historical records by Si Matsian (145–87 BC), it played an important role in food production. Moreover, it is widely recognized that China is the center of origin of naked oats, already known among the medieval farmers as "pillcorn" or "peelcorn," for example, in England. In China, oats are mostly grown in the north, particularly in Inner Mongolia, where wild species of oats are also found. In addition, European farmers distinguished "small oats" and "large oats." Variety names during the early eighteenth to nineteenth centuries were "Archangel," "Black Riga," "Blainslie," "Common Dun," "Common White Tartarian," "Cumberland," "Dun winter oat," "Dutch," "Early Angus," "Early Kent," "Friesland," "Georgian," "Hopetown," "Late Angus," "Magbiehill," "Old black," "Pilez," "Poland," "Red Essex," or "Black Tartarian" (*A. orientalis*) (MOORE-COLYER 1995).

7.3.1.4 Rye (*Secale cereale*)

Rye is a cereal that played a major role in the feeding of European populations throughout the middle Ages. Basically, there are two centers of rye origin: (1) a primary region around Tabriz (Iran) and toward the Black Sea and (2) a second region east of Iran and toward Afghanistan. They coincide with the recent distribution of weedy ryes. From here, cereal rye was distributed to the western and middle Europe via Anatolia and along the Danube River, to Scandinavia via Russia, and to China, Korea, and Japan. One of the earliest depictions of rye is found in the book of plants by Jacobus THEODORUS (1950), which shows compact spikes with many awns (SCHLEGEL 2013).

First targeted selections were made in the north of Germany by the Probsteier Seed Cooperative around 1850. Before numerous landraces were described typical for some regions of Europe and characters of the population. Among those early landraces are "Pico Gentario," "Trenelense" (Argentina), "Kärtner," "Marchfelder" (Austria), "Chlumecké," "Krmne Zito," "Pudmericke" (Czech Rep.), "Ging Zhouzao," "Kye-shan hye mai" (China), "Brattingsborg," "Sejet Kaernerrug" (Denmark), "Myttääla," "Slarma" (Finland), "Champagner rye" (France), "Biesenroggen," "Schlägler Alt," "Stauden," "Wollny's schlaffährigen Roggen" (Germany), "Tiszaközi feher" (Hungary), "Tetti Bartola" (Italy), "JKN" (Japan), "Kazimierskie," "Wojcieszyckie" (Poland), "Dolinskaya," "Pamirskaya," "Vyatka" (Russia), "Gotlands," "Kungsrug," and "Värnerag" (Sweden).

The Schlanstedter Roggen is the first modern rye variety with higher lodging resistance. It was bred by W. RIMPAU (1842–1903), who started winter rye breeding at Schlanstedt (Germany) in 1875. Most successful were the varieties of F. von LOCHOW (1849–1924) at Petkus (Germany), derived from the Probsteier gene pool. The village of Probstei is close to the Baltic Sea in the north of Germany.

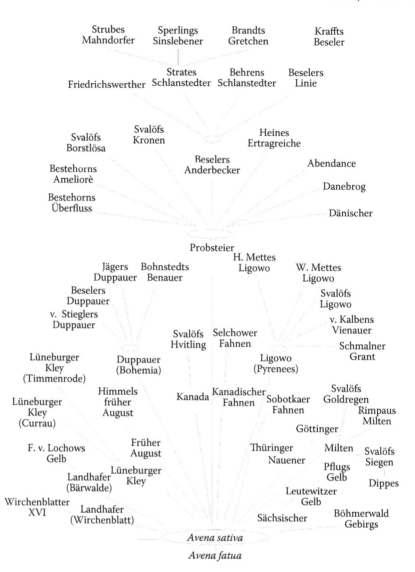

FIGURE 7.2 Oat varieties selected from European landraces at the beginning of the twentieth century. (Compiled and redrawn after ZADE, A., *Der Hafer – eine Monographie auf wissenschaftlicher und praktischer Grundlage*, Jena, Germany, G. Fischer VERL., 1918. With Permission.)

Since 1880, directed crosses led to the variety "Bestehorn's Riesenroggen" (G. BESTEHORN, 1836–1889) at Biberitz (near Könnern), Germany. Nickolay RUDNITSKY began rye breeding at Vyatka station (central Russia near Kirov) in 1894. The Vyatka variety is still used to some extent in the agricultural areas to the northwest of Russia. Friedrich Georg Magnus von BERG (1845–1938) was the breeder at Sangaste (Sagnitz) mansion in Estonia. In 1875, a local rye population "Sangaste" was developed, which was characterized by winter hardiness and large kernels. He started formation of an initial material for plant breeding in 1868. Over 40 crossing combinations between several landraces of rye were made. Mainly Probstei rye contributed to the first release. Currently, the variety "Sangaste" is considered as the oldest cultivated rye variety in the world. At the World's Fair held in Paris in 1889, "Sangaste rye" was awarded a gold medal, and at the world exhibition in Chicago in 1893, first prize was given to that variety.

7.3.1.5 Maize (*Zea mays*)

A study by MATSUOKA et al. (2002) has demonstrated that, rather than the multiple independent domestications model, all maize arose from a single domestication in southern Mexico about 9000 years ago. The study also demonstrated that the oldest surviving maize types are those of the Mexican highlands. Later, maize spread from this region over the Americas. The "Chapalote" maize seems to be the oldest form in North America. It is one of the most distinctive races of maize in Mexico. It is primitive in being not only popcorn but also a weak pod maize. One of the most distinctive characteristics of "Chapalote" is its brown pericarp (kernel) color (Figure 7.3). It performed well even during relatively dry years, because it was early maturing. Sometime between 1492 and 1952, "Chapalote" had fallen out of cultivation in the Southwestern Deserts, perhaps because of its susceptibility to rust disease. Nevertheless, it persisted in the Central Sonoran foothills of the Sierra Madre. To this day, Sonoran towns where "Chapalote" was grown still pride themselves on making the best pinole[4] in all of Mexico. Prehistoric cobs of a maize that is equivalent or close kin to today's chocolate-brown kernels of "Chapalote" date to 4100–4200 years ago.

Furthermore, "Chapalote" was prehistorically grown in fields along some of the oldest and most extensive irrigation canals anywhere in the New World. Recently, maize husks, stalks, cobs, and tassels dating from 6700 to 3000 years ago were unearthed at Paredones and Huaca Prieta, two sites on Peru's northern coast.

FIGURE 7.3 Cops of chapalote, the oldest cultivar of maize. (Courtesy of Gary P. NABHAN, Patagonia, Arizona, USA.)

---

[4] It is made principally of a unique roasted ground maize, which is then mixed with a combination of cocoa, agave, cinnamon, chia seeds, vanilla, or other spices. The resulting powder is then used as a nutrient-dense ingredient to make different foods, such as cereals, baked goods, tortillas, and beverages. The name comes from the Nahuatl word *pinolli*, meaning maizemeal. Today, pinole is generally made by hand using wood-burning adobe ovens and a stone and pestle, and is still consumed in certain, often rural, parts of Latin America. In fact, pinole is considered the national beverage of Nicaragua and Honduras.

"Choclero" refers to maize used for its "choclos" or green roasting ears. "Choclero" is probably the result of the introgression of dent germplasm from the USA into conical eared South American maize. Beside "Camelia," "Curagua," "Araucano," "Marcame," or "Chutucuno," it is one of the oldest landraces in Chile.

Already in 1525, the first fields were planted with maize in Spain, after Christopher COLUMBUS had discovered the plant in the Caribbean and brought it to Europe. From there he also brought the word "mays." This is derived from "mahiz," the word for maize and/or corn on Taino, the language of the Arawak (native the north coast of South America). The oldest specimen for maize is found in the herbarium of the Italian cardinal and archbishop Innocenzo CIBO (1491–1550) of Genoa from 1532, and maize plants are already depicted in Leonhart FUCHS's (1501–1566) herbal book in 1543. However, maize cultivation first spread in the east of the Mediterranean. In 1574, fields were already found in Turkey and on the upper Euphrates, on which maize was cultivated.

7.3.1.6 Rice (*Oryza sativa*)

Preliterate cultures could record knowledge about crop varieties only orally. But the beginning of writing opens us a window into the past. The Sanskrit—as one of the oldest Indo-European languages—is the primary sacred language of Hinduism and Mahāyāna Buddhism, a philosophical language in Hinduism, Jainism, Buddhism, and Sikhism of India. About 200 BC, one of its medical scholars and a native of Kashmir, Maharishi CHARAKA (Figure 7.4) wrote a list of various rice varieties, including "Raktashali" (reddish straw, glumes), "Mahashali" (large, fragant grain), "Kalam" (thick stem), "Shakunarhita" (curved grain), "Turnaka" (quick maturing), "Tapiniya"

FIGURE 7.4   The Indian "wandering physicians" H. CHARAKA (ca. 200 BC) monument of the Pantanjali Yogpeeth Campus, Haridwar, India. (From Wikipedia 2017.)

(maturing in hot weather), or "Kardamaka" (for slimy soil). He did this because some of the varieties as being good for particular medical therapies (NENE 2005).

"Bhutanese" rice is medium-grain rice grown since long in the Kingdom of Bhutan in the eastern Himalayas. "Bhutanese" is red japonica rice. Black rice belongs to the species *Oryza sativa*. Varieties include "Indonesian black" rice and "Thai jasmine black" rice. In China, Black rice is claimed to be good for the kidney, stomach, and liver. There is also the "Chakhao" black rice of Manipur (India) that came with the people of Manipur, particularly the Meiteis when they first settled on this land (chak—rice; hao—delicious). The words "amubi" and "angouba" stand for black and white, respectively, in Manipuri. "Chakhao amubi" and "Chakhao angouba" were named after the black and white color glutinous aromatic rice endosperm of these cultivars (OIKAWA et al. 2015). Other cultivars are "Chakhao poireiton" and "Chakhao sempa."

In China, SHOUSHITONGKAO (1742) recorded 3429 farmer cultivars of indica rices (TANG et al. 2010). In 1934, the early-maturing indica rice "Nantehao" was developed from the landrace "Boyangzao" through pure line selection at the Jiangxi Agricultural Experiment Farm, Jiangxi.

7.3.1.7 Millets

The origin of millet from Neolithic China has generally been accepted, but it remains unknown whether common millet (*Panicum miliaceum*) or foxtail millet (*Setaria italica*) was the first species domesticated. They were staple foods in the semiarid regions of East Asia (China, Japan, Russia, India, and Korea) and even in the entire Eurasian continent before the popularity of rice and wheat. About 30 years ago, the world's oldest millet remains, dating to nearly 8200 years BC, were discovered at the Early Neolithic site of Cishan, northern China. The site contained around 50,000 kg of grain crops of *S. italica* stored in the storage pits. Over hundreds of years, local Chinese cultivars became well established in different regions under names such as "Shishizhun," "Golden Black," "Maoti," "Banyelai," "Esiniu," "Huang Bianxiao," "Xifuxiao," or "Qisifeng." Many ancient millet cultivars are still popular among local farmers. For example, in the Taihang Mountain area in She County, Hebei Province, ancient cultivars such as "Laiwugu" and "Dabaigu" are still grown because of their unique adaptation to the local climate.

Beside finger (*Eleusine coracana*, East Africa) and foxtail millet (*S. italica*, China), other common millets exists, such as kodo (*Paspalum scrobiculatum*, India) and proso millet (*P. miliaceum*) as well as different barnyard millets (*Echinochloa* spp., Japan), pearl millet (*Pennisetum glaucum*, Sahel region), fonio millets (*Digitaria exilis, D. iburua*, West Africa), and little millet exist (*Panicum sumatrense*, Southeast Asia). Browntop millet (*P. ramosum*) is a native of India. The Mongolian "Buryat millet" (*P. miliaceum*) was already grown in Siberia during the seventeenth century. The two oldest finds of *P. miliaceum* in Germany (near Leipzig) are from the period of the line-band ceramics (old Neolithicum 5500–4900 BC).

Previous research recognized five different types of millets in South India: samai (*P. sumatrense*), thina (*Setaria* sp.), varagu (*Paspalum scrobiculatum*), kezhvaragu (*Eleusine coracana*), and solam (*Sorghum bicolor*). The etymology of these terms provides some historical evidence of a long history of these crops' use in the region. "Samai" and "Varagu" are featured in the oldest Tamil literature—the Sangam poems that date from 300 BC to 300 AD—as crops of pastoralists of the Deccan interior. These poems were composed by Dravidian Tamil poets. In Sangam literature, "Thina" and "Varagu" have been mentioned as the staple food for the people. Archaeological evidence suggests that the millet landraces such as "Samai" have been cultivated in Neolithic south India, such as at Hallur after 2000 BC (FULLER 2006). "Kezhvaragu" millet is not mentioned in Sangam literature (2500–1800 BC), suggesting it was introduced from China and Korea, after the Sangam age. In China, HU (1981) pointed out that the word for sorghum was frequently seen in documents of the Xian Qin period (2100–221 BC) and suggested the ancient "Liang" millet was today's sorghum. Early, it was introduced from Africa via India into China about 3000 BC.

Setaria millet occurs in a large number of varieties in China as well. Some of them contain a glutinous endosperm, which can be even used for wine making (NEEDHAM and BRAY 1984).

This is caused by a monogenic recessive mutation (cf. Section 3.5). That millet wine was used in SHANG era ceremonbials (about 1766–1122 BC). Thus, it can be supposed that the millet is grown since about 4000 years.

7.3.1.8 Quinoa (*Chenopodium quinoa*)

Quinoa is a domesticated staple food in Andean South America. It was domesticated by ancient Andean civilizations in the region surrounding the Bolivian and Peruvian Altiplano. Some 400 landraces of quinoa are known around that region. The oldest archeological remains of domesticated quinoa date to 5000 BC. Anciently, quinoa was known by a number of names in local languages. People of the Chibcha (Bogota) culture called quinoa "suba" or "supha," the Tiahuancotas (Bolivia) called it "jupha," and the inhabitants of the Atacama Desert (currently in Chile) knew it by the name "dahue." The names "quinua" and "quinoa" were used in Bolivia, Peru, Ecuador, Argentina, and Chile. "Huazontle," another chenopod (*C. berlandieri* subsp. *nuttalliae*) was domesticated in Mexico, where it served anciently as a seed crop and potherb, but is currently cultivated principally for its leaves and immature inflorescence.

7.3.1.9 Soybean (*Glycine max*)

"Flowery eyebrow," "Cuild of the white crane," "White spirit of the wind," or "Large jewel" are some of the names given to the indigenous Chinese soybean. The protein-rich soybean or "dadou" (means "great bean" in Chinese versa "xiaodou" small been that refers to the azuki bean) has been cultivated in China and Japan at least since the days of China's Western Chou dynasty (until 771 BC). The Japanese word for soybeans, "*daizu*," had been derived from the characters for the Chinese word "*ta-tou*," rather than from the Manchurian "*tu-ri*."

The bean of great antiquity, its origin can be traced back more than 3000 years. Its domestication dates back to the eleventh century BC (HYMOWITZ 1970). The first written record of the plant is contained in the books "*Pen Ts'ao Kang Mu*," describing the plants of China by Emperor Shen NUNG in 2838 BC. The earliest mention of the soybean in Japan occurs in the "*Kojiki*" (Record of Ancient Matters of Japan, 712 AD) (CHAMBERLAIN 1932). Dutch missionaries reached Japan in the 1700s, they discovered the attraction of "shoyu," the fermented soy sauce, but mistook its name for that of the bean itself. When they sent samples home to Europe, they described them as "shoyu" or "soya."

In the sixth century, the oldest variety names "Huang luo dou," "Chang shao," "Niu jian" were recorded in the agricultural book "Qui Min Yao Sho." "Huang luo dou" referred to the abscission of leaves during maturity, "Chang shao" indicated tall plants, and "Niu jian" described the seed shape. The variety names "Wu yue huang," "Liu yiu bao," or "Dong huang" from the MING dynasty (1368–1644 AD) can still be found in Chinese germplasm collections (SHURTLEFF et al. 2014).

7.3.2 Root Crops

7.3.2.1 Potato (*Solanum tuberosum*)

The potatoes cultivated today are derived from different landscapes that occur in the Andes from western Venezuela to Argentina and the island of Chiloé and the Chonos archipelago in the south of Chile. The oldest known traces of wild potatoes were found on Chiloé, their age is estimated to be 13,000 years. The Chilean landraces are presumably derived from the Peruvian species (*Solanum tuberosum* ssp. *andigena*), probably after hybridization with the wild *Solanum tarijense*. In Peru, which has long been known as the land of the potato, there are more than 3000 endemic potato cultivars. Most can only be grown in the Peruvian Andes due to their geological and climatic claims. The main difference between the Andean potatoes and the varieties cultivated in other cultivation areas is that they are adapted to the short-day light conditions. Early Spanish chroniclers—who misused the Indian word "batata" (sweet potato) as the name for the potato—noted the importance of the tuber to the Incan Empire. The enormous variety (more than 3800 types) of the grown potatoes (Figure 7.5) obviously had catchy names. "Papa Pitiquiña" (*S. stenotomum*) is widely considered the most primitive of the domestic potatoes. The highly frost-resistant "Papa Ajanhuiri" is extensively cultivated at altitudes up to 4100 m. The word "papa" is originally Quechua and simply means tuber, a general name of potato.

When, how, and through whom the potato came to Europe has not been clarified to this day. On their way from South America to Spain, the potato station on the Spanish Canary Islands. This is known because in 1567 three barrels containing potatoes, oranges, and green lemons were shipped

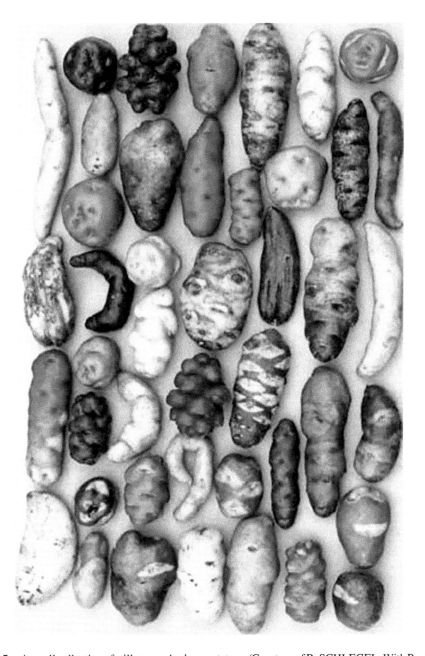

FIGURE 7.5   A small collection of still grown Andean potatoes. (Courtesy of R. SCHLEGEL. With Permission.)

from Gran Canaria to Antwerp, and in 1574 two barrels of potatoes were shipped from Tenerife via Gran Canaria to Rouen. If it is assumed that at least 5 years were needed to obtain so many potatoes that they could become an export item, the naturalization of the plant in the Canaries took place by 1562 at the latest.

The earliest proof for potatoes in Spain can be found in the books of the Hospital de la Sangre in Seville, which purchased potatoes in 1573. It is assumed that the potato reached Spain at the earliest 1564/1565 and at the latest 1570, since otherwise the botanist C. CLUSIUS (1525–1609), who visited the country in 1564 in search of new plants, would probably have noticed them. From Spain, the potato came to Italy and spread slowly on the European mainland.

For the first time, the potato in England was published in London in 1596 in the catalog of the plants bred by the botanist John GERARD (1545–1612) in his garden in Holborn. In Germany, the first potatoes were grown during the reign of FERDINAND III, around 1647 in Pilgramsreuth (Upper Franconia). In the monastery Seitenstetten of Lower Austria, the Benedictine abbot Caspar PLAUTZ wrote a cookbook with potato recipes that appeared in Linz in 1621. Large-scale cultivation began in 1684 in Lancashire (UK), 1728 in Scotland, 1716 in Saxony (Germany), 1738 in Prussia (Germany), and 1783 in France. First cultivars are described as "Rosa Tannenzapfen" and "Bamberger Hörnchen" (Germany 1850 and 1870, resp.), "Asparges" (Denmark 1872), "Belle de Fontenay" (France 1885), "Lumper" (Scotland ~1850), and "Duke of York" (England 1891). The first potatoes arrived in North America in 1621.

7.3.2.2 Sugarbeet (*Beta vulgaris* ssp. *sacharifera*)

In 1747, the Berlin pharmacists and chemist Andreas Sigismund MARGGRAF showed the sugar content of the beetroots (German: Runkelrübe) with concentrations of 1.3%–1.6%. In 1801, after the successful selection of the "Weisse schlesische Rübe" (White Silesian beet) as first sugarbeet variety with about 6% sugar content, the student of MARGGRAF and physico-chemist ▶ Franz Carl ACHARD also laid the foundations for industrial sugar production (cf. Section 3.2). This selection is the progenitor of all modern sugar beets. After crosses, already between 1786 and 1800, he grew about 22 varieties of beet on an experimental field at Kaulsdorf near Berlin (Germany) for testing the sugar content. Around 1885, the "Klein Wanzlebener Original" became outstanding species of its time, followed by "Ziemann-Quedlinburg" (~1900) or "Breustedts Elite" (1907). Luis VILMORIN, Olivier LECQ, and Fouquier d'HEROUEL (France) marketed their own varieties.

7.3.3 VEGETABLES

7.3.3.1 Carrot (*Daucus carota*)

It is believed that the carrot originated some 5000 years ago in Middle Asia around Afghanistan, and it is thought that Afghanistan is also the center of diversity of the anthocyanin carrot. The "redness" feature is thought to have emerged in varieties developed in post classical times, after hybridization with a central Asian species in the early middle ages. *Most wild carrots are white or muddy yellow.* They had edible leaves and thin, strong tasting white roots, which were prescribed for medicinal purposes. Names include Greek *keras, philtron, sisaron, staphylinos agrios, elaphoboscum, daukos,* and Latin *daucus, pastinaca rustica.* THEOPHRASTUS (371–287 BC) stated in the ninth book of his "History of Plants" that carrots grow in Arcadia, but that the best are found in Sparta. *Among other, the seeds of "daucus creticus" was mentioned in an ingredient of an antidote, that is, as pills found on board a second century BC (CELSUS 1935). This might be the first cultivar name of carrot. Carrots appeared in Asia Minor (modern Turkey) in the tenth century, in Arab Spain in the twelfth, in China in the fourteenth, and in Japan in the seventeenth. European carrots before the sixteenth century were purple or yellow.* In the German language, the "yellow turnip" or "horse turnip" is still a term for the original form of the carrot.

The first European author who mentions red and yellow carrots is the Jewish Byzantine doctor and dietician Simeon SETH, in the eleventh century. The word "carrot" was first recorded in English around 1530 and was borrowed from Middle French *carotte*, itself from Late Latin *carōta*, from Greek *karōton*, originally from the Indo-European root *ker* (horn), due to its horn-like shape. In Old English, carrots (typically white at the time) were not clearly distinguished from parsnips, the two being collectively called *moru* or *more* (from Proto-Indo-European *mork* (edible root), German for carrot is "Möhre."

7.3.3.2 Tomato (*Lypersicon* ssp.)

The tomato is the edible, red fruit of *Solanum lycopersicum*. The species originated in Central and South America. The Nahuatl (Aztec language) word "xitomatl" gave rise to the Spanish word "tomate," from which the English word "tomato" originates. "Calabacito Rojo" is very old American heirloom tomato that was documented in Philadelphia gardens as early as 1795. The bushy indeterminate plants can grow quite large and the small red fruit is flattened, slightly ribbed, and disease resistant. The "Yellow Pearshaped" appeared on the US market around 1800. The fruit has a sweet unique flavor. "Crimson cushion" (syn. "Henderson's crimson cushion," "Red ponderosa," "Beefsteak tomato") is known since 1892. It is said to be the original strain of "Red Beefsteak tomato." The large fruit can weigh up to 1 kg. As old is the "Brandywine tomato" that was introduced by the Amish in 1885. It comes from the collection of the late Ben Quisenberry, who collected hundreds of tomato strains from 1910 to the 1960s. Another Mennonite community in the Shennandoah Valley of Virginia (USA) has kept the old German varieties since the 1800. Tomatoes were the first vegetable in America to earn the trendy prefix "heirloom." What not commonly known is that "Landreth Seeds Co." was the first seed company to offer tomato seeds commercially to the general public—in 1820.

"Lukullus" is perhaps the oldest German tomato variety. It was said to have emerged at the beginning of the twentieth century from a cross between Danish "Export" and "Jewel." It helped to the breakthrough of tomato, which was largely unknown in Germany as a foodstuff. The "Goldene Königin" (Golden Queen) is a yellow-round tomato, which has been cultivated since 1871. Previously there was already the "Gelbe Johannisbeertomate" (*Lycopersicum pimpinellifolium*). The salad tomato "Moneymaker" is already over 100 years old. This high-yielding variety was offered for the first time in England in 1913.

The first descriptions of the plant come from the first half of the sixteenth century, mainly from Italy. In 1543, the German Joachim KREICH, a pharmacist in Torgau, founder of a botanical garden, was one of only four tomato owners in the then Germany. In 1544, Pietro Andrea MATTIOLI first described the plant as "Pomi d'oro" (Golden apple) and introduced the Latin term *Mala aurea*. The French name, "*pomme d'amore*," or "apple of love," suggests that they agreed, though some experts suspect that the name was a misunderstanding of the Spanish "pome dei Moro," or "apple of the Moors." The botanist Joseph Pitton de TOURNEFORT (1656–1708) at France grouped the tomato within the *Solanaceae* family and called it *Lycopersicon*. This Greek term appears to follow an old German word for tomato, "Wolfspfirsich," which was also translated into English as "wolf peach." It received today's common name "tomato" only in the nineteenth century. The public was wary. As potatoes, tomatoes had long been considered poisonous due to their ancestry as members of the Nightshade family.

7.3.3.3 Cabbage

In the Roman and Greek literature, a leafy kale is mentioned from the northwest Iberia that dates back to the Iberian farm activities of 961 AC. The earliest Medieval drawings of leafy kale are contained in the handbook "Tacuina sanitatis" of the fourteenth century, based on an Arab manual "Taqwin al-shha" of the eleventh century. However, C. PLINIUS, the Elder (23–79 AD) in the first century described already "*cumanum*" cabbage with a white head (*capitum patulum*), "*lacuturense*" with a wide head (capite preagrandes), and "*ariccinum*" that reminds of broccoli sprouts, the variety

"Pompeianum" as a marrow-stem kale. In the "Capitulare de villis," a "ravacaulos" is mentioned that was a turnip-like kale, probably kohlrabi. Hildegard von BINGEN (1098–1179) recognized a red cabbage. Cauliflower was first classified as "Syrian cole" in the book "Kitab-al-falaha" (Book of Agriculture), written in the twelfth century by AL AWAM (CUBERO 2002).

7.3.4 FRUITS

7.3.4.1 Apple (*Malus* ssp.)

Already in HOMER's "Odyssey," composed near the end of the eighth century BC, the apple is mentioned, namely, when, after the Trojan War, ODYSSEUS longed for the home garden full of apple trees. The apple, as "Forbidden Fruit" according to Old Testament's Book of Genesis, seems to have appeared in Western Europe at least by the twelfth century. The Latin word *malus* means both "apple" and "evil." The latter is probably due to the early Christians. Therefore, a 1504 engraving by Albrecht DÜRER (1471–1528) shows "Adam and Eve" with apples.

The tree (*Malus pumila*, commonly and erroneously called *Malus domestica*) originated in Central Asia, where its wild ancestor, *Malus sieversii*, is still found today. Apples have been grown for thousands of years in Asia and Europe, and were brought to North America by European colonists. The apple tree was perhaps the earliest tree to be cultivated. There are more than 7500 known cultivars of apples.

Apples were introduced to North America by colonists in the seventeenth century, and the first apple orchard on the North American continent was planted in Boston by Reverend William BLAXTON in 1625. The only apples native to North America are crab apples, which were once called "common apples." Apple varieties brought as seed from Europe were spread along Native American trade routes. An 1845 US apples nursery catalogue sold 350 of the "best" varieties. In Britain, old cultivars such as "Cox's Orange Pippin" (first grown in 1830) and "Egremont Russet" (first recorded in 1872) are still commercially important. "Ashmead's Kernel" was described at the beginning of eighteenth century in Gloucester (UK) and "Catshead" already 100 years before. In 1613, "Court Pendu Plat" was mentioned in France—an extremely old variety that may date from as early as Roman times. The Swiss "Glockenapfel" is reported from the same time. The famous "Gravensteiner" has is origin in Italy or Germany of seventeenth century.

In 1170, Cistercian monks from Burgundy mentioned the "Borsdorfer Apfel" as a very early European variety. The "Roter Kalvill" (syn. "Roter Himbeerapfel") is a very old variety of unknown origin. In 1565, it stood as a true red apple in the princely "Pleasure garden" at Stuttgart. In America, John CHAPMAN (1774–1845) grew in his nurseries apples from seed. A high degree of recombinations occurred from which CHAPMAN selected various cultivars. "Newtown Pippin," found as a sport in Flushing, New York, had already spread to Europe by 1781. Flushing was the first plant and tree nursery to be established in the USA. During the 1860s, Peter GIDEON (1820–1899) tried breeding with apples of Russian origin and discovered "Wealthy," a variety that survived the strong Minnesota winters and tasted good.

7.3.4.2 Pear (*Pyrus pyrifolia*)

It is thought that pears originated in the Caucasus from where they spread to Europe and Asia and that they were first cultivated more than 4000 years ago. By the thirteenth century, many varieties of pears had been distributed from France across Europe. In England, toward the end of the fourteenth century, the "Warden pear" had been bred and became famous for its use in pies. The variety is mentioned in Shakespeare's "*The Winter's Tale*." "Bon Chretien d'Hiver" was imported into France from Italy in 1495 by CHARLES VIII. About 1500, the "Red pear" was largely grown in Herefordshire (UK). "Messire Jean" is of uncertain origin. The reference to the variety in French pomological literature dates back to 1540/1550. "Besi d'Hery" originated in Brittany or France, about 1598. "Black Worcester" is possibly of French origin. British references to this cultivar date back to 1575. "Spina carpi" originated in Italy about 1575. By 1640, at least 64 varieties were being cultivated in England and at about this time grafting onto quince rootstock began to replace pear and crab apples rootstocks. Early in the nineteenth century, the renowned horticulturist ▶ Thomas Andrew KNIGHT began to develop pear varieties. The Royal Horticultural Society encouraged pear growing and in 1826 there were 622 varieties in their gardens at Chiswick. The "Williams" Bon Chretien' pear was mentioned in the "Transaction of the Horticultural Society of London" by the eminent botanist William HOOKER in 1816. It was marketed in Slovakia as "Williamsovka" and in the USA as "Bartlett."

A famous ancient pear "Shandong pear" from northeastern China dates back over 3000 years! It is still in production in the Chucheng, Laiyang, and Penglai districts.

7.3.4.3   Banana (*Ensete ventricosa, Musa* ssp.)

The Ari people of Ethiopia, whose staple food is the banana relative, *Ensete ventricosa*, have sacred areas, called "kaiduma," where wild ensete plants grow. Modern genome research came to the conclusion that the Ari population came to Africa at least 3000 years ago.

Archeologists have discovered in the Kuk valley of New Guinea an area where human first domesticated the banana around 8000 BC. Kuk is the first known instance of banana domestication, but it is probably not the cradle from which all other domesticated species sprang. But Kuk is probably not the cradle from which all other domesticated species sprang. Bananas originated also in Southeast Asia and the South Pacific around 8000–5000 BC. Plantains may have been grown in eastern Africa as early as 3000 AD, and in Madagascar by 1000 AD. The plantain had certainly reached the African continent between 500 BC and 500 AD. Buddhist literature notes the existence of the banana in 600 BC, and when Alexander the Great's expeditions led him to India in 327 BC, he stumbled across the fruit. By the third century AD, plantains were being cultivated on plantations in China. By the fifteenth and sixteenth century, Portuguese sailors were establishing the crop throughout Brazil.

The name of today's most important banana variety is "CAVENDISH." "Cavendish" bananas were named after William CAVENDISH, sixth Duke of Devonshire. Though not the first known banana specimens in Europe, at around 1834, Cavendish had received a shipment of bananas courtesy of the chaplain of Alton Towers. His gardener, Sir Joseph PAXTON cultivated them in the greenhouses of Chatsworth House. The plants were botanically described by Paxton as *Musa cavendishii*, after the Duke. Almost all bananas traded worldwide originate from the "Cavendish" cultivar from south China, which was developed in 1953. The "Cavendish" banana replaced the "Grand Michel" (syn. "Gran Michel" or "Jamaica banana") from the world market in the first half of the twentieth century. French naturalist Nicolas BAUDIN carried a few corms of "Gran Michel" from Southeast Asia, depositing them at a botanical garden on the Caribbean island of Martinique. In 1835, French botanist Jean François POUYAT carried Baudin's fruit from Martinique to Jamaica.

British Banana supplier, Thomas FYFFES, received its first consignment of bananas from Canary Islands 124 years ago, in September 1888. These bananas belong to the "Dwarf Cavendish cultivar." Derived clones are "Grande Naine," "Lacatan," "Poyo," "Valéry," and "Williams." "Grande Naine" is the most important clone in international trade, also known as Chiquita banana. Most cultivated bananas belong to the species *Musa acuminata*, *M. balbisiana*, and *Musa × paradisiaca* for the hybrid *Musa acuminata × M. balbisiana*.

7.3.4.4   Grapes (*Vitis vinifera*)

The cultivation of the domesticated grape began 6000–8000 years ago in the Near East. The earliest archeological evidence for a dominant position of wine-making in human culture dates from 8000 years ago in Georgia. Wine was grown during Bronze Age at Santorin, of Mino culture, and during the Sumerian kingdom before 3500 years. Greek colonists then introduced the Minoan practices in their colonies, especially in southern Italy (Magna Grecia). The "Black Corinth" grape is a seedless mutant. It may date back to age of classical Greece. Later, the Romans contributed to further domestication and distribution. Grapes followed European colonies around the world, coming to North America around the seventeenth century and to Africa (in 1655 by the Dutch), South America (by Spanish invaders; in 1700, to California), and Australia (by different immigrants). The first documentation of grapevines growing in the Americas was discovered in researching the logbook of navigator Giovanni de VERAZZANO, who reported in 1504 that a large "white grape" was vigorously growing at Cape Fear, North Carolina. The English explorer of the New World, Sir Walter RALEIGH, confirmed in a letter to Arthur BARLOWE in 1585, the discovery of a white grape, when he landed in coastal North Carolina.

The first written accounts of grapes and wine can be found in the Epic of GILGAMESH, the king of Uruk, an ancient Sumerian text from the third millennium BC. There are also numerous hieroglyphic references from ancient Egypt, according to which wine was reserved exclusively for priests, state functionaries, and the pharaoh.

When we are listing oldest grape cultivars, we do not mean the oldest 100 bottles of grapevine of the Slovenian varieties "Žametovka" (syn. "Ametovka," "Kavčina črna," "Modra kavčina"). That vine was planted at least 400 years ago, probably somewhat earlier. Its Methuselah state is confirmed in paintings of Maribor dating from the years 1657 and 1681, which are kept in the Štajerska Provincial Museum in Graz. In the pictures, the frontage of the house at Vojašniška ulica 8, built in the sixteenth century, can be clearly seen, and even then, it was already lushly overgrown with today's venerated variety (Figure 7.6). Just as in those days, the plant type still today clings to the trelliswork (ROBINSON et al. 2013).

Note, we are trying to compile those about 2000 cultivars that are grown worldwide. The ancient origin of "Pinot" was explained by the fact that it was identified as one of the parents of 46 modern varieties, including "Gamay," "Chardonnay," and "Melon," which are offspring of crosses between "Pinot" and "Gouais" of the Merlot group. "Clairette" variety was attested in Languedoc since the beginning of the twelfth century, still widely cultivated some kilometers north of the Valros area, and suggested an autochthonous and ancient origin.

FIGURE 7.6 A more than 400-year-old "Žametovka" grapevine plant growing outside the Old Vine House in Maribor, Slovenia. (From Maribor - Pohorje Tourist Board (Slovenia), www.maribor-pohorje.si, 2017; Courtesy of Vesna MALE. With Permission.)

The European cultivar "Heunisch Weiss" (syn. "Gouais blanc") contributed to the development of renowned varieties like "Chardonnay" and "Riesling Weiss." There are three phenotypic variants of "Heunisch Weiss:" (a) "Heunisch Dreifarbig" expressing intense anthocyanin coloration on shoots, inflorescences, and leaf petioles before fruit set and even red berry skin at fruit set stage, (b) "Pekasore" or "Heunisch Rotgestreift" with rose to red stripes on white berries, and (c) "Heunisch Weiss Seedless" or "Aspirant" (MAUL et al. 2015).

Although the geographic and genetic origin of "Heunisch Weiss" is still uncertain, AEBERHARD (2005) and KRÄMER (2006) believe in the introduction by the Huns in the fifth century, by the Magyars after the ninth century, or importation by order of Charlemagne around 800 AD.

First citation of "Huniscedruben" (= Heunisch grapes) was discovered in *"Summarium Heinrici"* from Lorsch-monastery (Germany), a scholar compendium of the eleventh century (STAAB 1997).

The first documented evidence for the "Riesling" variety was a sale that took place on March 13, 1435. The cellar log of Count Katzenelnbogen at Rüsselsheim (Germany) shows that Klaus KLEINFISH sold the six wines of "Riesling" for 22 solidi. "Malvasia" (syn. "Malvazia") is a group of wine grape varieties grown historically in the Mediterranean region, Balearic Islands, Canary Islands, and the island of Madeira. Most ampelographers believe that the Malvasia family of grapes is of ancient origin, most likely originating in Crete. The name "Malvasia" is generally thought to derive from Monemvasia, a medieval and early Renaissance Byzantine fortress on the coast of Laconia, known in Italian as "Malvasia." "Traminer" (syn. "Gewürztraminer," "Red Traminer," "Yellow Traminer," "Klevner," "Clevner," and about 150 other co-cultivars) is a white wine variety with yellow-reddish berries. As genetic studies showed, the origin is probably Southeastern Europe. Its name comes from Tramin (Lake Karer) in South Tyrol, where wines have been documented since the eleventh century under this name. In Germany, the cultivation of the grape variety around the year 1500 in the mixing set, for example, with Riesling, is fixed.

The variety "Tempranillo" (syn. "Ull de Llebre," "Cencibel," "Tinta del Pais," "Aragonez," "Tinta Roriz," and several other synonyms) is a black grape variety widely grown to make full-bodied red wines in its native Spain. Genetic studies revealed that "Tempranillo" is a spontaneous crossing of the white grape variety "Albillo Mayor" with the red "Benedicto." The first written mention of "Albillo" grape is from the fifteenth century in "Agricultura General" by Gabriel Alonso de HERRERA in which he says that the wine made from this variety is described as "very clear, with gentle color and taste." The "Tempranillo" genepool seems to be related to ancient Phoenician species in Lebanon. In addition, the "Tempranillo Blanco" was also discovered as a purely autochthonous white mutation of the "Tempranillo" (ROBINSON et al. 2013).

The "Koshu" grape of Japan is also more than 1000 years old and all the "Muscat" (syn. "Moscato," "Misket") grapes, including more than 100 co-cultivars, are known since 3000 years.

7.3.4.5 Olive (*Olea* ssp.)

The wild-olive (*Olea. europae* ssp. *oleaster*) has been considered by various botanists a valid species and a subspecies of the cultivated olive tree (*Olea europaea* ssp. *europaea*) that was domesticated at various places during the fourth and third millennia BC, in selections drawn from varying local populations. The Old Testament writers distinguished two trees: *"zayit"* designates the cultivated olive, the wild-olive being designated in the seventh century BC, Nehemiah 8:15 as *"eẓ shemen."* Some modern botanists take this latter term to apply to *Elaeagnus angustifolia*, the "Russian-olive." The word derives from Latin *ŏlīva* a borrowing from the Greek and means oil. Thus, the word "oil" in multiple languages ultimately derives from the name of this tree and its fruit.

It seems certain that the olive tree as we know it today had its origin approximately 6000–7000 years ago in the region corresponding to ancient Persia and Mesopotamia. The olive plant later spread from these areas to the Levant. The edible olive seems to have coexisted with humans for about 5000–6000 years, going back to the early Bronze Age (3150–1200 BC). As far back as 3000 BC, olives were grown commercially in Crete. They may have been the source of the wealth of the Minoan civilization. The first seedlings from Spain were planted in Lima (America) by Antonio de RIVERA in 1560.

There are close to 2000 of olive cultivars in the world. And just like grape varieties express certain flavors in wine, olive cultivars exhibit certain tastes in extra virgin olive oils. Olives, like the "Salonika" variety were likely first domesticated in the Levant around 6000 years ago. The "Kalamata" olive is a large purple olive with a smooth, meaty texture named after the city of Kalamata in the southern Peloponnese (Greece). The "Maltija" cultivar is probably the most popular Maltese cultivar and can give a high productivity; the "Bidnija" is believed to be the oldest Maltese olive cultivar and thought to date back to Roman times. "Nabali" is one of the oldest olive cultivars in the Middle East that originated on the banks of the River Jordan. The most ancient olives in Spain are "Gordal Sevillana," "Lechin de Granada," and "Verdial de Velez Malaga." The former cultivar is one of the most widely known table olive cultivars and is typically found in the western region of Andalusia, specifically the Sevilla province (RODRIGUEZ-ARIZA and MOYA 2005). According to the AFLP results, three cultivars "Vasilicada," "Throumbolia," and "Lianolia Kerkiras" were considered as ancient Greek cultivars.

7.3.4.6 Fig (*Ficus carica*)

Common fig (or just the fig) is an Asian species. Native to the Middle East and western Asia, it has been sought out and cultivated since ancient times, and is now widely grown throughout the world, both for its fruit and as an ornamental plant. The edible fig is one of the first plants that were cultivated by humans. Nine subfossil figs of a parthenocarpic (and therefore sterile) type dating to about 9400–9200 BC were found in the early Neolithic village Gilgal I (Jordan Valley). The find predates the domestication of wheat, barley, and legumes, and may thus be the first known instance of agriculture. Historical Sumerian tablets record the use and consumption of figs in 2500 BC. The most famous Biblical reference to figs is that, in which JESUS cursed a fig tree for not producing any fruit for him as he passed by, a curse that killed the fig tree (Matt 21:18). Figs were also a common food source for the Romans. CATO the Elder, in his "De Agri Cultura" (Chapter 8), lists several strains of figs grown at the time he wrote his handbook: the "Mariscan," "African," "Herculanean," "Saguntine," and the "Black Tellanian." The "Sar Lop" fig is possibly 2000 years old (STOREY 1995). Another "Dotatto" was mentioned by C. PLINIUS (23–79 AD). Figs probably first traveled east to China along the Silk Road after the Islamic conquests, as the first time we hear about figs in China is about 700 AD, during Tang Dynasty, and then people in China call figs by their Arabic name, "tin." In 1769, Spanish missionaries led by Junipero SERRA brought the first figs to California. The "Mission" variety was named after the Californian Franciscan Mission of California dating back to 1770. It was planted there and cultivated on a commercial scale. It is still popular there. The "Kadota" cultivar is even older, being mentioned by the Roman naturalist C. PLINIUS, the Elder (23–79 AD) in the first century AD. There are more than 100 cultivars of fig worldwide.

7.3.4.7 Strawberry (*Fragaria* ssp.)

C. PLINIUS, the Elder (23–79 AD) in the first century AD described the strawberry ground plant that was being grown for food to be used as a medicinal tonic in the first century AD. Strawberry fruits were depicted in European paintings during medieval times, for example, *The Garden of Earthly Delights* (or *The Millennium*); a triptych painted by the early Netherlandish master Hieronymus BOSCH (1450–1516) and housed in the Museo del Prado in Madrid since 1939. Strawberry was cultivated in gardens during the 1300s in Europe. HENRY the VIII, King of England, purchased some strawberry fruits to eat in the year 1530. During the 1600s, a strawberry plant shipment was received in England from the American colonies and planted in backyard gardens. These Virginia strawberries, *Fragaria virginiana*, were tasty and delicious growing larger in size than the European strawberries. The most widespread attempts of getting two strawberry crops a year stimulated systematic hybridization experiments, most of all in Germany, France, and England. Genes for this came from California strawberries, a race of *F. chiloense* that was introduced to Europe in the middle of the nineteenth century. Of these, "Vicomtesse Hericart de Thury" was the first widely cultivated variety to produce two full crops (KINGSBURY 2009). One of its parents was "Elton," bred in England in 1819 by T. A. KNIGHT.

7.3.5 INDUSTRIAL CROPS

7.3.5.1 Cotton (*Gossypium* ssp.)

Old World diploid cottons *Gossypium arboreum* and *G. herbvaceum* traditionally grown in Asia since as far back as 5000 BC were largely displaced by Spanish and Portuguese introductions of New World tetraploids (*G. hirsutum, G. barbardense*) during the colonial period. Already before 1820, Muhammad Ali PASHA (1769–1849) initiated the cotton breeding in Egypt, leading to the varieties "Jumel" and "Mako." By 1823, exports started and attracted some of the highest prices in Europe. This sales boom was based on a random find of Louis Alexis JUMEL (1785–1823), a French textile engineer, who around 1817 discovered unfamiliar cotton in a garden of Cairo. It was bush bearing and showed fiber superior in length and strength as compared to common cultivars. In 1869, in India, the variety "Khandish Hingunghat" was declared the best out of 300 varieties for its cleanness, strength, length, brightness, silkiness, and evenness.

In the Old South (USA), Henry W. VICK used to be a successful seed trader and cotton breeder. During the early 1840s, he spent winter evenings sorting through his cotton plants. One day he realized that "I plainly saw that what I supposed to be a homogenous stock of cotton seed, consisted in fact of ten or a dozen different varieties." This discovery led him to select out and grow out the best one (MOORE 1956). In its first year, it yielded exactly 100 seeds. In naming so, "100 seeds" the variety became one of the most widely grown cotton before 1860 at the southern cotton belt (USA).

8 Future Developments

Plant breeding remains a vibrant, multidisciplinary science characterized by its ability to re-invent itself by absorbing and utilizing novel scientific findings and technical approaches. Modern breeding programs should include hands-on experience with the inheritance and selection of complex traits in actual plant populations, basic biology of plants (reproductive biology, Mendelian genetics), principles of quantitative genetics and selection theory, principles and practice of plant breeding, and related sciences such as genomics, applied statistics, experimental design, and pest sciences. With regard to specialty crops, increased support for research and education may result from a focus on the unique features of these crops. Finally, it is important to cultivate public awareness of the accomplishments of plant breeding.

In developing countries, periurban agriculture and home gardening are becoming more important for national food security as a result of rapid urban expansion. After the urban gardening, indoor farming becomes the latest hype in urban centers. Of course, it is limited to horticulture until now. Nevertheless, plant breeding is particularly challenged. Hydroponics, light regimes, climatic champers, and so on require specific adaptation of plant varieties grown under those artificial conditions.

Hence, new cultivars will be needed to fit into intensive production systems, which may provide the food required to satisfy urban world demands of this century. Specific plant architecture, tolerance to urban pollution, efficient nutrient uptake, and crop acclimatization to new substrates for growing are, among others, the plant characteristics required for this kind of agriculture. Actually, the first step toward new type of agronomy is already made. The production of rapeseed for biodiesel reached a level in several European countries that it became a particular branch of agriculture. Alcohol production from sugarcane in Brazil is another indication for chances in agriculture.

Food crops with low fats, and high in specific amino acids may be another need to satisfy people who wish to change their eating habits. If genes controlling these characteristics do not exist in a specific crop pool or genebanks they may be incorporated genetic engineering.

Some publications anticipated that in the 21st millennium food would not need to be harvested from farmer's fields (ANDERSON 1996). Tissue culture of certain parts of the plant may provide a means to achieve success in this endeavor. For example, edible portions of fruit crops could be grown in vitro.

A steady and cheap supply of these edible plant parts will be required in this new agribusiness. It will take some time before such a process can be scaled up for commercial output. Nonetheless, a biotech company for producing a vanilla extract through cell culture got it patented (RAO and RAVISHANKAR 2000). Of course, this technique will not replace farming. This biotechnique, as well as other new farming methods, offers a means for new ways of producing food, feed or other products, such as medicines, solvents, dyes, and noncooking oils for many years. Hence, it would not be surprising to see, in few years from now, entire farms without food crops but growing transgenic plants to produce new products, for example, edible plastic from peas, maize, or potatoes and plant oils to manufacture hydraulic fluids or nylon (GRACE 1997). This new rural activity may result in important changes in the national economic sector. "Pharming" has been added to the dictionary of agricultural biotechnology, indicating a new kind of production system.

The private sector will play an increased role in the plant breeding. In 2017, major international breeders are Monsanto USA, DuPont USA, Syngenta Switzerland, Bayer Crop Science Germany, Sakata Japan, BASF Germany, Group Limagrain France to name a few.

The global players are focused on partnership and collaboration with other companies in order to increase its product portfolio, industry offering, and global presence. The global giants in genetically modified food[1] are actively involved in collaboration with the advanced research institutes in the industrial countries such as Brazil, Argentina, China, India, Malaysia, and the Philippines for significant R&D program in biotechnology and transgenic crops. Companies are also increasing their R&D in some of the African countries such as South Africa, Kenya, Zimbabwe, Mali, Nigeria, Egypt, and Uganda in order to cater the growing demand for genetically modified food in Africa.

Contrary to versatile criticism of the genetically modified (GM) plants, a recent study shows the positive effect over a longer period. The meta-analysis of the impacts of GM crops of KLÜMPER and QAIM (2014) revealed the following facts: on average, GM technology adoption has reduced chemical pesticide use by 37%, increased crop yields by 22%, and increased farmer profits by 68%. Yield gains and pesticide reductions are larger for insect-resistant crops than for herbicide-tolerant crops. Yield and profit gains are higher in developing countries than in developed countries. The analysis reveals robust evidence of GM crop benefits for farmers in developed and developing countries. Such evidence may help to gradually increase public trust in this technology.

In order to effect improvement in any plant's productivity or performance, the breeder must be fully aware of the attributes and breeding history of the plant in question. As well as more conventional concerns, such as disease resistance and day-length control of flowering, some destabilized environmental factors must now be given more attention. The predominant genotypes for each crop may in future need to be reassessed and modified in light of their responses to increasing mean global atmospheric CO_2 concentration, climatic changes, or increasing penetration of damaging UV light and deficient stratospheric ozone. Within the next 10 or 20 years, few research areas may become very important for crop improvement.

8.1 INCREASED YIELD AND INCREASED RELIABILITY OF PERFORMANCE, INCLUDING PHOTOSYNTHETIC EFFICIENCY

The increase of leaf photosynthetic rates seems to be a straightforward way. Conventional breeding already demonstrated genotypes with superior photosynthetic rates in maize, wheat, and soybean. One current target for molecular modification of photosynthesis is to introduce the precursor pathway for organic acid fixation of CO_2 (C4 pathway) into C3 species. Transgenic lines of rice have already been developed that have higher levels of phosphoenolpyruvate carboxylase activity, but their leaf photosynthetic rates were still comparably to untransformed lines. KROMDIJK et al. (2016) genetically engineered tobacco plants to react more rapidly to sudden switches between light and shade and report an approximately 15% improvement in the modified plants' productivity. It is a breakthrough and the first instance where it has been possible to demonstrate that, by improving the efficiency of photosynthesis, there is an increase in yield under field conditions.

It is to be expected that TILLING methodology identifies genes causal for quantitative trait loci (QTLs) of agronomic interest. The number of genes that need to be assessed in this way depends upon marker density and the decay of linkage disequilibrium and may be large.

---

[1] Genetically modified foods are foods that are derived from organisms whose genetic material has been modified in such a way that it does not occur naturally, for example, through the introduction of a gene from a different organism. Foods produced from by using GM organisms are often referred to as GM foods. The major advantages of genetically modified foods are better texture, flavor, and high nutritional value along with longer shelf life. However, unusual taste over non-GM food and safety concerns as it might interfere with the body normal functioning are some of the disadvantages of genetically modified food.

8.2 CHANGES IN PLANT ARCHITECTURE MODIFYING BALANCED PROPORTIONS OF TUBER, SEED, LEAVES, OR INTERNAL CHARACTERS

Novel gene variants affecting, for example, spikelet arrangement in barley, tendril development and inflorescence development in pea can be produced by TILLING approaches in accordance with the plant phenotype concept.

SOYK et al. (2016) identifies an array of new gene mutations that allow, for the first time, a way to fine-tune the balance of florigen to antiflorigen enzymes in tomato plants. This maximizes fruit production without compromising the energy from leaves needed to support those fruits. They mixed and matched all of the mutations to produce plants with a broad range of architectures. The collection of mutations forms a powerful toolkit for breeders to pinpoint a new optimum in flowering and architecture that can achieve previously unattainable yield gains. The breakthrough benefit of the toolkit is that it allows farmers to customize genetic variations for particular varieties and growing conditions (cf. Section 3.5.1.1).

8.3 IMPROVEMENT OF PEST AND DISEASE RESISTANCE

Although crop plants belong to certain plant families, most of their diseases are specific to a particular species. Therefore, resistance genes are often not usable across the genus barrier despite the cross-ability and incompatibility problems. However, recent transgenic methods demonstrate the feasibility of transferring nonhost resistance genes between distantly related grasses to control specific diseases. For example, a maize resistance gene, *Rxo1*, that recognizes a rice pathogen, *Xanthomonas oryzae* pv. *oryzicola*, which causes bacterial streak disease, conditioned a resistance reaction to a diverse collection of pathogen strains of rice (ZHAO et al. 2005). A brilliant approach to using CRISPR in plants is to edit the family of genes that confers susceptibility to bacterial blight in rice. Bacterial blight in rice, caused by *Xanthomonas oryzae* pv. *oryzae*, is a huge problem in Asia and Africa. Using a similar approach, disease-resistant citrus trees can also been developed. In Florida, the citrus industry faces disease challenges from citrus canker and citrus greening disease caused by two bacteria, *Xanthomonas citri* and *Candidatus Liberibacter asiaticus*, respectively. By application of TILLING methodology, variants alleles could be generated that have the potential to overcome dominant resistance because strong resistance genes generate a strong selective pressure for the evolution of resistance in the pathogen.

8.4 IMPROVED TOLERANCE TO ABIOTIC STRESS, INCLUDING WATER-USE EFFICIENCY

The increase of growth rate of individual seeds seems to be one approach. Transformation of ADP-glucose pyrophosphorylase in seed resulted in less sensitivity to phosphorus inhibition of wheat and rice.

Inadequate water availability is another crucial limitation to crop yield in most environments, and has been the focus for genetic improvement of crops for many years. The ratio between plant mass accumulation and transpiration, that is, water-use efficiency, has limited flexibility owing to the physics and physiology of leaf gas exchange. One molecular approach that has received considerable attention is the possibility that solute accumulation in plants, or osmoprotection, might confer drought tolerance (SINCLAIR et al. 2004). Transformation is being attempted for osmolyte accumulation and increased production of compatible solutes under water-deficit conditions.

Because many crop-breeding pools are limited by genetic bottlenecks during domestication and adaptation to intensive agriculture, which reduces the variation available in environmental response, TILLING by Sequencing could identify novel alleles to help maintain yield under abiotic stress.

8.5 APOMIXIS TO FIX HYBRID VIGOR

The introduction of apomixis-asexual reproduction through seeds-into crop plants is considered the holy grail of agriculture, as it would provide a mechanism to maintain agriculturally important phenotypes. Apomicts produce clonal offspring, such that apomixis could be used to transgenerationally fix any genotype, including that of F1 hybrids, which are used in agriculture due to their superior vigor and yield. At least in mouse-ear hawkweed (*Hieracium pilosella*) hybrids a step forward was made. SAILER et al. (2016) could fix F1 hybrids across generations by apomixis. Using a natural apomict, they created 11 hybrid lines. In these and a parental line, they analyzed 20 phenotypic traits that are related to plant growth and reproduction. Of the 20 traits, 18 (90%) were stably inherited over two apomictic generations, grown at the same time in a randomized design, in 11 of the 12 lines. Although one hybrid line showed phenotypic instability, the results provide a fundamental proof of principle, demonstrating that apomixis can indeed be used in plant breeding and seed production to fix complex, quantitative phenotypes across generations.

8.6 MALE STERILITY SYSTEMS WITH TRANSGENICS FOR HYBRID SEED IN SELF-POLLINATING CROPS

From all five male sterility systems—abolition restoration system, abolition reversible system, constitutive reversible system, complementary gene system, and gametocide-targeted system—the genic male sterility could be modified by transgenesis. It was done through transformation of male sterility causing gene constructs inside the nuclear genome. Such transgenic male sterility systems are currently being employed for variety development and seed production of rapeseed, chicory, and maize. Many others are under investigation.

8.7 PARTHENOCARPY FOR SEEDLESS VEGETABLES AND FRUIT TREES

In plants able to form fruits without fertilization can increase fruit acceptance by consumers. To achieve parthenocarpic development, it is common practice to treat flower buds with synthetic auxins. To mimic the hormonal effects by genetic engineering, the expression of a gene able to increase auxin content and activity should be induced in the ovule. In some experiments, the gene *iaaM* from *Pseudomonas syringae* under the control of an ovule-specific promoter from *Anthirrhinum majus* was utilized to induce parthenocarpic development in transgenic tobacco and eggplants. This approach could also be valuable in other horticultural crops, such as pepper, tomatoes, or melons.

8.8 SHORT-CYCLING FOR RAPID IMPROVEMENT OF FOREST AND FRUIT TREES AS WELL AS TUBER CROPS

Examples of single amino acid changes altering the time to vegetative–reproductive transition have been described, suggesting that screens for induced variation could identify novel means of controlling vernalization, flowering, anthesis, and grain-filling in many crops.

The ability to control sprouting time in potato tubers is of considerable economic importance. In order to prevent chemical treatment for stimulating tuber control, the transformation of the pyrophosphatase gene from *Escherichia coli* under the control of the tuber-specific patatin promoter was realized. Transformed potatoes displayed a significantly accelerated sprouting by 6–7 weeks.

In citrus trees, the generation time was reduced. Citrus trees have a long juvenile phase that delays their reproductive development by between 6 and 20 years. Therefore, juvenile citrus seedlings were transformed with *Arabidopsis thaliana Lfy* and *Ap1* genes, which promote flower initiation. Both types of transformed seedlings produced fertile flowers as early as the first year. These traits are submitted to the offspring as dominant alleles, generating trees with a generation time of 1 year from seed to seed. Constitutive expression of the *Lfy* gene also promoted flower initiation in transgenic rice.

8.9 NUTRITIONAL AND MICRONUTRITIONAL EFFICIENCY OF CEREAL AND TUBER CROPS

Increase nitrogen accumulation by crop has been a crucial feature of past yield increases. Increased nitrogen accumulation has usually resulted from applications of nitrogen to the soil and genetic improvement of plants to accumulate and store greater quantities. The enzyme pathways involved in uptake and/or biosynthesis of micronutrients lacking in many human diets are typically well known, but subtle changes induced by TILLING (TbyS) may alter their regulation or activity and so change flux through pathways.

The goal of molecular changes is to increase nitrogen-use efficiency. The activities of specific enzymes involved in nitrogen metabolism have been targeted for transformation, for example, the overexpression of a glutamine synthetase gene.

The detrimental effects of salt on plants are a consequence of both water deficit and the effects of excess sodium ions on key biochemical processes. To tolerate high levels of salts, plants should be able to use ions for osmotic adjustment and internally to distribute these ions to keep sodium away from the cytosol. The first transgenic approach introduced genes that modulate cation transport systems. Hence, transgenic tomato plants overexpressing a vacuolar Na^+/H^+ antiport were able to grow, flower, and produce fruits in the presence of 200 mM sodium chloride. Another overexpression of the *Hal1* gene from yeast in transgenic tomato plants had also a positive effect on salt tolerance by reducing K^+ loss and decreasing intracellular Na^+ from the cells under salt stress (HELLER 2003).

In arid and semi-arid regions of the world, the soils are alkaline in nature and therefore crop yields are limited by the lack of available iron (cf. Section 3.7.4). Under iron stress, some plants release specific Fe(III)-binding compounds, known as siderophores, which bind the otherwise insoluble Fe(III) and transport it to the root surface. To increase the quantity of siderophores released under conditions of low iron availability, two barley genes coding for the enzyme nicotianamine aminotransferase together with the endogenous promoters into rice plants. Transformed rice plants withstand iron deprivation remarkably better, resulting in a fourfold increase in grain yield as compared to the control plants. Alternately, increasing the rate-limiting step of Fe(III) chelate reduction, which reduces iron to the more soluble Fe(II) form, might enhance iron uptake in alkaline soils.

In 2004, "HarvestPlus" was implemented by CGIAR's Global Challenge Programs for breeding crops with better nutritional value. Initially, six crops and three nutrients are targeted (beans, cassava, maize, wheat, rice, sweet potato). Since this time, the so-called biofortification became a strategy that seeks to reduce human micronutrient deficiencies—vitamin A, zinc, or iron—by developing and disseminating food crops that contain high levels of micronutrients. As more evidence on the efficacy and cost-effectiveness of biofortification becomes available, stakeholders—including international NGOs and multilateral donor agencies—are increasingly interested in investing in biofortification. "HarvestPlus" has emerged as a global leader in developing biofortified crops and now works with more than 200 agricultural and nutrition scientists around the world. It is co-convened by the International Center for Tropical Agriculture (CIAT) and the International Food Policy Research Institute (IFPRI).

8.10 CONVERTING ANNUAL INTO PERENNIAL CROPS FOR SUSTAINABLE AGRICULTURAL SYSTEMS

The development of perennial crops will be especially important to protect soils from erosion, that is, sustainable agriculture. Perennial plants conserve freshwater better than annuals plants. Annual crops lose up to five times more water than perennials. Most farmland is devoted to annual agriculture. Cereals, oilseeds, and legumes—all annuals—occupy about 70% of global croplands. Many of these staple crops can be replaced by perennials by hybridization and other genetic engineering techniques. According to GLOVER and REGANOLD (2010), 10 of the 13 most grown cereals and oilseeds can be hybridized with perennial plants. However, perennial crops are not ready to replace annuals. Genetic development is needed for perennials to compete with the output of annual crops. Breeders need to modify perennial strains to express certain characteristics such as large seed size, palatability, strong stems, and high seed yield.

Initial cycles of hybridization, propagation, and selection in wheat, wheatgrasses, sorghum, sunflower, and Illinois bundleflower have produced perennial progenies with phenotypes intermediate between wild and cultivated species, along with improved grain production.

8.11 DNA REPAIR AND GENE EDITING IN PLANTS

The expected increase in UV-B radiation at the earth's surface by the early twenty-first century is of considerable concern because UV causes damage to genetic material (DNA). In the case of crop plants, unrepaired DNA damage may have penalty in yield losses. From *Arabidopsis*, a gene coding for nucleotide repair endonuclease, ERCC1, was already identified and isolated; its protein plays a pivotal role in the DNA nucleotide excision repair process. It is the first plant gene involved in DNA excision repair so far reported (XU et al. 1998, 1999).

Plant biotechnology will play, of course, an important role in achieving research and development success in all these areas. Banning transgenic crops in the farming system will be ineffective because the potential benefits are so great. Whatever scientists do to develop crops that eliminate or reduce the utilization of polluting agrochemicals in the farming systems must be welcome by farmers and consumers. For example, one interesting approach for developing resistant transgenic crops may be through the improvement of the plant's own defense system. Inducible and tissue-specific promoters could assist in this endeavor.

Collective approval may lead to new partnerships, cooperation, or joint ventures in research and development between scientists in the public and private sectors that will benefit farmers and consumers with profits and high-quality products, respectively. Any potential risk in human development associated with biotechnology applications in agriculture will be easily resolved in a democratic society. The public needs to choose between being safely self-regulated or to follow safety regulations as agreed by lawmakers after listening to the views of scientists, producers, and consumers.

In 2005, a study shows no maize, soybeans, or rapeseed crops in the USA that do not have traces of genetically engineered counterparts. An environmental advocacy group, tested nontransgenic seed from major seed suppliers and found low but detectable levels of DNA from two transgenic varieties in 50%–83% of maize and soybean and between 83% and 100% of rapeseed. In 2010, more than 2 million ha of GM rapeseed was commercially grown in the USA and 86% of test plants along a 5400 km distance of North Dakota highway ditches were polluted.

Eight percent of the US soybeans and 38% of the maize crop are genetically engineered to be herbicide-resistant, to manufacture their own pesticides, or both. But such mixing, that is, seed production and production of pesticides, is an economic threat to organic growers, who risk losing customers if their products are not 100% nontransgenic. Europe and Japan have long refused to purchase transgenic products. Damage to beneficial insects and soil fertility coupled with genetic pollution, absence of any centralized government regulation, and lack of awareness about the benefits of GM plants are factors that are restraining the market potential growth. Nevertheless, rising demand for healthy food products and year-round and easy availability of genetically modified food are main trends in the genetically modified food market globally.

The risk is manifold: extensive use of the same herbicide for long time in many different crops may hasten the appearance of tolerant weeds, and the herbicide-tolerant gene may introgress into a wild or weed wild relative of the transgenic crop. MIKKELSON et al. (1996) have demonstrated the introgression of the BASTA-tolerant gene from transgenic rapeseed (*Brassica napus*) to a weed (*B. campestris*). Transgenic crops could decrease the biodiversity. Concerning the latter, some contrary examples are already in the fields:

1. Herbicide-resistant transgenic crops have made new crop rotations possible due to more flexible weed control and less persistent herbicides (e.g., grain legumes in Canadian rapeseed rotations).
2. More birds and beneficial insects are seen in GM cotton fields.
3. Still in the research phase is use of transgenic fungi to slow down a devastating disease that has almost eliminated the native chestnut tree from North American forests.
4. A long-run global impact of biotechnology may be increased space for wilderness because the greatest threat to the earth's biodiversity is habitat loss through conversion of natural ecosystems to agriculture. Conservation of biodiversity in situ will continue to be possible if high-yield practices allow the same amount of land to support a higher standard of living, more people, or both. The deployment of transgenic crops could dramatically reduce use of fungicides and pesticides that are still more heavily used in Europe than in North America.

The general public should see biotechnology as a safe tool for scientific crop improvement because it helps in the fight against hunger and poverty. Therefore, research funding should be allocated accordingly to long-term plant-breeding programs, which include biotechnology as one of its tools. In this way, we may effectively face the serious challenge of feeding the rapidly growing world population in the next millennium. However, alternative interests should not be ignored. The European Consortium for Organic Plant Breeding (ECO-PB), founded in 2001 at Driebergen (the Netherlands) is one of them. It provides a platform for discussion and exchange of knowledge and experiences, international support of organic plant breeding programs, and the development of scientific concepts of an organic plant breeding.

In 2016, in North America the Organic Seed Growers and Trade Association (OSGATA) approved its policy on organic plant breeding as follows (ANONYMOUS 2016b):

- Seed is the foundation of crop agriculture. Therefore, establishing principles of organic plant breeding is essential to the integrity of organic systems.
- Organic plant breeding is a process that occurs at the plant level and respects the integrity of plants. It is free of genetic manipulation or other techniques at the subcellular level.
- Organic plant breeding maintains or enhances on-farm genetic diversity as opposed to fostering crop monoculture.
- Organic plant breeding must be farm-centric in that a given genetic technique must be of such a nature that it could be performed in a farm setting as opposed to a laboratory.
- The ownership of seed resulting from organic plant breeding resides within the public domain.

9 In the Service of CERES—A Gallery of Breeders, Geneticists, and Persons Associated with Crop Improvement and Plant Breeding

AARONSOHN, A. (1876–1919): He was an agronomist born in Bacau, Romania, and was very well known in the Jewish community in Ottoman Palestine, both as botanical explorer and agricultural expert on semiarid regions of the Mediterranean basin. At the age of 6 years, he was brought to Haifa (Palestine) in 1882 by his parents. He studied gardening and agriculture in France, and on his return, he was employed as an agronomist by Baron Edmond de Rothschild at Metullah in 1895. He made extensive explorations in Eretz (Israel) and neighboring countries in 1906 and discovered specimens of wild emmer wheat (*Triticum dicoccoides*) at Rosh Pinah, a discovery that made him famous among botanists throughout the world. He recognized wild emmer as the tetraploid ancestor of durum and bread wheats. Additionally, he claimed that wild emmer, as well as other wild forms of domesticated plants, may serve as "gene pool" for crop improvement. With the support of American Jews, Aaronsohn founded an agricultural research station in Atlit (Israel), where he built a rich library, collected geological and botanical samples, and inspected crops.

ACHARD, F. C. (1753–1821): A German scientist and chemist, he was the first person who selected beet plants successfully for sugar production. He is considered to be the founder of the beet sugar industry. By crossing of the "Weisse Mangoldrübe" (Zuckerwurzel = sugar root) and the "Roter Mangold" (Rübenmangold or Futterrübe = fodder beet), he was able to select genotypes showing high sugar content; it was the birth of a new crop plant. The first name of sugar beet was "Schlesische Rübe." First cropping of sugar beet started in Central Germany around Magdeburg—Halberstadt—Klein Wanzleben, where heavy, very fertile loess soil was available (it is still the center of sugar beet production of Germany). Later, Achard founded the first sugar beet factory of the world at Cunnern (Silesian, Germany/Poland).

ÅKERBERG, E. (1906–1991): Prof. ÅKERBERG, honorary member of the Swedish Seed Association and former director of the association during 1956–1971, was known to many plant breeders all over the world. His name was synonymous with Svalöf and its plant breeding and research during this 15-year period. He was born in 1906 and grew up on an agricultural experimental station, Flahult, Småland (Sweden), where his father was the farm manager. He was trained as an agronomist at Alnarp and qualified in 1931. Parallel to his studies at Alnarp, he also studied science at Lund University. He completed his BSc degree in botany and genetics in 1928. While continuing his postgraduate studies at Lund, he took up a position as plant breeder at the Weibullsholm plant breeding station, Landskrona (Sweden) during 1932–1938. He bred many crops, such as oats, peas, and beans; clover; grasses; and potato. His PhD thesis work was devoted to the analysis of the grass species *Poa pratensis* and *P. alpina* and crosses between them. In 1938, when he left Weibullsholm to be head at one of the Svalöf branch stations in Norrland (Lännäs), he was able to present a new red clover variety, "Resistenta." After 6 years of breeding work at Lännäs, he moved to Uppsala as head of the Swedish Seed Association's Ultuna branch station. In early 1956, ÅKERBERG again moved to Svalöf as the new director of the Swedish Seed Association. He succeeded ▶ Å. ÅKERMAN. In the same year, that is, 1956, he joined a group of leading European plant breeding professors who founded the European Plant Breeders Union (EUCARPIA), still existing in present days. He became president of EUCARPIA during 1965–1968, and later, he became vice-chairman of the Swedish State Agricultural Research Council. In finding new alternatives in crop production, he introduced rape as a new silage crop in northern Sweden and broad bean (*Vicia faba*) as another seed crop for the southern part of the country.

ÅKERMAN, Å. (1887–1955): He was a Swedish botanist and breeder working at Svalöf. In 1904, under the guidance of ▶ Prof. H. NILSSON-EHLE, he started genetic studies on quantitative traits of wheat and oats. During the 1920s, he carried out experiments on cold tolerance of wheats and other cereals; since the 1930s, he began studies on baking quality of wheat. In 1931, he established the first Swedish laboratory for baking and oil quality of wheat and/or rapeseed. His wheat varieties, derived from Swedish landraces and English square head strains, contributed to an increase in yield in Sweden by 50% during the first half of the twentieth century.

ALLARD, R. W. (1919–2003): The American breeder Robert W. ALLARD was born in the San Fernando Valley. When he was 10 years old, he already decided that he wanted to be a plant breeder. He graduated from the College of Agriculture at the University of California, Davis (United States). He earned his doctorate (PhD) from the University of Wisconsin in Madison. After 1945, he started his academic career at the University of California, where he served until his retirement (1946–1986). From 1952 to 1957, he worked as an associate professor and an associate agronomist in Experiment Station, and from 1957 to 1971, he worked as a professor. While in his early research ALLARD focused primarily on the genetic processes basic to plant breeding practices, his later research objectives were threefold: to derive theoretical models of population dynamics, to obtain experimental estimates of genetic and ecological parameters to enter into these models, and to utilize the information so gained to achieve an improved understanding of the biology of populations. He strongly contributed to the plant population genetics, predominantly self-fertilizing plants. He was also keenly aware of ecological settings and integrated genetics with ecology in the 1960s and 1970s. He wrote the major textbook on plant breeding (*Principles of Plant Breeding*, originally published in 1960, with three editions) and served as president of several scientific societies, including the Genetics Society of America, the American Genetic Association, and the American Society of Naturalists.

AMOS, H. R. ▶ BACKHOUSE, W. O.

ANDERSON, R. G. (1924–1981): Born in Ontario (Canada), he was working on genetics of rust resistance in bread wheat at the Canada Department of Agriculture. In 1964, he was appointed by the Rockefeller Foundation to work in India with the skilled assistance of Dr. M. S. SWAMINATHAN and Dr. S. P. KOHLI. He contributed to world leaders, wheat scientists, and farmers concerns in feeding a hungry world.

BACKHOUSE, W. O. (1779–1844): He was an English geneticist and breeder and graduated from Cambridge, United Kingdom, under ▶ J. BIFFEN. He came to Argentina along with two assistants, J. WILLIAMSON and H. R. AMOS. Their work was extremely valuable for Argentina, so genetic research was incorporated into breeding and new direction was given to agricultural experimentation. BACKHOUSE made extensive surveys throughout the wheat area, studying the characteristics of different populations and making single selection, mainly on "Barleta" and "Ruso" populations. He established many comparative trials in several locations, incorporating foreign varieties to determine their degree of adaptability and usefulness in Argentina. In 1925, he released a bread wheat cultivar "38 M.A." it was the predominant cultivar in the wheat production until 1944 and also a remarkable progenitor for a great number of new cultivars. The cultivar "Lin Calel M.A.," suitable for early sowing and grazing, was released in 1927. This cultivar was extensively grown in the center, south, and west of the Province of Buenos Aires, as well as in the Province of La Pampa. Its peak of popularity was in 1935 (21% of the total production).

BAILEY, L. H. (1858–1954): He was born in South Haven, Michigan (United States) and was the youngest child of a hardworking family that worked on a fruit farm. In 1877, BAILEY left South Haven and began his secondary education at the Michigan Agricultural College (now Michigan State University). While in Michigan, he worked with W. BEAL, a botanist with whom he became acquainted through participation in the Michigan State Pomological Society, and developed interest in plant breeding. He got involved in the Natural History Society, and to publish his first articles on the identification of local flora in *Botanical Gazette*; he began his long-term involvement with the classification of the genus *Rubus*. After leaving Michigan Agricultural College, BAILEY went to Harvard University to work for the botanist A. GRAY, where he was responsible for sorting and classifying plant specimens received from Kew Gardens (United Kingdom). By 1888, dynamic and prolific BAILEY had a well-established reputation as a scientist and an innovator in the area of horticultural research and education. Consequently, he was recruited to fill a new position as professor of horticulture and botany at Cornell University (United States) and dean of the Agriculture College. There, for many years, he was one of the most influential botanists and horticulturists of the twentieth century, having written countless authoritative articles and books. His *Manual of Cultivated Plants*, *Cyclopedia of Horticulture*, and *Hortus* I–III series are widely known. The term "cultivar" was coined by him. Among his horticultural monographs are works on *Dianthus*, *Campanula*, garden palms, and conifers.

BAKER, R. F. (1906–1999): He was the first maize breeder with PIONEER HI-BRED, one of the preeminent plant breeders of the twentieth century, and a graduate of the Iowa State University, Agronomy Department (United States).

BARABAS, Z. (1926–1993): He was a Hungarian agronomist and plant breeder born in Budapest (Hungary). After receiving his PhD degree at the University of Agricultural Sciences, he started his career at the place as teaching assistant. From 1951, he started flax and sorghum breeding. In 1960, he became the head of the Hungarian sorghum breeding program for 9 years in Martonvasar. He released four hybrid varieties during this time. As the first breeder, Barabas introduced genetically male sterility in a series of sorghum strains. In 1969, he moved to Szeged and became the leader of the wheat breeding program at the Cereal Research Institute of the Hungarian Ministry of Agriculture. He essentially contributed to the development of 18 wheat varieties. During the same time, he began a breeding program for durum wheat in Hungary. By use of cytokinins, he was able to accelerate the vernalization process. As a breeder and director, he successfully established the periodical *Cereal Research Communications* (still ongoing) and edited it for more than 20 years.

BATESON, W. (1861–1926): Born in Whitby (United Kingdom) and educated at St. John's College Cambridge (England), he was an English geneticist who recognized the importance of MENDEL's pioneering work. He translated and introduced MENDEL's papers to Britain and coined the word "genetics" in a paper of 1907. he was the founder of the "Cambridge School" of genetics. In 1899, the Royal Horticultural Society organized an international conference on hybridization. Corresponding to MENDEL (30 years before), ▶ VRIES, ▶ CORRENS, and ▶ TSCHERMAK, BATESON stated in his paper: "What we first require is to know what happens when a variety is crossed with its nearest allies. If the result is to have a scientific value, it is almost absolutely necessary that the offspring of such crossing should then be examined statistically. It must be recorded how many of the offspring resembled each parent, and how many showed characters intermediate between those of the parents. If the parents differ in several characters, the offspring must be examined statistically, and marshaled, as it is called, in respect to each of those characters separately... ." One year later, he introduced the term "allelomorph" for the opposite characters of a pair of characters, the term "heterozygote" for nuclei that resemble the opposite "allelomorphs" (later abbreviated to "allele"), and the term "homozygote" for identical "allelomorphs." He became the first director of the John Innes Horticultural Institute in Merton (United Kingdom).

BAUR, E. (1875–1933): A German botanist who contributed very much to the utilization of genetic knowledge in breeding. In 1894, he began to study medicine at the universities of Heidelberg, Freiburg, and Strasbourg. In 1897, he continued medical studies at the University of Kiel. At the same time, he attended lectures in botany and biology. In 1900, he received M.D. from the University of Kiel. In 1903, he accepted a doctorate with a study of the developmental aspects of fructification in lichens. In 1911, he received a professorship in botany at the Landwirtschaftliche Hochschule in Berlin, and in 1928, he became the director of the new "Kaiser Wilhelm Institut für Züchtungsforschung" in Müncheberg. Since 1931, he held lectures on evolution, applied genetics, and eugenics in London, Sweden, and South America. He was the editor of *Zeitschrift für induktive Abstammungs- und Vererbungslehre* (from 1908 onward; now: *Molecular and General Genetics*), founder of *Der Züchter* (1929; now: *Theoretical and Applied Genetics*), and cofounder of the "Deutsche Gesellschaft für Vererbungswissenschaft" with ▶ C. CORRENS and R. GOLDSCHMIDT. He was elected as president of the 5th International Congress of Genetics in 1927.

BEADLE, G. W. (1903–1989): He is credited as "the man who laid the foundation for the field of biotechnology." He studied agronomy under F. KEIM at the University of Nebraska Lincoln (United States) and earned a BSc degree in 1926 and an MSc degree in 1927. His interest in genetics began at Cornell, where he began his graduate work and earned his PhD. In 1931, BEADLE was a distinguished National Research Council post-doctoral fellow and instructed at the California Institute of Technology through 1935. He served on the faculties of Harvard University, Stanford University, and the Institut de Biologie in Paris (France). During these years, he conducted his prize-winning research, that is, subjecting the *Neurospora* (red bread mold) to X-rays in order to induce mutations. In 1941, from studies of mold, he recognized the genetic control of biochemical reactions in *Neurospora*. In 1946, he returned to Caltech as professor and chair of the Division of Biology. While serving as the George Eastman professor at Oxford University, Beadle received the 1958 Nobel Prize in Physiology or Medicine along with J. LEDERBERG and E. TATUM for their discovery that genes control chemical reactions by their formation of specific enzymes. Thirty years before he was awarded the Nobel Prize, BEADLE was working on teosinte, a wild grass that is closely related to maize. This early work on teosinte began a lifelong fascination with the origin of maize and set him on a mission to confirm a hypothesis that he had settled in his own mind as a graduate student—teosinte is the progenitor of cultivated maize. Although the path of events led him away from this mission during the most productive years of his career, upon retirement from the presidency of the University of Chicago in 1968, he again took up research on teosinte, employing experimental genetics and organizing an expedition to Mexico in search of naturally occurring mutants in teosinte populations that might shed some light on the steps that transformed teosinte into maize. From the time of its discovery until 1896, teosinte was known principally to a few botanists who had preserved a few dried specimens in European herbaria and bestowed upon it the Latin name, *Euchlaena mexicana*. Teosinte was placed in the genus *Euchlaena* rather than in *Zea* with maize (*Z. mays*), because the structure of its ear is so profoundly different from that of maize that nineteenth-century botanists did not appreciate the close relationship between these plants. When the first maize–teosinte hybrids were discovered in the late 1800s, they were not recognized as hybrids but considered a new and distinct species, that is, *Zea canina*. A Mexican agronomist, J. SEGURA, who made the first experimental maize–teosinte crosses, demonstrated that *Zea canina* was a maize–teosinte hybrid, thereby implying that maize and teosinte were much more closely related than previously thought.

BEAL, W. ▶ BAILEY, L. H.

BEKE, F. (1914–1988): He was born in Bana (Hungary), pursued his studies at Gyoer and later on at the Agricultural High School in Mosonmagyarovar. From 1949, he continued his research work at the Plant Breeding Institute, Fertoed (Hungary) on clover, rape, and wheat. One of the best results achieved by him was the improvement of the winter wheat variety "F293." This variety, showing excellent leaf rust resistance, has been a leading variety in Hungary over almost 29 years. The investigation of the relationships between plant development, growth, and productivity was his main concern.

BENARY, E. (1819–1893): After his apprenticeship with Friedrich A. HAAGE in Erfurt (Germany) and his traveling years, which imparted him further knowledge and connections with seed companies in France, Belgium, and the United Kingdom, in 1843, he founded an art and trade nursery. Its aim was to build up a garden seed company. Benefited by the expedient climate of Erfurt for maturity of seeds, he succeeded with diligence, know-how, and reliability to gain ground on German and foreign markets. His first flower breeding of campion (*Lychnis haageana*) started in 1859 by crossing *L. sieboldii* × *L. fulgens*. A variety of *Lobelia erinus* "Crystal Palace Compacta" is still grown in the present day. The F1 hybrid of *Begonia gracilis* "Primadona" was created in 1909. His dwarf petunia "Erfurter Zwerg" and "Weisse Flocke" became famous. Since 1843 until 1951, more than 500 varieties of ornamental plants were bred from species such as *Abutilon, Ageratum, Alonsoa, Angelonia, Anthirrhinum, Aquilegia, Arabis, Armeria, Calistephus, Begonia, Bellis, Browallia, Calceolalria, Calendula, Campanula, Celosia, Centaurea, Cheiranthus, Chrysanthemum, Cineraria, Clintonia, Coleus, Collinsia, Coreopsis, Cuphea, Cyclamus, Cynoglossum, Delphinium, Dianthus, Echium, Erigeron, Exacum, Gaillardia, Gloxinia, Sinningia, Godetia, Heuchera, Iberis, Impatiens, Lathyrus, Leucanthemum, Linaria, Lobelia, Lupinus, Lynchis, Mimulus, Myosotis, Papaver, Penstemon, Petunia, Phacelia, Phlox, Primula, Pyrethrum, Rudbeckia, Saintpaulia, Salvia, Saponaria, Scabiosa, Schizanthus, Sedum, Silena, Statice, Tagetes, Tropaeolum, Verbena, Viola,* and *Zinnia.* In about 50 years, he established a company that was one of the most important global suppliers of flower and vegetable seeds. For his employees, he established a health fund and a company pension scheme, and for the citizens of his Erfurt hometown, he donated a public park for recreation.

BERZSENYI-JANOSITS, L. (1903–1982): He was an outstanding personality and father figure of Hungarian plant breeding. He was mainly concerned with Hungarian maize breeding and organization of the Research Institute at Marosvasarhely (today Romania) over more than 20 years. He re-organized Hungarian plant breeding after 1945, and with the help of the Food and Agriculture Organization of the United Nations (FAO), he provided the first foreign (American) inbred lines to Hungary.

BESSEY, C. A. ▶ EMERSON, R. A.

BIFFEN, R. H. (1874–1949): He initiated the wheat breeding at the Plant Breeding Institute, Cambridge, United Kingdom, and was its first director in 1896. He introduced Mendelian genetics into wheat breeding. In 1905, he was the first who bred crops for disease (rust) resistance (vertical resistance) and recognized the single-gene resistance. The variability in the pathogen was not fully appreciated until the work of STAKMAN and PIEMEISAL in 1917, who reported that stem rust in cereals and grasses comprised six biological forms; these forms were distinguished from each other morphologically and parasitically and were differentiated on selected cereal and grass hosts. This led to research on variation and variability of plant pathogens and the breeding of plants for resistance to specific races. Moreover, BIFFEN bred the famous varieties, including "Little Joss," released in 1908, from a cross of the old English Wheat, "Squarehead's Master," and a Russian variety "Ghirka," with good resistance to yellow rust.

BLAKESLEE, A. F. ▶ MÜNTZING, A.

BOHR, N. ▶ DELBRÜCK, M.

BOHUTINSKY, G. (1877–1914): He was head of the Plant Breeding Station and professor at the Royal Farming School in Krizevci (Croatia) during the period 1903–1912. During that time, he conducted an extensive research work, mainly on wheat and maize. His paper "The Crossing of Squarehead × Banatska Brkilja" (1911), published in *Gospodarska Smorta*, represents one of the earliest significant works on wheat genetics in the territory of former Yugoslavia. By method of individual selection from autochthonous population such as "White Wheat from Srijem," "Somogy Wheat," and "Sirban Prolific," he created outstanding new varieties for Croatia, Istria, Dalmatia, Slovenia, and Bosnia. They were grown under the name of "Bohutinsky's wheats" until 1925.

BOJANOWSKI, J. S. (1921–2004): He was born in Warsaw (Poland). He graduated from the Agricultural University in Warsaw in 1949, getting his MSc degree in agronomy, with specialization in genetics and plant breeding, and received his PhD from Agricultural Academy, Poznan, in 1963. He started the work in plant breeding in a private company in 1943 and continued it till 1953 as its chief breeder. During this period, he was working with several crops, such as cereals, fodder beets, fodder carrots, forage grasses, and alfalfa. In 1953, he joined the Department of Plant Genetics of Polish Academy of Sciences, led by Prof. E. MALINOWSKI. From this time on, he concentrated his research on heterosis and hybrid maize breeding. In 1959, he moved to the Department of Genetics of the Agricultural University in Warsaw, where he remained until 1969. The years 1953–1969 were very important for hybrid maize breeding in Poland. By the end of the decade, first Polish hybrids were registered. During this period, BOJANOWSKI passed a 1-year training at the Iowa State College in Ames, Iowa, under the leadership of Prof. G. F. SPRAGUE and W. A. RUSSELL. In 1969, he was appointed as deputy director of the Plant Breeding and Acclimatization Institute (IHAR) in Radzików, which played an important role in the maize breeding program of Poland. In 1976, BOJANOWSKI, as the head of the Maize Department of IHAR, initiated work on breeding of high-lysine maize, which is continued up to the present. In 1960, BOJANOWSKI became a member of the Maize and Sorghum section of EUCARPIA. He was a member of the EUCARPIA Board in the years 1975–1981 and the president of EUCARPIA between 1983 and 1986. He published 20 scientific papers and developed 7 varieties of agricultural crops: 1 of spring barley, 1 of oats, and 5 of maize. He also translated two important textbooks on plant breeding to Polish: F. C. ELLIOTT's *Plant Breeding and Cytogenetics* and ▶ R. W. ALLARD's *Principles of Plant Breeding*.

BOLIN, P. ▶ NILSSON-EHLE, H.

BOOM, B. K. (1903–1980): He was a horticultural taxonomist in Wageningen (the Netherlands), a conifer expert, and co-author of the landmark *Manual of Cultivated Conifers*. He did much to stabilize garden conifer nomenclature.

BORLAUG, N. E. (1914–2009): Norman E. BORLAUG was born in Cresco, United States, and was an American agricultural scientist, plant pathologist, and winner of the Nobel Prize for Peace in 1970 (Figure 3.9). He studied plant biology and forestry at the University of Minnesota (United States), and in 1941, he earned PhD in plant pathology. In 1943, BORLAUG joined the staff of the Cooperative Mexican Agricultural Program as an employee of the Rockefeller Foundation. This program resulted from a trip to Mexico by Vice President H. A. WALLACE and a request from Mexico for technical assistance to improve its agricultural research. In 1944, he arrived to Mexico. BORLAUG labored for 13 years before he and his team of agricultural scientists developed a disease-resistant wheat. Native wheats produced low yields and toppled in wind and rain. In an attempt to solve this problem, BORLAUG turned to several dwarf strains, which he crossed with varieties raised in the hot, dry fields of Northern Mexico as well as the cool highlands near Mexico City. He began hybridizing with short-statured wheats in the mid-1950s, based on the productive germplasm provided to him by O. VOGEL (United States), the father of the wheat revolution in the United States. BORLAUG's first varieties were named in 1962 and were displayed at the 1st Wheat Genetics Symposium in Winnipeg (Canada) in 1968. First named varieties were "Pitic 62" and "Penjamo 62," followed by numerous other varieties. As a result, he produced hard spring wheat that resisted rust, tolerated the climatic and soil variations across Mexico, and reduced toppling. It produced large yields with the use of nitrogen fertilizer and irrigation. As a result, in 1956, Mexico achieved self-sufficiency in wheat production. He also created new wheat–rye hybrids for cereal production known as triticale. He served as director of the Inter-American Food Crop Program (1960–1963) and as director of the International Maize and Wheat Improvement Center in Mexico City, Mexico (1964–1979). He was one of those who laid the ground work of the so-called Green Revolution, the agricultural technological advance that promised to alleviate the world hunger.

BORODIN, I. P. ▶ NAVASHIN, S. G.

BOTHA, L. ▶ NEETHLING, J. H.

BOVERI, T. ▶ SUTTON, W.

BREMER, G. (1891–1969): He was born in Rotterdam (the Netherlands) and attended the primary and secondary school. In 1910, he registered as a student at the College of Agriculture, which became later the Agricultural University, Wageningen (the Netherlands). After graduating in 1914, majoring in tropical agriculture, BREMER left for the former Dutch East Indies in 1915 and joined the research staff of the Experiment Station for the "Java Sugar Cane Industry" in Pasuruan as a cytologist. At that time, sugarcane breeding had already made good progress and the new developments of hybrid varieties brought up problems of cytology and reproduction biology. His pioneering research brought order in the controversial statements on chromosome numbers of *Saccharum* species. During his home leave in 1921, BREMER was among the first Wageningen graduates to gain a doctorate in agricultural science.

BRENNAN, P. (1944–2013): Dr. BRENNAN was a graduate from the University of Queensland (Australia, BAgrSc, 1966; MAgrSc, 1972; PhD, 1976) and the University of Saskatchewan (Canada) in genetics and plant breeding (Figure 9.1). He spent 33 years

FIGURE 9.1 Paul BRENNAN, Australian wheat breeder, 2010. (Courtesy of Meg KUMMEROW, Grains Research Foundation Limited, Australia. With Permission.)

as a wheat breeder and held appointment as the director of the Queensland Wheat Research Institute (1994–1999), Toowoomba. While he was working at the Wagga Wagga Agricultural Institute of New South Wales Department of Primary Industries, 23 wheat varieties had been released as a result of his activities. These varieties, over an extended period, occupied between 20% and 25% of Australian wheat plantings. He is also the author of a number of scientific papers and presentations at both national and international conferences and held appointments as director of the Sugar Research and Development Corporation and the northern panel for the Grains Research and Development Corporation.

BREZHNEV, D. D. (1905–1982): He was a Russian plant scientist and geneticist and an expert in tomato. From 1937–1941 he was in charge of the Department of Vegetable Crops at N. I. Vavilov Institute of Plant Genetic Resources (VIR), and from 1965–1978 its director (Saint Petersburg, Russia).

BRIDGES, C. B. ▶ RHOADES, M. M.

BRIGGS, F. N. (1896–1965): Fred N. BRIGGS was born in Center, Missouri (United States). While his family moved to Hope, Arkansas, he returned to get admission in the University of Missouri, from which he graduated in 1918. His choice of college and career was influenced by his uncle, Henry Jackson WATERS, dean of Agriculture at the University of Missouri, the then president of Kansas State University. His first professional interest was plant pathology, and he worked in this field for 11 years, from 1919 to 1930, with the Division of Cereal Crops and Diseases, Bureau of Plant Industry, the United States Department of Agriculture (USDA), with headquarters at the University of California in Berkeley. During these years, he obtained his MSc degree in 1922 and his PhD in 1925 from the University of California. In 1930, he joined the faculty at Davis as assistant professor of agronomy. Assigned to the problems of smut and bunt in cereals, he concerned himself first with chemical control. Later, his interest turned to the genetics of resistance to these and other diseases. He successfully transferred resistance to the major diseases into several varieties of wheat and barley. In 1936, BRIGGS was promoted to associate professor in the College of Agriculture and Associate Agronomist in the Agricultural Experiment Station, and in 1942, he became professor and agronomist. He became acting chairman of the Department of Agronomy in 1947 and chairman in 1948. In 1952, he became dean of the College of Agriculture at Davis, assistant director of the Agricultural Experiment Station, and assistant dean of the Graduate Division. He was recalled to active service after his retirement (1963) to serve as the University's representative and the co-director of a cotton breeding program in the San Joaquin Valley to meet a potential disaster of a serious plant disease. His broad experience as a plant breeder was utilized to strengthen the program for improvement of disease tolerance.

BRINK, R. A. (1897–1984): The Canadian geneticist Royal Alexander BRINK had an important role in showing how plant genetics could affect agriculture for the better. He is credited with creating a hybrid maize variety that profited Wisconsin farmers. As a native of Woodstock, Ontario, he grew up on a dairy farm and was educated first at a local school and then at the Collegiate Institute in Woodstock. He enrolled at the Ontario Agricultural College in 1914 and graduated in 1919 in chemistry and physics. However, this degree failed to qualify him, as he wished for postgraduate admission in soil sciences at Cornell University, United States. So, after doing a milling and baking course at the Ontario Agricultural College, he worked for some time at the Western Canada Flour Mills in Winnipeg and then returned to academia, enrolled at the University of Illinois to study agronomy. His genetics studies at Cornell University under J. A. DETLEFESEN led him to consider a career in plant genetics. On DETLEFESEN's recommendation, in 1921, he received an Emerson Fellowship to work on a DSc degree under E. M. East at the Bussey Institute, Harvard. Here, he carried out research on pollen cultivation and on the maize waxy gene.

After receiving his DSc degree in 1923, he joined the University of Wisconsin–Madison as an assistant professor of genetics. He was to remain at Wisconsin University until his retirement in 1968. His first step after arriving at Wisconsin University was to initiate a hybrid maize breeding program, which was followed by an alfalfa breeding program in 1926. By applying genetic principles, BRINK and his research team created a cultivar called "Vernal" that could withstand winter colds and was resistant to diseases such as bacterial wilt. Around the 1930s, BRINK were also engaged in creating a non-bitter strain of sweet clover for use as forage, by crossing a forage quality bitter variety with a non-bitter strain from China, crossing the progeny with a hardier strain and then re-crossing with the forage quality bitter variety. During this experiment, he also noted that eating spoiled bitter sweet clover produced a blood-clotting deficiency in animals. In 1938, he found that the chemical toxin coumarin caused the sweet clover's bitterness and spoiling. This research later helped K.P. LINK to derive the anticoagulant dicumarol from sweet clover.

BRUUN von NEERGAARD, T. ▶ NILSSON-EHLE, H.

BUKASOV, S. M. (1891–1983): He was a Russian botanist and plant breeder and one of most prominent experts of potato breeding. From 1918, he worked at the Department of Applied Botany, and later, he became head of Potato Department of VIR (Saint Petersburg, Russia).

BURBANK, L. (1849–1926): Born in Lancaster, Massachusetts (United States), this famed horticulturist made his home in Santa Rosa for more than 50 years, and it was here that he conducted plant breeding experiments. BURBANK started to improve a spineless cactus, which should provide forage for livestock in desert regions. Later, he developed over 800 new strains of plants, including many popular varieties of potato, plums, prunes, berries, trees, and flowers. One of his greatest inventions was the "Russet Burbank" potato (also called the "Idaho potato"), which he developed in 1871. This blight-resistant potato helped Ireland recover from its devastating potato famine of 1840–1860. He also developed the "Flaming Gold" nectarine, the "Santa Rosa" plum, and the "Shasta" daisy. He was raised on a farm and only went to elementary school. He was self-educated. He applied Mendelian genetics to his breeding approaches. According to DARWIN's *The Variation of Animals and Plants under Domestication*, BURBANK said: "It opened up a new world to me."

BUSTARRET, J. (1904–1988): Born in Bordeaux (France), he enjoyed a brilliant school career at the Lycée Montaigne. Poor health led him to take the direction of the Institut National Agronomique. He was admitted in 1924 and graduated in 1926. In 1930, he joined the plant breeding station that had been set up at Epoisses/Dijon (France). In 1938, BUSTARRET joined the Agricultural Research Center at Versailles. His team worked on resistance to cold and bunt in wheat and resistance to smut in oats and resistance to bunt in wheat. From 1950, the variety "Etoile de Choisy" played an important role in the renewal of southwestern France. From 1937, BUSTARRET also carried out resistance breeding to viruses and blight in potato. He created an excellent variety that is still highly appreciated today, that is, "BF 15." One of the great concerns of Bustarret, in charge of the Central Plant Breeding Station from 1944, was to train young scientists at the National School of Horticulture (Ecole Nationale Supérieure d'Horticulture [ENSH]), beginning from 1941 until 1959. Since 1942, Bustarret inspired the policy of the Standing Technical Committee on Plant Breeding (Comité Technique Permanent de la Sélection, CTPS) that promoted standardization of breeding and propagation methods. Before the national field experiment network of Europe was established. The CTPS was followed by the International Convention for the Protection of New Varieties of Plants ("so-called" Paris Convention, adopted in 1961). The setting up of EUCARPIA (European Association for Research on Plant Breeding) crowned BUSTARRET's international activities because it was created on the initiative of Bustarret, J. C. DORST (Wageningen, Netherlands) and of ▶ W. RUDORF (Köln, Germany). From 1961 to 1964, BUSTARRET was chairman of this association that is today the most important body, well beyond the frontiers of Europe, for world-wide concertation in the plant breeding sector.

CAMARA, A. ▶ MELLO-SAMPAYO, T.

CARRIERE, E. A. (1818–1896): He was a French conifer expert and author of the well-known *Traite Generale des Coniferes* and of numerous Latin cultivar names still in widespread use today. He was one of the most important scholars of cultivated varieties of his time.

CATCHESIDE, D. G. (1907–1994): He was one of the seminal persons in the post-war development of genetics, both in the United Kingdom and Australia. He made distinguished contributions to plant genetics and cytology, to genetic effects of radiation, to fungal biochemical genetics, to control of genetic recombination, and, in his retirement, to bryology. As a professor and administrator, he was responsible for several new institutional developments, including the first Australian Department of Genetics, the first Department of Microbiology in Birmingham (United Kingdom), and, perhaps, most importantly, the Research School of Biological Sciences of the Australian National University.

CORRENS, C. E. (1864–1934): He was a German botanist and geneticist who, in 1900, independent of, but simultaneously with, the biologists ▶ TSCHERMAK and ▶ VRIES, rediscovered ▶ G. MENDEL's historic paper on principles of heredity. He was born in München (Germany) and was raised by his aunt in Switzerland. In 1885, he entered the University of München to study botany. C. NÄGELI, the botanist to whom MENDEL wrote to about his pea plant experiments, was no longer lecturing at München. However, NÄGELI knew CORRENS' parents and took an interest in him. NÄGELI was the one who encouraged CORRENS' interest in botany and advised him on his thesis subject. Later, CORRENS became tutor at the University of Tübingen (Germany), where he began to deal with trait inheritance in plants in 1893. He already knew about some of MENDEL's hawkweed plant experiments from NÄGELI. However, NÄGELI never talked about MENDEL's key pea plant results, so he was initially unaware of MENDEL's conclusions of heredity. By 1900, when CORRENS submitted his own results for publication, the paper was called "Gregor Mendels Regel über das Verhalten der Nachkommenschaft der Bastarde" (Gregor Mendel's Law Concerning the Behavior of the Progeny of Hybrids). CORRENS was active in genetic research in Germany and was modest enough to never have a problem with scientific creditor recognition. He believed that his other work was more important, and the rediscovery of MENDEL's laws only helped him with his other work. He was supposedly indignant that H. de VRIES did not mention G. MENDEL in his first printing. In 1913, CORRENS became the first director of the newly founded "Kaiser Wilhelm Institut für Biologie" in Berlin/Dahlem (Germany). Unfortunately, most of his work was unpublished and destroyed when Berlin was bombed in 1945.

COTTA, H. (1763–1844): He was a German forester and founder of scientific and practical forestry in Europe, on the grounds of the Tharandt School of Forestry in Germany.

CRICK, F. H. C. (1916–2004): He was an English scientist, born in Northampton (England). He attended the University College in London and received a BSc in physics in 1937. In 1947, he began to study biology. He joined the Cavendish Laboratory (Cambridge) in 1949. There, he started work on his PhD thesis, studying proteins by X-ray diffraction. It was at the Cavendish Laboratory where he and ▶ J. WATSON met each other. They soon realized their common interest in the nature of the "genetic material." There was still a debate among scientists as to whether proteins or DNA was the genetic material. Even without knowing the identity of the molecule, they were among those who thought nucleic acids were the key to hereditary transfer. With the experiments in 1952, they, and others, became more sure of this hunch. They wanted to determine the structure of DNA and hoped that this would lead to a deeper understanding of how life is propagated.

CURTIS, W. (1746–1799): He was the founder of *Botanical Magazine* in 1787 in the United States—one of the first periodicals devoted to ornamental and exotic plants. The 176 volumes contained over 10,000 high plates and added much to the early literature of new ornamentals.

DARLINGTON, C. D. (1903–1981): He was a British cytogeneticist born in Chorley, Lancashire, and educated at Mercer's School, Holborn (1912–1917), St Paul's School (1917–1920), and Wye College, Ashford (1920–1923). In 1923, he began an association of more than 30 years with the John Innes Horticultural Institution. In 1937, he became head of the Cytology Department and director of the institute in 1939. Much of his work on cytology and chromosome theory was done augmented by expeditions and work abroad. In 1953, he resigned from the institution and accepted the Sherardian Professorship of Botany at Oxford. In addition to his research, teaching, and publication, he took a keen interest in the botanic garden and created the "Genetic Garden."

DARWIN, C. (1809–1882): He was an English biologist who wrote the book *The origin of Species* in 1859. He first presented the idea in a scientifically plausible way that biological species can change over time. He was appointed to the post of naturalist on the scientific expedition of HMS Beagle (1831–1836). In 1842, he bought Down House, in Kent, where he lived for the rest of his life, apparently suffering from Chagas disease, which he had contracted in South America. Having a private income, he could investigate as he pleased and at his own slow pace. By 1844, he had developed his theory of evolution, but he delayed publication until a note from Alfred WALLACE revealed his independent discovery of the same idea. In 1858, their joint paper was read to the Linnaean Society in 1859. His book was widely and quickly recognized, but opposition came from religious groups who preferred a literal interpretation of the Bible.

DELBRÜCK, M. (1906–1981): He was a German biologist, born in Berlin. His father was H. DELBRÜCK, a professor of history at the University of Berlin, and his mother was the granddaughter of Justus von LIEBIG, a famous German chemist and plant physiologist. DELBRÜCK studied astrophysics, shifting toward theoretical physics, at the University of Göttingen. After receiving his Ph.D., he traveled through England, Denmark, and Switzerland. where he met W. PAULI and N. BOHR, who got him interested in biology. DELBRÜCK went back to Berlin in 1932 as an assistant to L. MEITNER. In 1937, he moved to the United States, taking up research at Caltech on *Drosophila* genetics. In 1942, he and S. LURIA demonstrated that bacterial resistance to virus infection is caused by random mutation and not by adaptive change. For that, they were awarded the Nobel Prize in Physiology or Medicine in 1969. From the 1950s onward, DELBRÜCK worked on physiology rather than genetics. He also set up the Institute for Molecular Genetics at the University of Köln (Germany).

DICKSON, A. D. (1900–1997): He was the first director of the USDA, Agricultural Research Service, Barley and Malt Laboratory, serving from its establishment in 1948 until his retirement in 1968. He was born in Moxee City, Washington. He earned his BSc degree in 1919 his PhD in biochemistry in 1929 at the University of Wisconsin, Madison. After a brief period in Washington, he returned to Wisconsin in 1931, where he took a position with the USDA. Together with his colleagues, he initiated a barley testing and research program directed toward two goals: (1) developing procedures and equipment for evaluating the malting quality of barley on a laboratory scale and (2) comparing the new hybrid barleys with the older varieties, for example, "Oderbrucher" and "Manchuria." He conducted research on physical and chemical variables of malt, particularly the amylases, as influenced by variations in the malting process. His major accomplishments were the establishment of the Agricultural Research Service (ARS) and barley quality testing program and the development of the close relationship with the barley breeders and the malting and brewing industry. He also did some research on *Fusarium* toxins in scab-infested barley.

DIPPE, G. A. (1824–1890): He was a German agronomist and breeder who founded a breeding and seed company in Quedlinburg (Germany) (Figures 9.2 and 9.3). It was one of the biggest seed producer for sugar beet between 1870 and 1920. Founded in 1784, it was the later called Henry METTE GmbH. In 1850, the family of Gustav Adolf DIPPE established a second enterprise, which later evolved into the largest of its kind in the city (Figures 9.2

FIGURE 9.2 Gustav Adolf DIPPE at his former headquarters, Neuer Weg, Quedlinburg (Germany). (Courtesy of Dr. M. STEIN, Quedlinburg, Germany. With Permission.)

FIGURE 9.3 Monument commemorating 150 years' professional plant breeding and seed marketing at Quedlinburg (Germany), in 1934. (From R. SCHLEGEL. With Permission.)

and 9.3). The two companies were merged in 1946 in the German seed company, Deutsche Saatzucht Gesellschaft (DSG), and in 1971 in the VEB Saat- und Pflanzgut, producing, handling, and trading seedlings for horticultural crops, small-seed plants, and legumes. In 1990, the company was privatized as Quedlinburger Saatgut GmbH. In 1998, it was acquired by Julius WAGNER GmbH in Heidelberg as a subsidiary and merged in 2002 with Julius WAGNER GmbH as Quedlinburger Saatzucht GmbH. Until 1989, seeds and plants of 2250 registered cultivars of 370 crop species were annually multiplied and marketed. The company VEB Saat- und Pflanzgut handeled about 150.000 ha land for cereals and pulses, 150.00 ha for potato and beets, 145.000 ha for forage and vegetable plants as well as

5.000 ha for horticultural and ornamental plants. A total of 57 breeding stations, including 2500 employees, provided 4000 ha for selection.

DIPPE, F. C. von (1855–1934): He was a German horticulturist in Quedlinburg (Germany). After a gardener apprenticeship in Stendal, he attended the family enterprise "Gebrüder Dippe" (Dippe Brothers), where he became manager of agriculture and breeding of flowers and vegetables. Since the 1880s, he became manager of crop breeding program. In 1890, he started with individual plant selection and examination of the progeny.

DOBZHANSKY, T. ▶ RHOADES, M. M.

DOROFEEV, V. F. (1919–1987): He was a Russian agronomist, botanist, and expert in wheat breeding. He was head of the Department of Wheat (1978–1987) and director of VIR (Saint Petersburg, Russia).

DUDLEY, J. W. (1931–): He was a quantitative geneticist and plant breeder at the Department of Crop Sciences, University of Illinois, Urbana (United States). For more than 50 years, he had been a consistent contributor to developing theoretical models, gathering, analyzing, and interpreting data relative to specific problems; analyzing long-term selection studies; and integrating molecular marker data with the expression of quantitative trait loci; DUDLEY started his career working on sugar beet and alfalfa, but in the majority of his career he conducted basic research on the genetics and breeding of maize. Because of his earlier interests in autotetraploids, he developed mating designs to study the inheritance of quantitative traits in autotetraploid crop species. The longest continuous selection study for oil and protein content of maize has been thoroughly evaluated by him and his colleagues to measure rates of response, relative importance of types of genetic effects responding to selection, and estimates of the number of effective factors involved in the expression of oil and protein contents (Section 3.10.2).

EAST, E. M. (1879–1938): He was born in Du Quoin, Illinois (United States). From 1897, he attended the Case School of Applied Science in Cleveland, and from 1898, he studied chemistry at the University of Illinois (BSc in 1900, MSc in 1904, and PhD in 1907). From 1905 to 1909, he became an assistant at the Connecticut Agricultural Experiment Station and moved to Harvard University in 1914. There, he became a professor of biochemistry. Most recognized was his book, *The Role of Selection in Plant Breeding*.

EIHFELD, J. G. (1893–1989): He was a Russian biologist and plant breeder. From 1923 to 1940, he was in charge of the Polar Branch of VIR (Saint Petersburg, Russia), and from 1940 to 1951, he was director of VIR headquarters.

ELLIOTT, F. C. ▶ BOJANOWSKI, J. S.

ELLISON, F. (1941–2002): He had a long association with the University of Sydney Plant Breeding Institute (Australia). He received his BSc in agriculture in 1967, MSc in 1971, and PhD in 1977. In 1975, he was appointed as assistant wheat breeder and undertook postdoctoral research in wheat breeding at the University of Manitoba, Winnipeg (Canada). In 1976, he was appointed as plant breeder and was promoted to the position of senior wheat breeder in 1988. His efforts were particularly directed at the release of Prime Hard wheats for northern New South Wales and Queensland. He bred the wheat cultivars "Sunkota," "Shortim," "Suneca," "Sunstar," "Sundor," "Sunbird," "Sunelg," "Sunco," "Sunfield," "Miskle," "Sunbri," "Sunmist," "Sunstate," "Sunland," "Sunvale," "Sunbrook," "Sunlin," "Sunsoft 98," "Braewood," and "Marombi." ELLISON also assisted N. DARVEY and R. JESSOP (University of New England, Armadale) in the development of triticale cultivars "Ningadhu," "Samson," "Bejon," "Madonna," and "Maiden."

EMERSON, R. A. (1873–1947): He was born in Pillar Point (New York State, United States), but his early development and schooling took place in Nebraska. He spent 15 years of his professional career at the University of Nebraska, followed by 33 years at Cornell University. In 1893, he enrolled in the College of Agriculture at the University of Nebraska. He was greatly influenced by the ideas of the great teacher C. A. BESSEY. In 1897, he received his BSc degree. His first job was in the office of Experiment Stations

(USDA, Washington, D.C.). He served there for 2 years (1897–1899) as assistant editor of *Horticulture*. In 1899, he accepted an appointment at the University of Nebraska as assistant professor and chairman of the Horticulture Department and horticulturist in the Experiment Station. He held these positions for 15 years, with a promotion to full professor in 1905. He covered a wide range of horticultural projects, such as different culture methods for fruits and vegetables, domesticating native wild fruits, and winter hardiness of trees. Through the years in Nebraska, he became increasingly absorbed in a different and fundamental type of research, the nature of heredity in plants. He began to hybridize garden beans in 1898 while he was in Washington, to find out if there were any definite principles controlling heredity in plants. In 1902, he first became aware of MENDEL's laws because he referred to it in a paper published in the 15th Annual Report of the Nebraska Experiment Station. He quickly realized that his own studies on beans could be used to test MENDEL's principles, and he confirmed MENDEL's observation that some characters were dominant over the alternative forms. In 1909, EMERSON summarized his findings on "Inheritance of color in the seeds of the common bean, *Phaseolus vulgaris*" for the Annual Report of the Nebraska Experiment Station. EMERSON's interest in maize for studies of heredity began in around 1908, when he grew some plants from a cross between a rice popcorn and a sweet corn variety for a teaching demonstration. The segregation of starchy and sugary kernels deviated from the expected ratio based on a single factor pair. He collected a wide variety of genetic deviants in maize, many of which were used in later studies at Cornell University. At that time, he presented a report, when the relationship between genetic linkage and chromosomes had not jelled, and stated "that if genes were definitely located in chromosomes and that if parental chromosomes separated bodily at the reduction division, we should have an explanation not only of perfect genetic correlation and of allelomorphism but of independent inheritance as well." In 1908, by making some crosses between a Missouri dent corn and two dwarf types of popcorn, he initiated a study on quantitative inheritance in maize. In 1911, he spent a year at Bussey Institution of Harvard University, pursuing a graduate program for a D.Sc., which he was awarded in 1913. His advisor was a distinguished geneticist, ▶ Prof. E. M. EAST, whose special interest was quantitative inheritance. EAST, from studies on endosperm color in maize, and a Swedish geneticist, ▶ NILSSON-EHLE, from studies on color segregations in wheat and oats, independently proposed what came to be known as the "multiple-factor hypothesis" to explain the inheritance of quantitative characters. They assumed that the continuous variation in segregating progenies was governed by several to many genes, which were cumulative in their action. EMERSON and EAST collaborated in compiling Nebraska Research Bulletin 2 on "The inheritance of quantitative characters in maize," which was published in 1913. Because of his merits, in 1914, he accepted an offer from Cornell University in Ithaca to become professor and head of the Department of Plant Breeding, positions that he was to hold until his retirement in 1942. He recommended a number of proposals to develop the Plant Breeding Department, such as to inaugurate clonal selection with fruit and flower crops, scion selection with apples, to make biochemical-genetic investigations of color inheritance, and to develop a botanic-genetic garden to illustrate MENDEL's experiments. In 1924, he and F. D. RICHEY, USDA maize investigator, made a scientific expedition to Argentina, Bolivia, Chile, and Peru. They collected around 200 samples of maize. In 1935, "A summary of linkage studies in maize" was compiled by EMERSON, ▶ G. W. BEADLE, and A. C. FRASER. For the first time, all the known more than 300 genes in maize were cataloged alphabetically, with appropriate symbols, descriptions, and chromosomal locations, when known. EMERSON was cited for 41 of these genes. For a few years (1938–1942), he was associated with a project on breeding muskmelons for resistance to *Fusarium* wilt.

ENGLEDOW, F. L. ▶ HUNTER, H.

EYAL, Z. (1937–1999): He was a prominent leader of plant pathological research in cereals and its practical applications in combating fungal disease. He studied agronomy at Oklahoma State University, Stillwater (United States) and earned a PhD in plant pathology in 1966. He served twice as Head of the Department of Plant Sciences of Tel Aviv University (Israel) and was appointed as director of the University's Institute for Cereal Crops Improvement in 1996. He became one of the leading researchers in the field of wheat–*Septoria* interactions, initiating and guiding approaches toward understanding both plant and pathogen biology.

FENZL, E. ▶ TSCHERMAK-SEYSENEGG, E. von.

FIALA, J. L. (1924–1990): He was a Catholic priest in the United States and one of the century's leading hybridizers of *Malus* and *Syringa*. He wrote very remarkable books on both genera. His lilac breeding program spanned more than 50 years and produced great results. "Avalanche," for example, is one of top-rated white lilacs.

FIFE, D. (1805–1877): He was a Scottish farmer and wheat breeder. He moved to Canada with his parents in 1820 and settled in Otonabee/Peterborough (Ontario), where he grew up and married. FIFE brought the spring wheat to Canada. He grew it on his farm and harvested half a bushel of grain, which he shared, in part, with his friends. In 1842, this wheat came to Russia from Ukraine. The original strain was called "Halychanka" and was later known as "Red Fife" (Section 7.2.1). The strain was used widely after 1848, becoming the leader in Ontario by 1851 and virtually replacing all others there by 1860. From Fife's farm in Otonabee, it spread to Illinois and Ohio in the United States and then to Saskatchewan, Alberta, and Manitoba in about 1870, ranking as the leading variety there from 1882 to 1909. "Red Fife" also served as the male parent of the "Marquis" strains, which proved more frost tolerant and even less susceptible to rusts, allowing wheat farming in Manitoba to spread farther west and north.

FISHER, R. A. (1890–1962): He was the second of twins born in St. James, London (United Kingdom). In 1904, he entered Harrow School and won the Neeld Medal in 1906 in a mathematical essay competition. FISHER was awarded a scholarship from Caius and Gonville College, Cambridge, which was necessary to finance his studies. In 1909, he matriculated at Cambridge. Although he studied mathematics and astronomy, he was interested in biology. In his second year as an undergraduate, he began consulting senior members of the university about the possibility of forming a Cambridge University Eugenics Society. He graduated with distinction in the mathematical Tripos of 1912. After leaving Cambridge, FISHER had no means of financial support and worked for a few months on a farm in Canada. Later, he returned to London, taking up a post as a statistician in the Mercantile and General Investment Company. The interest in eugenics and his experiences in working on the Canadian farm made him interested in starting a farm of his own and gave up being a mathematics teacher in 1919. At that time, two posts were simultaneously offered to him—the first by K. PEARSON as chief statistician at the Galton Laboratories and the second as statistician at the Rothamsted Agricultural Experiment Station (Table 3.1). He accepted the post at Rothamsted, where he made many contributions, both to statistics, in particular the design and analysis of experiments, and to genetics. There, he studied the design of experiments by introducing the concept of randomization and the analysis of variance. FISHER's idea was to arrange an experiment as a set of partitioned sub-experiments that differed from each other in having one or several factors or treatments applied to them. The sub-experiments were designed in such a way as to permit differences in their outcome to be attributed to the different factors or combinations of factors by means of statistical analysis. This was a notable advance over the existing approach of varying only one factor at a time. In 1921, he introduced the concept of likelihood, that is, the likelihood of a parameter is proportional to the probability of the data and it gives a function that usually has a single maximum value, which he called the "maximum likelihood." In 1922, he gave a new definition of statistics as a method to reduce the amount of data. He identified three fundamental problems: (a) specification of the kind of population from which the data came, (b) data estimation, and (c) data distribution. FISHER published a number of important books, for example, *Statistical Methods for Research Workers* (1925), *The Design of Experiments* (1935), and *Statistical Tables* (1947). The first ran to many editions, which he extended throughout his life. It became a handbook of methods for design and analysis of experiments. During his time at Rothamsted, he had conducted breeding experiments with mice, snails, and poultry, and the results he obtained led to theories about gene dominance and fitness, which he published in *The Genetical Theory of Natural Selection* (1930). In 1933, K. PEARSON retired as professor of eugenics at University College, and FISHER was appointed to the chair as his successor. He held this post for 10 years, before being appointed as Arthur Balfour professor of genetics at the University of Cambridge in 1943. He retired from his Cambridge chair in 1957 but continued to carry out his duties there for another 2 years, until his successor could be appointed. He then moved to the University of Adelaide (Australia), where he continued his research for the final 3 years of his life.

FORREST, G. (1873–1932): He was an American plant collector, explorer, and *Rhododendron* expert of the United States. He was perhaps the leading introducer of new *Rhododendron* taxa to horticulture. His collections are still studied today.

FORTUNE, R. (1812–1880): He was curator of the "Chelsea Physic Garden" (United States) and a plant explorer, particular to China, beginning in 1852. Most plants bearing the epithet "fortune" are in his honor. He introduced *Camellia sinensis*, the tea plant, to the West.

FRANKEL, O. (1900–1998): Born in Vienna (Austria), he studied agriculture at the University of Berlin (Germany) and earned his doctorate in agriculture from this university in 1925. He was employed for 2 years (1925–1927) as a plant breeder on a large private estate in Dioseg/Bratislava (Slovakia). Later, he began wheat and barley breeding at Lincoln College/Christchurch (New Zealand), where he was to work until 1951. He began his breeding program with introducing quantitative assessments of grain yield, of milling and baking quality, which led to the release of the widely grown variety "Cross 7" (1934), "Taiaroa" and "Tainui" (1939), "Fife-Tuscan" (1941), and "WRI-Yielder" (1947). He put a considerable effort into optimizing the role of quality testing in the selection process. His post-retirement research was focused on the base-sterile mutants of speltoid wheats and examined the photoperiodic effects on floral initiation and floret sterility. He was one of the pioneer of genetic resources movement, that is, the conservation of biological diversity. From 1951 to 1962, he became chief of the Commonwealth Scientific and Industrial Research Organisation (CSIRO) Division of Plant Industry (Australia) and on the CSIRO Executive (1962–1966).

FRASER, A. C. ▶ EMERSON, R. A.

FREISLEBEN, R. (1906–1943): He was a German botanist working at the Institute of Plant Production of the University of Halle/S. (Germany). He died early in the World War II. During his short life as a cytogeneticist, he was able to select (together with ▶ A. LEIN) a mutant line of barley after X-ray treatment, showing stable mildew resistance (*mlo*); this so-called *mlo* locus is still present in many advanced barley variety around the world.

FRIEDRICH, B. (1899–1980): He studied plant breeding during the 1920s and assisted Prof. ▶ TSCHERMAK at the "Hochschule für Bodenkunde" in Vienna (Austria). From 1938 to 1948, he was working as barley and wheat breeder in Sladkovicovo (Czechoslovakia). After 1945, he was appointed in Martonvasar (Hungary) until his retirement. He was active in searching for new ways in plant breeding and applying new ideas in mechanization of nursery techniques. Since 1946, he successfully planted rust nurseries of wheat in naturally infected locations for detection of resistant genotypes against rusts.

FRUWIRTH, C. (1862–1930): He was professor of plant production at the University of Vienna (Austria). In 1905, he founded the "Königliche Württembergische Saatzuchtanstalt" in Hohenheim (Germany)—an institution that still contributes to successful breeding in

Germany. He returned to Austria and bought the estate Walddorf near Klagenfurt. There, he started cereal breeding. In addition, he received particular reputation as scientific author. He was the founder of the *Handbuch der Züchtung landwirtschaftlichen Kulturpflanzen* (P. Parey Verl., Berlin), with five volumes and seven editions. With the same publisher, he founded the journal *Zeitschrift für Pflanzenzüchtung* in 1913 (now: *Plant Breeding*).

GALLESIO, G. (1772–1839): He was an Italian naturalist and pomologist and conducted many hybridization experiments with carnations. He was born in Liguria and died in Florins (Italy). Although he studied jurisprudence and was active in several political positions, in 1811, he published his first book, *Traite du Citrus,* and, in 1816, the second book, *Teoria della riproduzione vegetale*, in which he described his observation, for example, "… I have crossed white with red flowering carnations and reciprocally… the seeds I produced brought carnations with mixed colors…the plants sometimes show the characters of or the other parent, depending which one is dominating…;" therefore, the term "dominate" was already used before it was applied by ▶ M. SAGERET in 1826 and ▶ T. A. KNIGHT.

GALTON, F. (1822–1911): He was a British geneticist and a cousin of ▶ C. DARWIN. His statistical analysis of genetic segregation patterns led him to the introduction of the "correlation coefficient;" he also coined the word "eugenics."

GOLDSCHMIDT, R. ▶ BAUR, E.

GOULDEN, C. H. (1897–1981): He was a geneticist born in Bridgend, Wales (United Kingdom). As son of a homesteader, he took the course for farmers at the University of Saskatoon (Canada) and went on to earn his PhD in plant breeding before becoming chief cereal breeder at the Dominion Rust Research Laboratory, Winnipeg (Canada). In 1925, he succeeded L. H. NEWMAN as dominion cerealist in 1948 and later became assistant deputy minister for research in the Department of Agriculture. As natural mathematician, he took up the new specialty of biostatistics and wrote the first North American textbook on this subject in 1937—for the students he taught—at the University of Manitoba. As head of cereal breeding in Winnipeg for 23 years, GOULDEN was responsible for creating "Renown," "Regent," and "Redman" wheats, suitable for the Canadian climate, and possessing various rust-resistant qualities. He also developed six varieties of rust-resistant oats.

GOVOROV, L. I. (1885–1941): He was a Russian agronomist and plant breeder. From 1915, he worked at Moscow Breeding Station. From 1923, he was the head of the Steppe Experiment Station and later became in charge of the Department of Leguminous Crops at VIR (Saint Petersburg, Russia).

GRABNER, E. (1878–1955): He is supposed to be doyen of Hungarian plant breeding. He started the breeding work in Hungary in the beginning of the last century. By influences and visits of European plant breeding institutes in Sweden and Austria, he created the prerequisites for professional plant breeding in Hungary, based on MENDEL's laws of inheritance. It is summarized in his book *Breeding of Agricultural Plants*, published in 1908. Soon after 1908, the first Hungarian institute of plant breeding was founded in Magyarovar in 1909. He directed this institute for 28 years. Owing to his inspiring work on 20 locations of the country, professional plant breeding was carried on several crops in 1911. Based on his activities, the first Hungarian order regulating the system of variety registration was passed in 1915.

GRAY, A. ▶ BAILEY, L. H.

GREBENSCIKOV, I. S. (1912–1986): He was born in St. Petersburg (Russia). He studied at the Agricultural Faculty of the University of Belgrade (Serbia). In 1938, he obtained the MSc degree, and from 1942, he continued his work at the Genetic Department of the "Kaiser Wilhelm Institut," Berlin (Germany), where he carried out human brain research, together with his Russian colleague N. TIMOFEEFF-RESSOVSKY (one of the early mutation researchers). In 1946, he was appointed by the Zentralinstitut für Genetik und Kulturpflanzenforschung, Gatersleben (Germany). He became a recognized specialist for genetic–taxonomic studies in maize and Cucurbitaceae. His particular interest was focused on the inheritance of quantitative traits, for example, the ontogenetic dominance variance. Moreover, he was one of the first taxonomists introducing the term "convariety" in maize taxonomy. Both in maize and Cucurbitaceae, he studied the phenomenon of heterosis.

GREW, N. (1641–1712): He was co-founder of the discipline of plant anatomy with ▶ MALPIGHI and was born in Coventry (United Kingdom). He was a practicing physician first in Coventry and then in London and became secretary of the Royal Society. His work on plant anatomy began in 1664, with the object of comparing plant and animal tissues. His essay was published by the Royal Society of London in 1670. MALPIGHI, working independently on the same subject in Italy, also sent his work to the Royal Society. GREW approached botany from the medical standpoint. His fundamental thesis was that every plant organ consists of two "organical parts essentially distinct," that is, a "pithy part" and a "ligneous part;" in the seed, the pithy part is composed of "parenchyma," a term first used by GREW. He described stages of seed germination; however, the underlying physiology was hopelessly confused. He used the term "radicle" for embryonic root and "plume" for "plumula." He called cotyledons "leaves" and recognized that they could appear above ground and turn into green. He observed monocot stems with scattered (vascular) bundles and a lack of distinct bark and pith, resin ducts in cortex pine stem, wings and "feathers" on seeds and fruit, protection and economy of space gained by overlapping of bud scales, folding and rolling of leaves in buds, and forming of buds months before they expand ("… a bulb is, as it were, a great bud underground…"). He described tulip flower in bulb in September and noted that pollen grains are "bee-bread." He believed that micropyle allowed water to enter the seed and cause germination. GREW initiated the study of tissues (histology). He made the first successful attempt to extract chlorophyll from leaves, using oil as a solvent. With his important work *Anatomy of Plants* (1682), he related anatomy with physiology.

GRUNDY, P. M. ▶ YATES, F.

GYORFFY, B. (1911–1970): He was a Hungarian geneticist who obtained his PhD at the University of Szeged (Hungary). In 1937, he obtained a postdoctoral position at the "Kaiser Wilhelm Institut" (Berlin, Germany). ▶ F. von WETTSTEIN, a former student of ▶ C. CORRENS, strongly influenced GYORFFY's scientific career. Inspired by the numerous scientific papers about induced polyploidization via colchicine treatment, he devoted to polyploidy research. From 1944, he became director of the National Institute of Plant Breeding in Magyarovar (Hungary). In several positions, he contributed very much to the development of the modern Hungarian plant breeding and agriculture.

HADJINOV, M. I. (1899–1988): He was an outstanding Russian maize breeder and geneticist working since 1940 at Research Institute of Agriculture, Krasnodar (Russia). From 1946 to 1948, he widely used inbred lines to obtain variety–line crosses that resulted in a number of commercial hybrids sown over large acreage in the USSR. Since 1954, he was attracted to studies and utilization of cytoplasmic male sterility (CMS) in maize breeding at almost the same time when in the United States the first publication on CMS for hybrid seed production appeared. He initiated studies on maize grain quality (opaque-2), polyploidy, distant hybridization with *Tripsacum* and/or *Teosinte*, induced mutagenesis, haploidy, and so on.

HAGBERG, A. (1919–2011): He was born in Gothenburg, on the west coast of Sweden. He started university studies in Lund, where he defended his doctoral thesis in 1953 on the subject "Studies in Heterosis." He continued the achievements of his teacher ▶ Arne MÜNTZING regarding genetic studies in the plant genus *Galeopsis*. During the time in the University of Lund, he became acquainted with plant breeding. In 1942, he got an assistant employment at the famous plant breeding station for sugar beets at Hilleshög in Landskrona. Later, in 1945, he was employed at the Swedish Seed Association in Svalöf and worked with several species, for example, rye, clover, lupines, and potatoes. By the director of the Swedish Seed Association, ▶ Åke ÅKERMAN, in 1951, Arne HAGBERG got the trust to build up the Chromosome Department. At that time, cytology and cytogenetics were very strong developing sciences, especially in Sweden, where the pioneers ▶ Arne MÜNTZING, ▶ Albert LEVAN, and ▶ Åke GUSTAFSSON made Sweden a leading nation in these areas. Aat the new department, scientists such as Sven ELLERSTRÖM, Diter von WETTSTEIN, Volkmar STOY, Nils NYBOM, Göran PERSSON, and Arne WIBERG contributed much to plant cytogenetics. HAGBERG spent 1 year (1955–1956) as Rockefeller scholar in the United States (St. Paul, Minnesota, and Pullman, Washington State), where he collaborated with Charlie BURNHAM, Tom RAMAGE, ▶ Gus WIEBE, Ernie SEARS, and ▶ Bob NILAN. Years later, Arne was also appointed as head of the Barley Breeding Department, a position that he held for about 10 years. Already in the beginning of the 1950s, he developed interest in barley as a model crop in genetic research. Induction of mutations for day-length neutrality (barley: "Mari" and "Mona") became important for global increase of barley cultivation; breeding of rye and flax were as important. many of the former plant breeders received their education at the University of Lund, where ▶ Professor Hermann NILSSON-EHLE was the first holder of the professorship in genetics as well as the director of the Swedish Seed Association. In 1979, Arne became professor at the newly established Department of Plant Breeding at the Swedish University of Agricultural Sciences. Here, he stayed until his retirement in 1985. He was president of many international organizations in research and development of plant breeding, that is, International Crops Research Institute for the Semi-arid Tropics (ICRISAT), European Association for Plant Breeding Research (EUCARPIA), and the European Brewery Convention. Together with Bob NILAN (United States) and Evald FAVRET (Argentina), who were visiting researchers in Svalöf in the beginning of the 1960s, Arne initiated "The International Barley Genetics Symposium" (IBGS). The first IBGS was held in Wageningen (the Netherlands) in 1963. The IBGS is still a leading organization regarding issues of barley research and breeding. He published several books, for example, *Plant breeding—Green Revolution* in Swedish (1977) and *Mutations and Polyploidy in Plant Breeding* (1961, together with Erik ÅKERBERG). In total, Arne published more than 300 articles. During the years 1972–1979, Arne HAGBERG was the director of the Swedish Seed Association. He became a member of several important societies: the Royal Swedish Academy of Sciences, the Royal Swedish Academy of Forestry and Agriculture, the Royal Physiographical Society in Lund, and the Mendelian Society of Lund. For his great and important efforts within Swedish genetic research, plant breeding, and the Swedish agriculture, he was awarded with the "Nilsson-Ehle" medal and "The Engström Medal."

HAKANSSON, A. ▶ NILSSON-EHLE, H.

HALDANE, J. B. S. (1892–1964): A native of Oxford (United Kingdom), he studied mathematics and biology at the University of Oxford and carried out several genetic studies in plants. After his time as lecturer on enzyme research at Cambridge University, he became the successor of ▶ W. BATESON at the John Innes Horticultural Institute in Merton near London. From 1927 to 1936, he carried out intensive genetic research, before he became head of the Chair of Genetics (later: Chair of Biochemistry) at University College London in 1933. From 1957, he was an active supervisor at the Indian Federal Office (Calcutta) and founded the Laboratory of Genetics and Biometrics in Bhubaneswar (Orissa), where he died.

HALES, S. (1671–1761): He was an English physiologist, chemist, inventor, and country vicar. In 1709, he resigned from a fellowship at Cambridge University to become a perpetual curate in Teddington. He studied physiology based on the foundation of ▶ GREW's work on plant anatomy. For 40 years, he devoted his leisure time to research in botany and zoology. His memoirs were published in a collected form as the "Statical Essays," dealing with problems of plant and animal physiology. "Vegetable Statics" (1727) became the classic work in plant physiology. HALES is considered the founder of experimental plant physiology. The great part of his work is a record of successive experiments. An attempt to stop "bleeding" in a badly pruned grape vine by means of a piece of bladder tied over the wound gave him the idea of manometer. He found that root pressure showed a daily periodicity and was affected by changes in temperature. He noticed that leaves gave off water, and thus, he proceeded to measure the amount of transpiration and to compare it with the amount absorbed by the root. He studied variations in the quantity of water transpired over a 24-hour period and demonstrated reduction in transpiration at night. He had definite notions of the part that the leaves played in plant nutrition; therefore, he studied leaf structure. He also contended that "plants very probably draw through their leaves some part of their nourishment from the air" and that leaves also absorbed light. HALES had a scientific mind of the highest order and is ranked along with ▶ GREW and ▶ MALPIGHI as outstanding leaders in botany and physiology up to the end of the eighteenth century. He was the first to use quantitative results in botanical experiments, such as movement of sap, transpiration, and flow of nutrients by girdling.

HAMMARLAND, C. ▶ NILSSON-EHLE, H.

HÄNSEL, H. (1918–2005): He was an Austrian breeder born in Vienna. He started his professional career after studying languages and agriculture at the University of Agricultural Sciences in Vienna. He received his PhD in plant breeding in 1948. After several postdoctoral studies in Cambridge (United Kingdom) and Wageningen (the Netherlands), he decided to work as a practical plant breeder in the "Probstdorfer Saatzucht Co." (Austria), located in the cereal growing area east of Vienna. Breeding bread and durum wheat and barley became his passion; he never lost contact with his former university and received a DSc degree in 1954 as a lecturer and later as an external professor of plant breeding. For more than 33 years, he offered lectures on mutation breeding, breeding methodology, and developmental physiology. More than 55 varieties were bred by him. In wheat breeding, he was able to combine high yielding performance with superior baking qualities in varieties such as "Probstdorfer Extrem," "Perlo," and "Capo." Some of his varieties sometimes covered 75% of the wheat production areas (of about 225.000 ha) in Austria. His spring barley "Adora" and "Viva" combined excellent yield stability with durable resistances. In durum wheat breeding, his varieties are now cultivated in Italy, France, and Spain. No other breeder personality has dominated for the last 45 years in the Austrian breeding.

HARLAN, J. R. (1917–1998): Jack Rodney HARLAN went to the University of California in Berkeley (United Kingdom) and was the first graduate student to receive the PhD, along with the great botanist and evolutionist G. L. STEBBINS. He began his professional career as forage and rangeland grass breeder for the USDA in 1942 in Woodward (Oklahoma), and in 1951, he transferred to Oklahoma State University (Stillwater). While holding a joint appointment as professor of genetics at the university, he left the USDA in 1961 and joined the faculty of Oklahoma State University as a full-time professor in 1966. Later, he moved to the University of Illinois as professor of plant genetics in the Department of Agronomy. A year later, with J. M. J. de WET, he founded the Crop Evolution Laboratory. In 1984, his formal professional career at the University of California (Davis) ended, where he completed the revision of the book *Crops and Man* and formulated the basic outline of *The Living Fields: Our Agricultural Heritage*. He developed a deeper understanding of the domestication of many crops through his extensive plant exploration work and astute observations in some 45 countries on all of the continents over a period of 35 year. HARLAN was a very keen student of VAVILOV's work and synthesized his observations in a classic paper "Agricultural Origins: Centers and Noncenters" (1971). He introduced the concept of "noncenters" as a complement and refinement of Vavilovian theories of crop origins and diversity. HARLAN is prominent among the founders Otto FRANKEL, Erna BENNETT, Jack HAWKES, Dieter BOMMER, M. S. SWAMINATHAN, and John CREECH of the modern movement, which established plant genetic resources as an interdisciplinary field for scientific study and for biological conservation (Figure 9.4).

HAVENER, R. D. (1930–2005): He was one of the pioneers in the global agricultural research system, working for the world's rural poor for more than five decades. He led The International Maize and Wheat Improvement Center (CIMMYT) (Mexico) from

FIGURE 9.4   J. R. HARLAN (1917–1998), 1990. (Courtesy of Dr. Theodore HYMOWITZ, Illinois. With Permission.)

1978 to 1985 as the center's third director general, bringing recognition as one of the leading international agricultural research organizations in the world. When he came to CIMMYT, ▶ N. BORLAUG was director of the wheat program and ▶ E. SPRAGUE was the director of the maize program. During his leadership, CIMMYT expanded its regional presence and strengthened the economics program. For 14 years, he worked as a senior agricultural program officer at the Ford Foundation. He served as interim director general at both International Center for Tropical Agriculture (CIAT), (Cali, Colombia) (1994) and International Rice Research Institute (IRRI), (Los Baños, Philippines) (1998) and was instrumental in the founding of International Center for Agricultural Research in the Dry Areas (ICARDA), (Beirut, Lebanon) and International Livestock Research Institute (ILRI), (Nairobi, Kenia). He served as chair of the ICARDA Board of Trustees from 1999 to 2003 and was the founding president of the Winrock International Institute for Agricultural Development, a fellow of the American Association for the Advancement of Science, an advisor for the World Food Prize, and also sat on the board of directors of Sasakawa Africa Association, whose president was N. BORLAUG.

HAYMAN, D. L. (1929–2006): He retired as reader in Genetics in 1992. He was closely associated with the University of Adelaide (Australia) for more than 50 years. In his final year for the BSc degree at the Waite Institute in 1952, he obtained a splendid grounding in genetics with Prof. D.G. CATCHESIDE, G. MAYO, and J. MATHIESON as lecturers in the new department. In 1953, he began a study of cross-incompatibility in the grass *Phalaris coerulescens,* which led to his PhD and a paper sent for publication early in 1956, with his discovery of a novel genetic system with two separate gene loci, both with multiple alleles, controlling incompatibility. A similar discovery was reported for rye in 1954 by A. LUNDQVIST in Sweden. Both HAYMAN and LUNDQVIST are now acknowledged as co-discoverers of the genetic basis for incompatibility found in all grass species that have been studied. After several years as a research officer in the CSIRO Division of Plant Industry, he went to Adelaide as a lecturer of genetics in 1959 and began a big project on marsupial cytogenetics. From studies of chromosome morphology and banding patterns in Australian and South American species, he identified two contrasting themes in marsupial evolution, that is, conservation of chromosome number and morphology, as well as chromosomal fission events leading to an increase in chromosome number. HAYMAN maintained his interest in incompatibility in *P. coerulescens*, supervising studies of mutations where the incompatibility had been completely or partially lost, and after retirement, he worked with Prof. P. LANGRIDGE at the Waite Institute (Australia) in an attempt to experimentally clone an incompatibility gene from pollen. He is remembered as an inspiring teacher of cytology and genetics and an excellent supervisor of research students.

HAYS, W. M. (1859–1927): Willet HAYS was born in 1859 in central Iowa on a farm near Eldora. He attended Oskaloosa College, founded by the Disciples Christian Church. It was the liberal fraction of the school that supported DARWIN's theory of evolution, and later, it was merged with Drake University. In 1885, he graduated from Drake University. As he was interested in scientific agriculture, he enrolled at Iowa State College in Ames, where he earned an MSc degree in 1886. Then, he became associate editor of *Prairie Farmer* magazine in Chicago. In 1888, HAYS was the first faculty member selected for the University of Minnesota's new Minnesota Agricultural Experiment Station in St. Paul. Later, he was appointed as instructor of agriculture in the School of Agriculture and assistant agriculturist in the Experiment Station under Director E. D. PORTER. By 1890, the young HAYS had postulated that "there are Shakespeares among plants," recognizing the individual plant as the unit of crop improvement; therefore, he began systematically breeding and testing large numbers of plants to find the outstanding individuals. By 1900, he had a crop nursery of several acres with millions of plants. HAYS left Minnesota late in 1891 and spent 1892 and much of 1893 as professor of agriculture at the North Dakota Agricultural College in Fargo. There, he was the first professor of agriculture and agriculturist for the agricultural experiment station and established the experiment station's field crop test plots and lasting, productive, plant breeding programs. There, a wheat plot and a flax plot have grown continuously since HAYS first planted them in 1892 (TROYER and STOEHR 2003). He returned to Minnesota in 1893 and served as professor and vice director of the agricultural experiment station through 1904. He was appointed as assistant U.S. Secretary of Agriculture in 1904, where he served from 1905 until 1913. He started the first systematic pure-line selection and progeny tests of oats in the United States at the Minnesota experiment station in 1888 and designed the centgener system of plant testing, including a special seed drill (Section 3.2). He was listed among the first four pioneers in barley breeding and tested barleys of hybrid origin as early as 1904. His 1889 selections of timothy plants at the Minnesota station are the earliest records of timothy improvement. In 1894, he started flax selection to develop "Minn 25" ("Primost"), the first pure-line flax variety developed and distributed in the United States, followed by "Minn 169" and "Blue Stem" wheat, "Minn 13" and "Minn 23" maize, "Minn 105" barley, and "Minn 281" and "Minn 295" oat varieties. As assistant secretary of agriculture, HAYS introduced the project system for agricultural research, which was later extended by the USDA to state experiment stations. In this position, HAYS wrote the protocol for the New International Institute of Agriculture, which was organized in Rome in 1913. It was the forerunner of the United Nations Food and Agriculture Organization. Later, in 1913, he went to Argentina, where he helped organize the Argentine Department of Agriculture along the lines of the USDA.

HEINE, F. (1840–1920): He was born in Halberstadt (Germany). After school from 1850 to 1859, he began to study agronomy in Ahlsdorf. As owner of the Emersleben Estate, he started with plant breeding in 1869. In 1885, he bought additional land in Hadmersleben and moved his seed company to this place. He was the founder of systematic plant breeding in Germany. He started with mass and single plant selection and with progeny testing in cereals, legumes, and root crops. His wheat varieties "Heines Squarehead" (1872), "Heines Teverson" (1893), "Heines Rivetts" (1896), "Heines Kolben" (1871), "Heines Bordeaux" (1891), "Heines Noe" (1900), and "Heines Japhet" (1903) were standards for long time in Germany. He also bred about 12 other varieties of spring barley, oats, winter rye, pea, broad bean, and sugar beet. The breeding station Hadmersleben remains a famous place for plant breeding until present time (now Syngenta®, Switzerland).

HELLRIEGEL, H. (1831–1895): A German from Mausitz near Pegau, he studied chemistry at Tharandt and founded the Agricultural Experiment Station, Dahme, where he was director since 1873. In 1886, together with H. WILFAHRT, he demonstrated the assimilation of atmospheric nitrogen by bacteria in the root nods of legumes.

HERBERT, W. (1778–1847): He has been a contemporary of ▶ T. A. KNIGHT and a dean of Manchester of the Anglican Church in England. He dealt with the question whether or not the fertility of interspecific hybrids is a measure for the distance between the species. He came to the conclusion that the fertility or sterility of hybrids is not an evidence for the taxonomic relationships and that there are no borders between varieties and species concerning hybridization. In his opinion, the environment under which a certain species grows up is more important for a successful hybridization than the systematic (genetic) borders between races "... the only thing certain is, that we are ignorant of the origin of races... that God has revealed nothing to us on the subject... and that we may amuse ourselves with speculating thereon... but we cannot obtain negative proof, that is, proof that two creatures or vegetables of the same family did not descend from one source. But can we prove the affirmative... and that is the use of hybridizing experiments, which I have invariably suggested... for if I can produce a fertile offspring between two plants that botanists have reckoned fundamentally distinct, I consider that I have shown them to be one kind, and indeed I am inclined to think that, if a well-formed and healthy offspring proceeds at all from their union, it would be rash, to hold them of distinct origin;" he was convinced that hybridization may improve the value of horticultural and agricultural crops. Many years later, he read the papers of ▶ KÖLREUTER, which confirmed his intention.

HOFFMANN, W. (1910–1974): He was professor of genetics and breeding research at the University of Berlin (Germany). He started his career as junior assistant at the Botanical Institute of Heidelberg (1933–1934) and later as assistant at the "Kaiser Wilhelm Institut für Züchtungsforschung" (Müncheberg) (1935–1936). From 1942 to 1946, he was appointed as Head of the Department of Genetics, Breeding and Agronomy in Schönberg (Moravia). From 1946 to 1949, he worked as scientific officer at the Institute of Plant Breeding, Hohenthurm (Germany), and later as full professor from 1950 to 1958. In 1958, he moved to Berlin, where he remained as professor and director of the Chair of Genetics and Breeding Research of the Technical University Berlin (from 1972: Institute of Applied Genetics of the Free University of Berlin) until 1974. His main interest was the systematic combination breeding in barley by utilization of induced mutants. In hemp, he developed pale-stalked mutants with good photoperiodic adaptability (released in 1940). Moreover, he developed several allopolyploids of cabbage and wheat. His synthetic tetraploid *Brassica campestris* ssp. *pekinensis* × *B. campestris* ssp. *oleiferea* hybrid received much attention as fodder and green manure crop; it was released in 1969.

HOLDEFLEISS, P. (1865–1940): He was a German breeder in Salzmünde (Germany). In 1894, he received his PhD from the University of Halle/S. (Germany) after apprenticeship in agriculture at the same university under ▶ Prof. J. KÜHN. After habilitation in 1897, he became assistant and professor at the Agricultural Institute, and, in 1931, he became dean of the agricultural faculty.

HOLTKAMP, H. (1904–1988): He was a master gardener and founder of the Holtkamp Greenhouses Co., the largest grower of African violets (*Saintpaulia* species) in the world. It is credited with a number of notable innovations, including multiflorescence and semper florescence. HOLTKAMP was responsible for many innovations, such as non-dropping flowers, which have contributed significantly to the worldwide popularity of African violets. He is remembered as one of the most influential pioneers in the African violet industry, creating many new varieties, each year, in both the United States and Germany.

HOOKE, R. (1634–1703): He was an English experimental physicist and had wide interests in science. He was the first to clearly state that the motion of heavenly bodies must be regarded as a mathematical problem, and he approached the discovery of universal gravitation in a remarkable manner. HOOKE had a strong personality and temper and made virulent attacks on I. NEWTON and other scientists, claiming that their published work was due to him. He is remembered in biology for the discovery and naming of "cells" as the units of plant structure. With the aid of a microscope, he examined a wide range of materials and substances, including feathers, lice, fleas, and cork. His observations were published in *Micrographia* (1665); he first recognized that charcoal, cork, and plant tissues were "... all perforated and porous, much like a honeycomb...;" to these pores, he gave the name cells; however, the cell walls were not considered as constituent parts.

HUNTER, H. (1882–1959): He was a British agricultural scientist and graduated in 1903 from the University of Leeds. He was appointed officer in charge of the barley investigations being conducted by the Department of Agriculture and Technical Instruction in Ireland. There, he developed the variety of barley "Spratt-Archer," which was, for many years, the most widely grown malting barley in Britain. In 1919, he was appointed head of the Plant Breeding Division of the Ministry of Agriculture for Northern Ireland, and, in 1923, he moved to Cambridge to join ▶ R. H. BIFFEN, T. B. WOOD, and F. L. ENGLEDOW in the Plant Breeding Institute of the University School of Agriculture. HUNTER became director of this Plant Breeding Institute in 1936. After his retirement in 1946, he served as president of the Council of the National Institute of Agricultural Botany for 3 years (1951–1953).

INNES, J. (1829–1904): He was the founder of John Innes Horticultural Institute. Later, it became the John Innes Institute of Plant Science, Norwich (United Kingdom). He was a city businessman working in partnership with his brother James INNES in a company that owned large sugarcane plantations in Jamaica and imported rum into England.

IVANOV, I. V. (1915–1998): He was professor of agronomy and a member of the Bulgarian Agricultural Academy of Sciences. He was born in Karnobat (Bulgaria) and received his MSc (agronomy) degree in 1946 from Sofia University and his PhD in 1974 from Agricultural Academy, dealing with wheat breeding and seed production techniques. He started work in 1946 as an agronomist at the Institute of Scientific and Applied Research in Karnobat. In 1948, he was appointed at the Agricultural Institute of Dobrich to assist Prof. T. SHARKOV, a wheat breeder. He returned to the institute in Karnobat from 1951 until 1962, when two wheat breeding centers were formed at Sadovo and Dobrich. Later, he served as chairman of the wheat breeding program at the Agricultural Experimental Station in Sadovo (from 1962 until his retirement in 1976). Between 1966 and 1969, he was deputy director of Agricultural Experiment Station in Proslav, near Plovdiv. He contributed greatly to the development and release of 12 Bulgarian wheat varieties and received the outstanding research award from the government in 1978. His bread wheat cultivars "Sadovo 1" and "Katya" were the second and first places finishers in 1977 and 1984 international field testing, respectively, organized by the Agronomy Department at the University of Nebraska, Lincoln (United States). Those cultivars are still grown in Bulgaria, Greece, and Turkey as good and productive bread wheats possessing a good balance of agronomic traits.

IVANOV, N. R. (1902–1978): He was a Russian plant scientist and a plant breeder. From 1926, he worked at the Institute of Applied Botany (Saint Petersburg, Russia). He had been the institute's director during the siege of Leningrad in the World War II. From 1967, he was the Scientific Secretary of the Commission on ▶ N. I. VAVILOV's scientific heritage. The commission belonged to the Academy of Sciences of USSR.

JENKINS, M. T. ▶ RHOADES, M. M.

JENNINGS, H. S. (1868–1947): He began his career after graduating from high school in his home town of Tonica (Illinois, United States) as a teacher. In 1889, he received a position as assistant professor of botany and horticulture in the Texas A&M College. During the decade in which he was pursuing study of protozoan behavior, ▶ MENDEL's laws were rediscovered. In a series of fundamental, exhaustive, pioneer papers on *Paramecium*, published between 1908 and 1913, he laid broad and deep foundations for all subsequent genetic work. He showed that heredity and its problems are essentially the same in microorganisms as in plants, and he formulated the "pure line theory" for vegetative reproduction shortly after ▶ JOHANNSEN's comparable theory for sexual reproduction. He also analyzed the phenomenon of assortative mating and pointed out its role in the isolation of races. He was struck by the continued production of hereditarily diverse clones at conjugation, even after many successive inbreedings that he undertook to examine the matter mathematically. As a result, general formulae for the diverse systems of mating were published in a series of papers between 1912 and 1917. He dealt with selection and mutation, multiple factors and multiple alleles, the demise of the unit factor and representative particle interpretation of Mendelism, the inheritance of acquired characters, the interaction of genes and environment in the determination of the phenotype, and the limitations of evolution by loss.

JENSEN, N. F. (1915–): He was born in Hazen (United States) and studied at Cornell University till 1939. In 1946, he returned to Cornell and became assistant professor in the small grains breeding program. During more than 30-year career, he marked major achievements in the breeding of spring barley, winter barley, winter wheat, spring oats, and winter oats. He viewed the breeding process as being composed of three stages. He first divided it into heterozygous and homozygous phases, where heterozygous phase deals with everything up to individual line selection (F5 or F6) and the homozygous phase deals with all subsequent line evaluation and variety release. He further subdivided the heterozygous stage, separating the planning and hybridization from the handling of the early-generation progenies.

JESENKO, F. (1875–1932): He was a research assistant of ▶ E. von TSCHERMAK in Vienna (Austria). He successfully crossed different wheat varieties with, for example, *Triticum dicoccoides*, *Secale cereal*, and *S. montanum*. By crossing "Mold Squarehead" wheat and "Petkus Rye," he produced one of the first perennial wheat–rye hybrids; after its backcrossing to wheat, a fertile perennial wheat plant was obtained. The first report of this extensive work was presented during the 4th International Genetic Conference, Paris (France) in 1911. In 1919, the Faculty of Agriculture and Forestry was established at Zagreb University (Croatia), and F. JESENKO was appointed as lecturer. Later, from 1921, he was professor of botany at Ljubljana University (Slovenia). His later research on interspecific hybridization, particularly studies on F1 plants from crosses between *Triticum aestivum* and *Aegilops geniculata,* were never published due to his incidental death.

JOHANNSEN, W. L. (1859–1927): He was a Danish biologist from Copenhagen who called the phenomenon of dominance and recessiveness "genes." He studied at the University of Copenhagen, where he became a professor of botany. He conducted numerous crossing experiments, particularly with Phaseolus *vulgaris*. He was able to isolate four "pure lines," leading him to the theory about populations and pure lines. He strongly promoted the genetics as science after 1900.

JOHNSON, R. (1935–2002): He was a British plant pathologist and wheat rust expert. His research was a groundbreaking work on durable resistance at the Plant Breeding Institute in Cambridge (United Kingdom), where he started his career in 1964. Following a brief interlude with UNILEVER Co. after the privatization of the Plant Breeding Institute, he continued research at the John Innes Centre in Norwich until his retirement in 1995. Until his death, he continued to serve the scientific community as senior editor of *Plant Pathology* and chairman of the U.K. Cereal Pathogen Virulence Survey.

JOHNSON, V. A. (1921–2001): He was an outstanding American wheat breeder and was born in Newman Grove, Nebraska. He graduated from Albion High School in 1939. He earned his BSc degree (1948) and PhD (1952) from the University of Nebraska in Lincoln (United States). He was employed by the USDA as part of the Agricultural Research Service.

He was with the University of Nebraska Department of Agronomy from 1952 to 1986 as a professor of agronomy and coordinator of the USDA–ARS Hard Red Winter Wheat Research Program. With longtime colleague Dr. J. SCHMIDT, he was co-leader of the internationally recognized Nebraska Wheat Research Team. In more than 30 years of active service, this team developed and released 28 new varieties of hard red winter wheat. Notable varieties released during their tenure included "Scout," "Centurk," and "Brule;" the variety "Scout," released in 1963, was grown on more than 3 million ha, making it the most extensively grown cultivar in the United States at that time. The team also carried out pioneering research on the enhancement of nutritional value of wheat, on selection for yield stability, and on the development of hybrid wheat. He was the principal organizer of five international wheat conferences, sponsored by the US Agency for International Development.

JOHNSTON, R. P. (1939–2001): He was a plant breeder of the Department of Primary Industries, Queensland (Australia). He studied his BSc at the University of Queensland, graduating in 1961, and was immediately employed by the Department of Primary Industries as a linseed breeder. In 1967, he initiated a barley variety testing program with the intention of identifying material suited to malting barley production in the South Queensland cropping region. His PhD thesis "Single Plant Selection for Yield in Barley" was conducted at the University of Adelaide and accepted in 1973. On his return to Queensland, he developed and led a full-scale barley breeding program. The barley varieties released from the program under his guidance include "Grimmett," "Tallon," "Gilbert," and "Lindwall."

JONES, A. N. (1843–): He was born in Cookham, situated on the Thames River, United Kingdom. At the age of 5 years, he came with his family to America and was located in Rochester, New York. Working at Batavia (United States), he began work on crop hybridization in 1869 at LeRoy, New York. By crossing, he produced many promising hybrids of the potato and strawberry. Among these were the "Early Gem," "Genesee County King," and "Tioga" potato, and the "Laural Leaf" strawberry. The "Amber Cream" sweet corn was originated in these early days, being cataloged first in 1879. In 1878, JONES began his work in cross-fertilization of wheat, and, in 1886, he introduced to the trade his first named wheat hybrid "Golden Cross," a cross between "Mediterranean" and "Clawson." This was followed in 1888 by two other varieties "New Early Red Clawson" (a cross of "Golden Cross" and "Clawson") and "Jones' Square Head" (a cross of "Landreth" and an unnamed hybrid), known in Canada as "Harvest Queen" after permission of JONES. In 1889, the first of JONES' hard gluten cultivars "Jones' Winter Fife" was sent out. This variety resulted from composite crossing of "No. 87" and "Mediterranean." In 1919, it was estimated by the USDA that nearly 250,000 ha of this variety were grown in the United States, principally in Washington, Illinois, Missouri, Indiana, Idaho, and Montana. In 1889, he wrote as follows: "My most successful cross-breeding has been from combination (composite) crossing, as in crossing Mediterranean 'Longberry' upon American wheat, progeny of which is crossed with Russian 'Velvet'; a smooth chaffed wheat is sometimes used, progeny of which is again crossed with American wheat. This cross gives a strong healthy growth, deep root, thick walled stocky straw, and grain of a fine milling quality in a compact head." In addition to his work on wheat, JONES did a great deal of work with beans and introduced many important varieties, such as "Jones' Ivory Pod Wax" in 1881, "Lemon Pod Wax" in 1881, "Jones' Round Pod Wax" in 1898, "Golden Crown White Seed Stringless Wax" in 1899, "Garden Pride Stringless Green Pod" in 1902, "Green Pod Stringless" in 1902, and "Jones' Marrow Pea" in 1909. The "Jones' Ivory Pod Wax" was a parent of many of the later cultivars.

JONES, D. (1890–1963): Donald JONES was a US maize geneticist and practical maize breeder at the Connecticut Agricultural Experiment Station, New Haven. Beginning at the station in 1914, he made high-yielding hybrid maize practical by his invention of the double-cross hybrid. In JONES' method, four inbred maize lines are used. The seeds from two initial crosses are used to grow up parental hybrids for the production fields; the production fields yield seeds in sufficient quantity to make the scheme practical. Until JONES invented the double-cross method, the yield from the parent lines (inbreds) was insufficient to allow practical production of hybrid maize seed. JONES' work received significant public attention and was used to make the first commercial hybrid maize in the 1920s. He was the sole geneticist at the Connecticut Station from 1915 until 1921, when Paul MANGELSDORF became his assistant. JONES was the president of the Genetics Society of America in 1935.

KEDAR, N. (1920–2015): Nachum KEDAR was a plant breeder and geneticist at the Hebrew University (Israel). He grew up and joined the Jewish Zionist Youth movement (Young Maccabees). In 1938, after the Nazi invasion of Austria, he was sent by the leaders of the movement to Denmark to practice farming and agriculture in preparation for his planned immigration to what was then a British mandate, Palestine. When Denmark became occupied by the Nazis in 1943, he and nine of his friends decided to find shelter in Sweden. NACHUM enrolled at Lund University, where he obtained his BSc degree, and later at Uppsala University, where he got his MSc degree in 1947. He also started working at the Swedish National Station for Plant Diseases in Stockholm. In 1950, he fulfilled his dreams and immigrated to Israel. A year later, he joined the Department of Plant Genetics at the Weizmann Institute of Science in Rehovot. He then assisted Oved SHIFRISS in studying castor bean plants (*Ricinus communis*), including breeding, production, evaluation of hybrids, and field technologies. In 1954, he started his studies toward a PhD degree at the Faculty of Agriculture of the Hebrew University of Jerusalem, in Rehovot, researching potato late blight and the role of polyphenol oxidases in the response of plants to infection. While carrying out this research, he was appointed as a research assistant and received his PhD in 1958. When hired by the Faculty of Agriculture, he initiated research on *Fusarium* wilt of tomatoes and applied his knowledge and results in developing the first *Fusarium* wilt-resistant tomato "Rehovot 13" for the Israeli market. From 1959 until his retirement in 1988, NACHUM served as senior teacher (full professor) on vegetable physiology, genetics, and breeding. In this capacity, he educated and trained thousands of students and supervised Masters and PhD theses of 20–30 students. The most significant work that he initiated was in the late 1960s to early 1970s, when reports on the discovery of random mutations *rin* and *nor* were published. He was intrigued by the potential of these mutations and was not discouraged by the recessive nature of these genes. In the early 1970s, numerous crosses were made in order to produce red-fruit tomatoes with considerable longer shelf life (Section 4.2). he received numerous awards, including the Rothschild Prize, the Israel Prize, the Kaye Prize, and the ASHS Award Best Vegetable Paper Award.

KEIM, F. ▶ BEADLE, G. W.

KAPPERT, H. (1890–1976): He was a German botanist, and between 1914 and 1920, he was an assistant of ▶ C. CORRENS in Berlin (Germany). As university professor, scientist, and author of numerous papers and books, he heavily contributed to the breeding research and plant breeding in Germany.

KARPECHENKO, G. D. (1899–1942): Born in Velsk (Russia), he was one of the leading geneticists in Russia during the 1930s. He was Head of the Department of Genetics at the Institute of Plant Industry (Saint Petersburg, Russia; before 1930, Russian Institute of Applied Botany and New Crops) and the Chair of Plant Genetics at Leningrad State University. In 1941, by political reasons, he was arrested, together with several leading scientists, following the arrest of ▶ N. I. Vavilov, director of VIR (Saint Petersburg, Russia). Later, he was executed by shooting. He was the first to produce polyploid hybrids between *Raphanus sativus* and *Brassica oleracea*.

KEMENESY, E. (1891–1981): He studied at the Agricultural Academy, Debrecen (Hungary), under Prof. K. KERPELY. As soil scientist, he emphasized the soil fertility for optimal plant production. He organized the Agricultural Research Institute in Keszthely and educated many of famous Hungarian agronomists and plant breeders.

KERPELY, K. ▶ KEMENESY, E.

KIHARA, H. (1904–1986): He did his graduation from Hokkaido University (Japan) in 1918. At about the same time, his interest became fixed, when T. SAKAMURA and K. SAX simultaneously discovered the ploidy evolution of wheat. As teacher in genetics at the Botany Department at the Kyoto University (Japan) since 1920, he developed his wheat research. After 2 years of study at the "Kaiser Wilhelm Institut für Biologie" in Berlin (Germany) under ▶ C. CORRENS, he returned to Kyoto. He revealed step by step the secrets of wheat evolution from genome to plasmon interrelationships. He was one of the discoverers of *Aegilops squarrosa* as the donor to the third wheat (D) genome. KIHARA established the cytogenetics of interploid hybrids, clarifying the meiotic chromosome behavior as well as the chromosome number and genome constitution of their progeny, based on which KIHARA formulated the concept of genome. He proposed the methodology for genome analysis and determined the genome constitution of all *Triticum* and *Aegilops* species. Later, Ohta re-evaluated the genome relationships among the diploid species, using the B chromosomes of *Ae. mutica*. After completing the genome analyses, KIHARA's interest was shifted to the genome–plasmon interaction that led to the discovery of cytoplasmic male sterility in wheat. He classified their plasmons into 17 major types and 5 subtypes and determined the maternal and paternal lineages of all polyploid species.

KISS, A. (1916–2001): He was born in Budapest, Hungary, and graduated from the Horticultural High School. At age of 18 years, he moved to the Agriculture University Mosonmagyarovar (West Hungary) to study agricultural sciences, where he earned a degree in agronomy. After working at a private farm (1936–1939), he began his research career at the Plant Breeding Institute, Mosonmagyarovar in 1941. From 1950 to 1957, he worked at the Plant Breeding Institute of the Hungarian Academy of Sciences. His interest was focused on the genetics and breeding of pea, watermelon, and small grain cereals, particularly wheat-rye hybrids. Working with G. REDEI, who emigrated to the United States in 1956, KISS observed a strong genotypic control of crossability of various wheat and rye genotypes. Their fundamental work was successful in developing primary octoploid triticale stocks. KISS discovered early that the hexaploid triticale is more useful in agriculture than octoploids. By 1962, he developed "No. 30," the first secondary hexaploid triticale, from a cross of "F481" (*T. aestivum* × rye // *T. turgidum* × rye). From his nursery, the world's first released triticale cultivars for commercial production appeared in 1968 as "T-No. 57" and "T-No. 64." He established a modern triticale breeding. Between 1967 and 1970, KISS developed more than 10 hexaploid lines and sent them to the Cereal Research Institute, Szeged, for further tests on various locations. In 1970, his semi-dwarf cultivar "Bokolo" was registered in Germany. KISS received his PhD on the topic of "Microevolution of wheat–rye hybrids" in 1964. In 1972, he earned his DSc degree from the Hungarian Academy of Sciences for his work entitled *Genetics and Breeding of Triticale*; he published more 138 scientific papers.

KNIGHT, T. A. (1759–1838): From 1811 to 1838, he was the president of Horticultural Society of London (United Kingdom)—as botanist, he displayed experience and scientific instinct. He was convinced that yield increase in plants and animals can be achieved by crossbreeding. In 1779, he emphasized the practical aspects of hybrids in grape, apple, pear, and plums, particularly to improve winter hardiness. His pea crosses (1799–1823) are of genetic interest. Just like ▶ G. MENDEL, he recognized the advantage of pea as a research and breeding subject. He often noted luxuriance of hybrids and the advantage of outcrossing to produce new forms. He was the first to describe the dominance of the gray seed color over the white one; however, he did not calculate the relation of different segregating fractions, as MENDEL did.

KNOWLES, P. F. (1916–1990): Paulden (Paul) KNOWLES was born in Saskatchewan, Canada, where he attended the University of Saskatchewan and received his BSc Agriculture and MSc degrees; he then went to the University of California, Davis, completing his PhD research in genetics in 1943. After only a year in the Department of Plant Science at the University of Alberta in 1946, he returned to the University of California, Davis, as assistant professor of agronomy, and it was here that he devoted the next 35 years until his retirement in 1982. After considerable work on flax and other oilseeds, KNOWLES turned his attention to safflower. His work stimulated commercial production by 1950, and the crop has continued to the present time. He made extensive plant collection expeditions in 1958 and in 1964–1965 through South Asia, the Middle East, and North Africa, in order to increase his safflower genepool (e.g., for oleic, linoleic, and other fatty acids). This led to the development of the safflower cultivar "UC-1" with high oleic acid, resulting in a vegetable oil similar in characteristics to olive oil. "UC-1" also became the basic germplasm for many breeding programs; moreover, he and his family gathered most of the germplasm of wild and cultivated species, now in the USDA world collection. KNOWLES and his associates authored 117 scientific papers, including his co-authorship on the successful textbook *Introduction to Plant Breeding*. From 1970 to 1975, he was chairperson of the Department of Agronomy and Range Science and, for many years, advisor to graduate students in both agronomy and genetics. In 1972, he was elected as a fellow in the American Society of Agronomy and the Crop Science Society of America. The California Chapter of the American Society of Agronomy gave him an Award of Honor, and the University of Saskatchewan recognized him as a Distinguished Graduate in Agriculture.

KOCH, K. H. E. (1809–1879): He was a German medical doctor and botanist who named and studied many of the plants known as cultivars. He brought great clarity and unity to the nomenclature of plants, cultivated and not cultivated.

KOHLI, S. P. ▶ ANDERSON, R. G.

KOL, A. K. ▶ LYSENKO, T. D.

KÖLREUTER, J. G. (1733–1806): Between 1760 and 1766, the German carried out the first series of systematic experiments in plant hybridization by using tobacco (*Nicotiana paniculata* × *N. rustica*). He demonstrated that the hybrid offspring generally resembled the parent as closely as the seed parent; thus, for the first time, he found that the pollen grain has an important part in determining the characters of the offspring; this was a novel idea, which was disbelieved by his contemporaries. He also observed accurately the different ways in which the pollen can be naturally conveyed to the stigma of the flower and discovered the function of nectar and the role of wind in flower pollination. He also observed that hybrid plants often exceed their parents in vigor of growth (hybrid vigor; now called "heterosis").

KÖNNECKE, G. (1908–1992): He was a German agronomist working as a professor at the University of Halle/S. (Germany). He was concerned about the utilization of crop rotations for increasing the agricultural productivity and choosing right crop varieties for suitable environments.

KORIC, M. (1894–1977): He was head of the Plant Breeding Station at Agricultural School in Krizevci (Serbia) from 1922 to 1929. By applying combination breeding and crossing domestic wheats with imported varieties, which carried genes for certain desirable traits (lodging resistance, earliness, rust resistance, and bread-making quality), he achieved important progress in local wheat breeding. He introduced earliness and shorter straw by introgression of Italian wheats carrying genes of the Japanese wheat "Akakomugi;" quality improvement was achieved by crossing with Canadian and Indian ("Calcutta Red") spring wheats. Among the best known and most spread new wheats were the cultivars "K6" and "K9." In Osijek (1929–1948), he continued his work and developed a number of additional varieties, such as "Koric's Awnless," "Osjecka Sisulja," and "U-1." Very soon, they spread over 50,000 ha and became leading cultivars in wheat production in the western parts of former Yugoslavia till the early sixties.

KOSAMBI, D. D. (1907–1966): Damodar Dharmanada KOSAMBI was not a geneticist by training and profession, but a mathematician. Actually, he was also a statistician, historian,

FIGURE 9.5   KOSAMBI, D. D. (1907–1966), honored on Indian stamp of 2008, including his invented statistic formula. (Courtesy of Indian Post, New Delhi, India.)

linguist, and writer—a multifaceted scholar. His famous mapping function was published in 1944. KOSAMBI's mapping function estimates the recombination fraction between two loci as a function of the map distance between the loci, by allowing some interference, as $c = (e^{4m} - 1)/2(e^{4m} + 1)$. The estimate of the map distance between two loci can be obtained from $m = \ln [(1 + 2c)/(1 - 2c)]/4$. KOSAMBI was born in Goa (India). After his early schooling, he moved to Cambridge, Massachusetts (United States) and studied grammar and Latin. In 1924, he joined Harvard University, where he studied mathematics. He discontinued studies for a brief period and returned to India in 1926, where he was awarded with Bachelor of Arts degree. After marriage in 1931, he joined Aligarh Muslim University as professor of mathematics, two years after he joined Fergusson College in Pune. It was during this period that his famous paper on mapping function was published in 1944. He published 127 papers and authored 9 books; however, many of his publications went unnoticed by Indians. No students of genetics were told that KOSAMBI was an Indian scientist. In 1964, KOSAMBI was appointed as a scientist emeritus of the Council of Scientific and Industrial Research (CSIR) in Pune, India. Posthumously, he was decorated with the Hari Om Ashram Award by the government of India's University Grant Commission in 1980 (Figure 9.5).

KOSTOV, D. (1897–1949): He was a Bulgarian geneticist and cytogeneticist. He studied agriculture and obtained PhD in agriculture at the University of Halle/S. (Germany). Under the guidance of ▶ E. M. EAST at Harvard University, he studied the ontogeny, genetics and cytogenetics of *Triticum* and *Helianthus* hybrids, as well as tumors and other malformation on certain *Nicotiana* hybrids. Later, in ▶ VAVILOV's laboratory of genetics, he continued the research work on polyploidy of crop plants.

KRISTEV, K. K. (1912–1986): He was an outstanding Bulgarian plant pathologist born in Khaskovo. He was educated at Sofia State Agro-Forestry Institute and trained as young agronomist in 1935 at the newly established Institute of Plant Protection (Sapareva Banya). Afterward, he served for about 11 years at the Department of Plant Pathology in the University of Sofia. He earned a PhD degree on smut and bunt diseases of wheat in 1943. Till 1976, he remained at different positions on this site; thereafter, he was appointed as head of the International Wheat Immunity Laboratory at the Dobroudja Wheat and Sunflower Institute, General Toshevo/Varna (1976–1979). Besides his teaching and organizing activities, his scientific activity was dedicated to research on smut and rust diseases; on the mechanism of enzyme activity, virology, and toxicology; and on several new diseases of crops cultivated in Bulgaria.

KROLOW, K.-D. (1926–2008): He was professor of genetics and breeding research at the Technical University Berlin (Germany) until 1989. He studied agriculture, promoted ▶ Prof. H. KAPPERT in 1957, and habilitated in 1969 under Prof. W. HOFFMANN (TU Berlin). He spent most of his time in the development of wheat–rye hybrids by sexual combination. Besides the predominant octoploid and hexaploid triticale lines, he was the first to produce a viable tetraploid triticale by subsequent introgression of D genome chromosomes of hexaploid wheat.

KRONSTAD, W. E. (1932–2000): He was born in Bellingham (United States). Following active military service from 1952 to 1954, he attended Washington State University (United States), from where he received a BSc degree in agronomy in 1957. In 1959, he was awarded an MSc degree in plant breeding and genetics from the same institution. He then joined the USDA wheat breeding program at Washington State University as a research assistant with Dr. O. A. VOGEL. From 1959 to 1963, KRONSTAD served as an instructor in the Farm Crops Department at Oregon State University (United States) and received his PhD degree in 1963. He remained at Oregon State University and was appointed project leader for cereal breeding and genetics in 1963. He continued to serve in this role, and many others, until his retirement in 1998.

KUCKUCK, H. (1903–1992): He was a German plant breeder born in Berlin. From 1925, he studied at the "Landwirtschaftliche Hochschule," Berlin, and received his PhD in 1929. His professional career started as assistant at the "Kaiser Wilhelm Institut für Züchtungsforschung" in Müncheberg. For political reasons, he was released in 1936. In 1946, he was appointed as head of the Chair of Plant Breeding at the University of Halle/S. (Germany) and returned to Müncheberg in 1948. He moved as a guest researcher to Sweden in 1950. From 1952 to 1954, he worked as a breeding expert and consultant in Iran on behalf of the FAO. In 1954, he received a call from the "Technische Hochschule," Hannover (Germany) and became full professor at the Institute of Horticultural Breeding, which he renamed as Institute of Applied Genetics. He retired from this institute in 1969. He was author of many scientific papers and several outstanding textbooks on plant breeding.

KÜHN, J. (1825–1910): He was a German agronomist and university teacher (Figure 9.6). He was the founder and organizer of the first university study of agricultural sciences in Germany. He was born in Pulsnitz (Germany) and studied at the Polytechnic School Dresden before he went as an agricultural volunteer to a manorial estate to practice agriculture. From 1848 to 1855, he became director of the estate Gross-Krausche/Bunzlau. There, he studied crop plant diseases by advanced techniques (microscopy) and published several papers. In 1855, he became a student at the Agricultural School in Bonn-Poppelsdorf. He received his PhD from the University of Leipzig in 1857 (thesis: "About the Smut Diseases in Cereals"). In 1858, he published the book *The Diseases of Crop Plants, Origin and Protection*, famous as textbook for many years. In 1862, KÜHN became full professor of agriculture at the University of Halle/S. (Germany). Soon, later in 1863, he received the governmental permission for the establishment of an independent Institute of Agriculture, including lecture halls, experimental fields, and laboratories, which was developed over 40 years as the most important place of agricultural research and teaching in Germany. In 1878, KÜHN began a permanent trial with rye in the experiment station of 1868 (founded by him just after the oldest in Rothamsted, United Kingdom, in 1842) that still continues until recent time ("Ewiger Roggenbau"). In 1889, he also founded the first experimental station for nematode removal in order to prevent the so-called "beet exhausting" of soils. He initiated ▶ K. von RÜMKER's basic studies on cereal breeding. His most famous students were ▶ E. von TSCHGERMAK-SEYSENEGG, ▶ F. von LOCHOW, and ▶ W. RIMPAU.

FIGURE 9.6 Julius KÜHN (1825–1910) in his office. (Courtesy of Agricultural Faculty, Martin Luther University Halle-Wittenberg. With Permission.)

KULPA, W. (1923–1984): He was born in Bialobrzegi (Poland). From 1945 to 1950, he studied at the University of Lublin. He started his career as seed scientist at the same university. In 1954, he became head of the Chair of Plant Breeding and Seed Science at the University of Lublin. He obtained his PhD degree in 1959 ("Biology and Germination of *Adonis vernalis* L."). After a post-doctoral study at the "Zentralinstitut für Kulturpflanzenforschung," Gatersleben (Germany), he devoted himself to the studies of systematics of *Linum usitatissimum* and later of *Veronica* species. After 22 years of teaching and scientific work at the University of Lublin, he was appointed as Head of the Department of Plant Collections at the "Institut Hodowli i Kalimatyzacji Roslin" in Radzikow (Poland). In that position, he became the main organizer of crop plant collections in Poland. He successfully developed the scientific exchange with EUCARPIA, FAO, and IBPGR. He organized and participated in several expeditions within Poland, the Tatra Mountain region, Bieszczaden, Caucasian region, and Turkey in order to collect wild species of crop plants and to safeguard plant genetic resources.

KHUSH, G. S. (1942–): Gurdev S. KHUSH began his research with rye and then continued with tomato, and for the last quarter of a century, he has focused on rice. He made remarkable contributions to the application of new (molecular) techniques in rice breeding. He established complete series of primary trisomics of rice, utilized for the first time, for associating linkage groups with cytologically identifiable chromosomes. From the progenies of primary trisomics, he selected secondary and telotrisomics,, used for determining the orientation of linkage groups, positions of centromeres, and arm location of genes and molecular markers. KHUSH was one of the global leaders on crop breeding and a major brain behind the development of productive rice varieties and the "Green Revolution" in plant breeding. Along with mentor Henry BEACHELL, KUSH received the 1996 "World Food Prize" for his achievements in enlarging and improving the global supply of rice during the time of exponential population growth. He was born in Rurkee village, Punjab (India). As son of a farmer, he finished his BSc degree from Punjab Agriculture University and went to the University of California, Davis (United States), to do his PhD at the age of 25 years. In 1967,

KHUSH joined the International Rice Research Institute (IRRI), Manila (Philippines), and was there till 2000. He joined the IRRI after postdoctoral studies on tomato breeding. He has been able to produce rice varieties resistant to several major insect pests, such as brown plant hopper (*Nilaparvata lugens*), green leafhopper (*Nephotettix virescens*), white backed plant hopper (*Sogatella furcifera*), and gall midge (*Orseolia oryzae*). In the beginning, during 35 years, he and his team introduced varieties such as "IR8," "IR36," "IR64," and "IR72." Those varieties and their progenies are planted in over 70% of the world's rice fields. He developed the so-called "miracle rice," "IR36," by using "IR8" as donor parent and crossed it with 13 other parental varieties from six countries. "IR36" is a semi-dwarf variety that proved highly resistant to a number of the major insect pests and diseases (Section 3.2.2). It matures rapidly in about 105 days as compared with 130 days for "IR8" and 150–170 days for traditional types. The combination of suitable characteristics made "IR36" as one of the most widely planted food crop varieties of the world ever known. In 1994, he announced a new type of "super rice," which has the potential to increase yields by 25%. KHUSH's latest work dealt with the so-called "new plant type" for irrigated rice fields.

LAUBSCHER, F. X. (1906–): He was born in Vredenburg (South Africa) and received his education there. He qualified for his BSc degree in 1928, MSc in 1942, and DSc in 1945. He was appointed to the Chair of Genetics at his *alma mater,* Stellenbosch University (South Africa), in 1950. From 1936 to 1949, while being stationed at the Potchefstroom College of Agriculture, he was involved in wheat breeding and released the cultivars "Spitskop," "Goudveld," "Magaliesburg," and "Flameks," a wheat with exceptional baking attributes. He published a monograph entitled *A Genetic Study of Sorghum Relationships*. In 1949, he was appointed as technical advisor to the Maize Board, in which, together with scientists from the United States, he was responsible for the successful initiation of the national maize hybrid-breeding program. His versatility can probably best be exemplified by the fact that he was requested to study sheep breeding in Australia, and this visit resulted in the founding of the Sheep Performance Testing Center in Middelburg (South Africa).

LEDERBERG, J. ▶ BEADLE, G. W.

LEEUWENHOEK, A. van (1632–1723): A Dutch microscopist, called "Father of Scientific Microscopy," was a cloth merchant and wine taster by trade. His works were published under title *Secrets of Nature* (1668), where he refers to "animalcules" (little animals). LEEUWENHOEK extended ▶ MALPIGHI's demonstration of blood capillaries, and 6 years later, he gave the first accurate description of red blood corpuscles, completing W. HARVEY's (1578–1657) discovery of the circulation of the blood in 1628. He discovered the effect of aphids on plant life and showed that they reproduced parthenogenically. He described different stem structures in monocots and dicots and observed polyembryony in citrus seed. He was responsible for the first representation of bacteria by a drawing in 1683 and constructed over 400 microscopes and bequeathed 26 to the Royal Society of London.

LEIN, A. (1912–1977): He was a distinguished German cereal breeder. He started his career at the Cytogenetic Department of the Institute of Plant Cultivation and Plant Breeding of the University of Halle/S. (Germany). In 1942, he pursued his PhD with the subject "The genetic basis of the ability of crossbreeding between wheat and rye." Since 1944, he evaluated wheats from collections of the German Hindu Kush expedition (1935–1936). In 1947, he continued his scientific work as the Head of Department of Self-fertilization at the "Max Planck Institut für Züchtungsforschung" in Voldagsen (Germany). In 1949, he became head breeder of the company ▶ Ferdinand HEINE in Schnega (Germany), and, in 1969, he was responsible as head breeder for barley and wheat at the F. von Lochow-Petkus Ltd. (Germany). Among his numerous successful varieties are the winter wheats "Kranich" and "Kormoran," the spring wheats "Kolibri" and "Selpek," and the spring barley "Oriol."

LELLEY, J. (1909–2003): Born in Nyitra (Hungary), he was one of the outstanding Hungarian wheat breeders (cf. Figure 9.7). He studied in Budapest. In 1931, he graduated

FIGURE 9.7   Janos LELLEY (1909–2003), retired in late age. (Courtesy of T. LELLEY, Tulln. With Permission.)

from the Agricultural High School in Mosonmagyarovar. After a short stay in Bratislava (Slovakia) in an agricultural office, he received a postgraduate training in plant breeding after 1946. He moved to the Plant Breeding Station in Kompolt. There, he developed an extensive wheat breeding program and conducted basic research on the methodology of breeding and on resistance to different stresses. He also organized a special network of research stations in the mountains for screening wheat for frost resistance. He achieved improvements of leaf rust and drought tolerance of wheat and worked out an effective method for artificial rust infection. He bred the spring and winter wheat "K169." In 1962, he started to organize a new wheat breeding center in Kiszombor. There, he improved the cultivars "Kiszombori 1" and "GK Tiszataj;" the latter was one of the best quality wheats in Europe, with a protein content of 16%–17%. In 1954, he wrote a handbook on wheat breeding, *Wheat Breeding—Theory and Practice*, for Hungarian students and breeders (Figure 9.7).

LEVAN, A. ▶ MÜNTZING, A.

LIEBIG, J. von ▶ DELBRÜCK, M.

LINDLEY, J. (1799–1865): He was one of the most remarkable horticultural scientists of the nineteenth century in Britain. His book *The Theory of Horticulture* (1840) is a classic and is still considered "one of the best books ever written on the physiological principles of horticulture." His formal education lasted only through age of 16 years, but his ability to work hard enabled him to become one of the most productive plant scientists of his era. LINDLEY had several careers, most of them simultaneously. He was the "mainspring" of the London Horticultural Society for 40 years, professor of botany at the University of London for 33 years, editor of the *Botanical Register* for 18 years, editor of the *Gardener's Chronicle* for 25 years, and professor of botany and director of the Physic Garden. He played a major role in saving Kew Gardens from being disbanded by the government as a budget-cutting measure. His pioneering works on orchid taxonomy earned him the title of "Father of Modern Orchidology," and he authored books on medical uses of plants, general botany, popular horticulture, and fossil plants. His botanical texts helped establish the natural system of plant classification as the system of choice. He named innumerable new species brought back by plant explorers and started the practice of ending plant family names in "-aceae." As the editor for the *Gardener's Chronicle*, he worked to improve the state of horticultural science for 25 years. His book *Theory of Horticulture* (1831) had a major impact on horticulture; it was translated into German, Dutch, and Russian and was published in an American edition (1841). LINDLEY received recognition from the University of München (Germany), which awarded him an honorary PhD in 1832; from the Royal Society, which awarded him their Royal Medal in 1857; from the Royal Horticultural Society, which named their Lindley Medal and Lindley Library in his honor; and from numerous taxonomists, who named six genera and numerous species in his honor.

LINNAEUS, C. (Carl von LINNÉ) (1707–1778): He was a Swedish botanist and physician. He began his education as a theology student, but at the age of 23 years, he became curator of the Gardens of the University of Lund (Sweden). From 1732 to 1738, he traveled through Lapland, Holland, England, and France and returned to Stockholm, where he practiced medicine. In 1741, he became Head of Botany at the University of Uppsala, where he remained until his death. His botanical contributions have earned him the title of "Father of Taxonomy;" LINNAEUS established groups of organisms, large and small, that depended on structural or morphological similarities and differences. The basic taxonomic criteria for grouping plants were based on the morphology of their reproductive parts—the plant organs least likely to be influenced by environmental conditions. His "sexual system" of classification used the number of stamens and carpels (styles) as a method for grouping plants (however, this was an artificial system that is no longer used). He described and assigned names to more than 1300 different plants. He is credited with the establishment of the binomial nomenclature, and with having replaced the long-winded and confused descriptions of the herbalist with clean and succinct descriptions, his works include *Systema Naturae* (1735), *Fundamenta Botanica* (1736), *Genera Plantarum* (1737), *Classes Plantarum* (1738), and *Philosophia Botanica* (1751).

LINSKENS, H. (1921–2007): Hans Ferdinand LINSKENS was a German botanist and geneticist and was born in Lahr/Baden. He was educated in Cologne, Bonn, Berlin, and Zurich, and in 1949, he received his doctorate from the University of Cologne. From 1957 to 1986, he was professor of botany at the Radboud University Nijmegen (the Netherlands). Main fields of his research were sexual plant reproduction, pollen physiology, and self-incompatibility. He was one of the first scientists worldwide to apply molecular biology to studies on reproductive biology in plants. LINSKENS authored and co-authored over 500 publications. He became the editor-in-chief of the journals *Theoretical and Applied Genetics* (1977–1987) and *Sexual Plant Reproduction*. Moreover, he was an influential editor of handbooks on plant breeding and related subject (BECKER 2007). LINSKENS was an elected member of the Deutsche Akademie der Naturforscher Leopoldina (Halle/S., Germany), Linnean Society of London (United Kingdom), Koninklijke Nederlandse Akademie van Wetenschappen (the Netherlands) in 1978 and the Academie Royale des Sciences de Belgique (Belgium).

LOCHOW, F. von (3rd of the LOCHOW dynasty, 1849–1924): He was a German agronomist on his estate at Petkus (Germany). When he selected the first seeds from the Austrian landrace "Probsteier Winterroggen" in 1880, he recognized the big variability between the single-seed progenies. This inspired him to start breeding with rye. The subsequent elite breeding improved the winter hardiness, fertility, and uniformity of seeds. In 1925, ▶ RÜMKER described LOCHOW's method as subsequent pedigree breeding. LOCHOW independently established the method from ▶ VILMORIN in France and ▶ FRUWIRTH in Germany (1908). Later it became known as "German selection method." First yield testing of the German Agricultural Society in 1891 showed that "Petkuser Winterroggen" (Petkus winter rye) outyielded standard rye by 8%–11%. From that time, "Petkus rye" (1891–1960) was distributed around the world and served as basic genepool for many other rye varieties. In 1911, a spring type of this rye was also released and multiplied until 1960 as "Petkuser Sommerroggen."

LUKYANENKO, P. P. (1901–1973): He was a very successful Russian wheat breeder working in Krasnodar (Russia). His name became known throughout the world in connection with outstanding varieties. The worldwide successful variety "Bezostaya 1" became famous in every wheat-producing country of the world. His varieties were grown on more than 10 million ha. New varieties followed, such as "Avrora," "Kavkaz," "Bezostaya 2," "Skorospelka," and "Rannaya 12;" the varieties "Avrora" and "Kavkaz" carried the so-called 1RS.1BL translocations that were transferred around the world; they are still used in many wheat improvement programs (Section 3.7.3).

LURIA, S. ▶ DELBRÜCK, M.

LYSENKO, T. D. (1898–1976): He was a Russian botanist and horticulturist. He was born in Karlowka (Ukraine). After study at the Horticultural School Belozersk (1921) and the Agricultural Institute of Kiev (1925), he was appointed at the breeding station Gandže (Azerbaidshan) from 1925 to 1929. LYSENKO was a young agronomist when he came in connection with an experiment in the winter planting of peas to precede the cotton crop in the Transcaucasia region. He subsequently became famous for the discovery of "vernalization," an agricultural technique that allowed winter crops to be obtained from summer planting by soaking and chilling the germinated seed for a determinate period of time. He was the first to use the term "vernalization" but not in fact the first to discover this technique. LYSENKO ignored previous studies of thermal factors in plant development. After being overshadowed by MAKSIMOV, his critic, at the All-union Congress of Genetics, Selection, Plant, and Animal Breeding, held in Leningrad (Russia) in 1929, LYSENKO organized a boisterous campaign around vernalization and made extravagant claims based on a modest experiment carried out by his peasant father. The Ukrainian Commission of Agriculture, in the hope of raising productivity after 2 years of famine, ordered massive use of the vernalization technique. LYSENKO was moved to a newly created Department for Vernalization at the All-Union Institute of Genetics and Plant Breeding in Odessa. Therem he began to publish the journal *Jarovizatsiya* (*Vernalization*), in which he disseminated his ideas on a wide scale and created a mass movement around vernalization. The next stage in LYSENKO's career came when, from 1931 to 1934, he began to advance a theory to explain his technique. According to the idea of the phasic development of plants, a plant underwent various stages of development, and during each stage, its environmental requirements differed sharply. The conclusion LYSENKO drew from this was that knowledge of the different phases of development opened the way for human direction of this development through control of the environment. It was a very vague theory, never to be spelt out fully, but it provided the link in the evolution of LYSENKO's platform from a simple agricultural technique to a full-scale biological theory; the underlying theme was the plasticity of the life cycle. LYSENKO came to believe that the crucial factor in determining the length of the vegetation period in a plant was not its genetic constitution but its interaction with its environment. His theory developed in a pragmatic and intuitive way as a rationalization of agronomic practice and a reflection of the ideological environment surrounding it and was not pursued according to rigorous scientific methods. Contemporaneous with LYSENKO's vernalization movement was a growing interest in the work of ▶ I. V. MICHURIN (1855–1935), the last in the line of an impoverished aristocratic family in central Russia, who cultivated fruit trees and began experimenting with grafting and hybridization. MICHURIN worked on the assumption that the environment exercised a crucial influence on the heredity of organisms, and he queried the relevance of MENDEL's so-called "peas laws" to fruit trees. Michurin's name was soon to be seized upon by LYSENKO to designate a whole new theory of biology in opposition to classical genetics, even though MICHURIN himself had no such theoretical pretensions, in 1931 and 1932, a number of geneticists were branded as "menshevising idealists" and lost their positions at the Communist Academy of Sciences. A particularly vicious article that appeared in the newspaper "Ekonomicheskaya zhizn" in 1931 was directed against Academician ▶ N. I. VAVILOV, founder and president of the Lenin Academy of Agricultural Sciences, director of its All-Union Institute of Plant Breeding, and director of the Institute of Genetics of the Academy of Sciences. VAVILOV was an internationally eminent plant geneticist and an ardent advocate of the unity of science and socialism. The article was written by a subordinate of VAVILOV's, A. K. KOL, who accused VAVILOV of a reactionary separation of theory and practice and advised him to stop collecting exotica and to concentrate on plants that could be introduced directly into farm production. Unrealizable goals were imposed on VAVILOV's All-Union Institute of Plant Breeding in 1931, and in 1934, he was called in by the Council of Peoples Commissars to account for the "separation between theory and practice" in the Lenin Academy of the Agricultural Sciences. LYSENKO was very much a part of this campaign, stirring up a negative attitude to basic research and virulently demanding immediate practical results. In 1940, VAVILOV was arrested, and LYSENKO replaced him as director of the Institute of Genetics of the Academy of Sciences. He believed in Lamarckianism, and by his positions (as president of Agricultural Academy, 1938–1956, and director of the Institute of Genetics of the Academy of Sciences, 1961–1962), he negatively influenced agricultural policy in the former USSR.

MACKEY, J. ▶ NILSSON-EHLE, H.

MALINOWSKI, E. ▶ BOJANOWSKI, J. S.

MALPIGHI, M. (1628–1694): He was an Italian physician, an anatomist, a physiologist, and a pioneer microscopist. He graduated in medicine in 1653, became lecturer in 1656, and was appointed to the Chair of Theoretical Medicine at the University of Pisa (Italy). He became the personal physician to Pope INNOCENT XII. His major contribution was *Anatome Plantarum* (1675). He was one of the first to utilize the microscope in the study of animal and vegetable structures and is considered the founder of microscopic anatomy. He applied himself to vegetable histology and became acquainted with spiral vessels of plants in 1662. He made the important discovery that the layers of tissues in leaves and young shoots are continuous with those of the main stem. He distinguished fibers, tubes, and other constituents of wood and was the first to understand the food functions of leaves. He observed stomata in leaves and nodules on legume roots and realized that the ovule developed into a seed and the carpel into a fruit or a portion thereof.

MARAIS, G. F. ▶ PIENAAR, R. V.

MÄRKER, M. (1842–1901): He was a German agronomist promoting plant breeding at the Agrochemical Research Station in Halle/S. (Germany). He introduced a system for testing crop varieties for defined environments.

MARSCHNER, H. (1929–1996): He was born in Zuckmantel (Czech Republic). He studied agriculture and chemistry at the University of Jena (Germany), obtained a PhD in agricultural chemistry in 1957, and then joined the Institute of Crop Plant Research, Gatersleben (Germany). During these years, he developed interest in modern techniques for studying plant nutrient uptake. In 1966, he became professor of plant nutrition at the Technical University of Berlin (Germany), and since 1977, he was director of the Institute of Plant Nutrition at the University of Hohenheim (Germany). In the beginning of his career, he mainly studied the uptake of mineral nutrients but then extended this to include nutrient transport and use within the plant. His later research greatly advanced the understanding of rhizosphere processes and iron uptake by plants. He also included environmental aspects of plant nutrition in his work, for example, on the side effects of high rates of agricultural fertilizer use, on heavy metal contamination of soils, and on the effect of changes in forest ecosystems on the uptake and use of nutrients by trees. He published extensively on the adaptation mechanisms of plants to adverse soil conditions and low nutrient supply. He was one of the first who related plant nutrition phenomena with genetic control and breeding approaches.

MARTYN, T. ▶ MILLER, P.

MAYSTRENKO, O. I. (1923–1999): Born in Orsk in a family of Ukrainian farmers, Olga finished school in Samarkand (Uzbekistan) and studied at the Moscow Timiryazev Agricultural Academy (Russia) during the hard years of the World War II (1942–1947). In 1947, after graduation from the Department of Breeding and Seed Multiplication, she began her scientific career at a breeding station in Kirgizia as spring and winter barley breeder, resulting in the cultivar "Nutans 45." From 1950 to 1954, she took a postgraduate course in the All-Union Institute of Plant Industry in Leningrad (VIR), where she gained her PhD She began to work with wheat firstly in Kirgizia and later in Sverdlovsk, where she became Head of the Cereal Laboratory (1951–1960). She developed breeding on fast ripening cultivars of oats and wheat with high bread-making quality. Her major scientific contributions followed after her appointment at the Institute of Cytology and Genetics of the Siberian Branch of the Russia Academy of Sciences in 1960 in Novosibirsk. Having been acquainted with the cytogenetic stocks of ▶ E. R. SEARS, she utilized them for the development of adapted wheats of Russia. She developed the monosomic series and later ditelosomics as well as monotelosomics of cultivars of Russian varieties "Diamant" and "Saratovskaya 29." She contributed to the chromosomal localization of genes *Vrnl* and *Vrn3* and a gene for resistance to race 20 of leaf rust as well as genetic factors determining physical properties of dough, resistance to lodging, and plant height. The choice of these two varieties is the evidence of her foresight. "Diamant" is an outstanding genotype of high-grain protein, and "Saratovskaya 29" was superior in bread-making quality and adaptive potential. She also established various sets of intervarietal substitution and near-isogenic lines (NILs).

McCLINTOCK, B. (1902–1992): Born in 1902 in Hartford (United States), she studied plant genetics at Cornell University in Ithaca, receiving her doctorate in botany in 1927. She took a research position at Carnegie's Cold Spring Harbor Laboratory (New York) for more than 40 years. By observing and experimenting with variations in the coloration of kernels of maize, she discovered that the genetic information is not stationary and suggested that the transposable elements are responsible for the diversity in cells during an organism's development. She won the Nobel Prize for Physiology in 1983 for discovery of mobile genetic elements, a discovery that heavily influenced molecular genetics during the last two decades of the twentieth century (Figure 9.8).

FIGURE 9.8 Barbara McCLINTOCK (1902–1992) together with Alfred HERSHEY (1908–1997), Nobel Laureate in Physiology/Medicine 1969, during a meeting of Nobel Prize winners, around 1985. (Courtesy of American Philosophical Society, Philadelphia. With Permission.)

McEWAN, M. (1931–2004): He was New Zealand's foremost wheat breeder of Palmerston North. His achievements helped change bread from the white, unsliced loaves of the 1960s and early 1970s to the convenient sliced, multigrain loaves of today. While working for the Department of Science and Industrial Research (DSIR), he bred several highly successful wheat cultivars, including "Otane," which had excellent milling qualities and produced exceptionally high-grade flours. These were important in the development of new bakery products and processes that became possible in New Zealand after deregulation of the wheat industry in the late 1980s. For a period in the early 1990s, "Otane" commanded over 80% of New Zealand's wheat production. Other successful wheat cultivars that he bred and named after areas in Manawatu were "Rongotea" (1979) and "Oroua" (1979). Others included "Karamu" (1972) and "Endeavour" (1994). All these cultivars resulted from semi-dwarf wheat germplasm that he brought to New Zealand. McEWAN worked on other cereals too, releasing the successful general-purpose feed oat "Awapuni," a forage oat "Enterprise" for Australia, and the black feed oat "Finlay;" a forage barley "Opiki;" and the triticale "Aranui." He retired in 1993. He was a fellow of the New Zealand Institute of Agricultural Science and received the New Zealand 1990 Commemoration Medal and the DSIR Ministerial Award in 1991.

McFADDEN, E. S. (1891–1956): He was born and raised in a homestead in Day County, South Dakota. After graduating in 1918 from South Dakota State, with a BSc degree in agriculture, he began a career with the USDA Bureau of Plant Industry and later became an enabling colleague with ▶ Norman BORLAUG during the Rockefeller Foundation wheat improvement program in Mexico from 1944 to 1955. Edgar S. McFADDEN was a remarkable scientist who made breakthroughs in wheat genetics in South Dakota and Texas. By 1913, he conducted an aggressive breeding program in wheat, oats, maize, and barley. As a student in South Dakota State in 1916, he envisioned and accomplished the first major breakthrough in conferring genetic resistance to stem rust. By the age of 34 years, he accomplished what became the centerpiece of his career. From a single seed, he developed a spring wheat variety that was immune to stem rust; he named it "Hope." This wheat was the first successful mating between common wheat and an ancestral wheat species.

McVEY, D. V. (1922–2010): He retired from the Agricultural Research Service (ARS), St. Paul (United States). His career began with ARS in 1960, working in Puerto Rico testing wheat accessions for resistance to race "15B" of stem rust, which had caused serious losses in wheat in 1953 and 1954. In 1965, he was transferred to the US Cooperative Rust Laboratory. Together with B. BUSCH, he was involved with the release of "Era" wheat in 1970. It was the first semi-dwarf spring wheat cultivar in the upper Midwest and was resistant to stem and leaf rust. The variety became used as a parent in the development of the cultivar "Marshall," which was released by Minnesota in 1982. McVEY helped develop the winter wheat "Siouxland" released by Nebraska. This cultivar was the first wheat to have two leaf and stem rust-resistance genes derived from wild relatives of wheat. It was widely adapted to the Great Plains region and was grown from Texas to South Dakota. A recent cultivar from Minnesota that McVEY helped develop is "McVEY," which was one of the first modern spring wheat cultivars with some resistance to *Fusarium* head blight.

MEITNER, L. ▶ DELBRÜCK, M.

MELCHERS, F. (1905–1997): He was a German geneticist who strongly contributed to the cell research and utilization of cell techniques in breeding. Beginning with mutation research, he was the first to develop a vital potato–tomato hybrid by somatic cell fusion of protoplasts, the so-called "Tomoffeln" or "Karmaten."

MELLO-SAMPAYO, T. (1923–1997): He was born in Pangim, Nova Goa, ex-Portuguese State of India. He graduated in 1949 in agronomy at the Technical University of Lisbon (Portugal). After a short period in Mozambique, he went to the National Agricultural Station, where he began his cytogenetic studies under the supervision of Prof. Antonio CAMARA—mainly on wheat cytogenetics, for example, aneuploidy, chromosome pairing regulation, and nucleolar organizer region (NOR) activity. He paid particular attention to aneuploids of tetraploid wheat and obtained two substitution lines ("Camara" and "Resende"). Moreover, he studied the dose effects of the *Ph* gene on chromosome pairing and achromatic fusion and on chromosome interlocking. With other collaborators, he developed the practical and theoretical concept of mixoploid genomes of wheat and triticale. Several papers were devoted to regulation of NORs and amphiplasty in interspecific hybrids.

MENDEL, G. J. (1822–1884): He was an Austrian Augustinian monk in the monastery of Brünn (now: Brno, Czech Republic, cf. Figure 3.1). By experiments, he discovered the underlying principles of heredity based on his work with pea plants, but his work was so brilliant and unprecedented at the time it appeared that it took 34 years for the rest of the scientific community to catch up to it. The short monograph *Experiments with Plant Hybrids*, in which he described how traits were inherited, has become one of the most enduring and influential publications in the history of science. MENDEL, the first person to trace the characteristics of successive generations of a living thing, was not a world-renowned scientist of his day. He was the second child of a farmer in Brünn, Moravia. MENDEL's performance at school as a youngster encouraged his family to support his pursuit of a higher education, but their resources were limited. So, MENDEL entered an Augustinian monastery, continuing his education and starting his teaching career. His attraction to research was based on his love of nature. He was not only interested in plants but also in meteorology and theories of evolution. MENDEL often wondered how plants obtained atypical characteristics. On one of his frequent walks around the monastery, he found an atypical variety of an ornamental plant; he took it and planted it next to the typical variety. He grew its progeny side by side to see if there would be any approximation of the traits passed on to the next generation. This experiment was "designed to support or to illustrate LAMARCK's views concerning the influence of environment upon plants." He found that the plant's respective offspring retained the essential traits of the parents and therefore were not influenced by the environment. This simple test gave birth to the idea of heredity. He saw that the traits were inherited in certain numerical ratios. Then, he came up with the idea of dominance and segregation of genes and set out to test it in peas. It took 7 years to cross and score the plants to the thousand to prove the laws of inheritance. From his studies, MENDEL derived certain basic laws of heredity, such as hereditary factors do not combine but are passed intact, each member of the parental generation transmits only half of its hereditary factors to each offspring (with certain factors "dominant" over others), and different offspring of the same parents receive different sets of hereditary factors.

MERKLE, O. G. (1929–1999): He was born in Meade (United States). He earned a BSc in agronomy in 1951 and an MSc in plant breeding in 1954, both degrees from Oklahoma State University. In 1957, he moved to College Station, Texas, where he worked as an agronomist with the USDA-ARS. He received the PhD degree from Texas A&M University in 1963, with a major in plant breeding and a minor in plant pathology. He continued with ARS at College Station until 1974, when he was transferred to Stillwater, Oklahoma. There, he worked as an ARS research agronomist until he retired in 1988. His work encompassed many facets of practical research, ranging from interactions of environment with fertilization and plant spacing to the inheritance of wheat flour quality and resistance to pests and drought. Improvement and development of small grain germplasm and cultivars were among his most important contributions to agriculture. As an agronomist–breeder on research teams in Texas and Oklahoma, he made significant contributions to the development of flax cultivars "Caldwell," "Dillman," and "Mac" and wheat cultivars "Caddo," "Milam," "Sturdy," "Fox," "Mit," and "Century." He and his co-workers released and registered more than 15 wheat germplasm lines with resistance to disease (rust) or insects (Hessian fly and the yellow sugarcane aphid) and with improved characters (large seed) and pearl millet germplasm lines with resistance to chinch bug. Prior to retirement, he was active in evaluating barley, wheat, and wild *Triticum* species for resistance to the Russian wheat aphid. He also evaluated winter and spring wheat genotypes for tolerance to drought in cooperative programs with colleagues in Lubbock (United States) and El Batan (Mexico).

METTIN, D. (1932–2004): He attended Berlin Eosander Junior College from 1942 to 1943, Eisleben Martin Luther Gymnasium from 1943 to 1950, Salzmünde Agricultural College from 1950 to 1952, and University of Halle/S. (Germany) from 1952 to 1955, where he obtained his MSc He obtained his PhD degree from the same university in 1961, with a major in plant breeding and genetics under the supervisor ▶ Prof. H. STUBBE (thesis: "Genetic and Cytological Studies in the Genus *Vicia*"). He was awarded a DSc degree from that institution in 1977 (thesis: "Selection, Identification and Genetic Utilization of Aneuploids in Hexaploid Winter Wheat, *Triticum aestivum* L."). In 1961, he moved to the Institute of Plant Breeding of University Halle/S., where, after graduation, he began his professional career as assistant professor at the Institute of Plant Breeding, Hohenthurm. In 1968, he was appointed as reader of cytogenetics in plant breeding, and in 1977, he became full professor of plant breeding. In 1983, METTIN accepted a call as director of the Central Institute of Genetics and Crop Plant Research Gatersleben, from which he retired in 1991. During his time at the University of Halle/S., he contributed significantly to the improvement of academic education in genetics, plant breeding, and seed production, as well as applied cytogenetic research. He had the opportunity to encourage and direct several graduate students. More than 40 BSc students and more than 60 MSc students studied under his supervision. Many of the research projects involved the participation of the 12 PhD and 2 DSc applicants. METTIN's research work encompassed many facets of basic and applied research, ranging from cytotaxonomic studies of *Vicia*; induced auto- and allopolyploidization in *Brassica* and *Secale*; wide hybridization in *Triticineae*, production and utilization of aneuploids in *Aegilops*, *Secale*, and *Triticum*; genetic mapping of resistance genes to leaf diseases; and quantitative traits in wheat and homoeologous chromosome pairing in cereals to first applications of molecular genetics and biotechnology in plant breeding. He was among the pioneers of wheat aneuploid research in the world. The creation of the complete series of monosomics of the German wheat cultivar "Poros" (winter type) and "Carola" (spring type), the production of the first complete series of primary rye trisomics, the co-discoveries of the 1B/1R wheat–rye translocations and substitutions in hexaploid wheat, and the spontaneous homologous recombination between wheat and rye chromosomes are four of his most essential merits.

MIKUZ, F. (1889–1978): He was a co-worker of ▶ F. JESENKO, and since 1921, he headed the newly established Plant Breeding Station in Beltinci (Yugoslavia). He was engaged in breeding of wheat, oats, rye, maize, and buckwheat. His wheat cultivars "Beltinska 227," "Beltinska 321," and "Beltinska 831" were known and spread for long in Slovenia.

MILLER, P. (1691–1771): He was a British gardener to the Worshipful Company of Apothecaries at their Botanic Garden in Chelsea (United Kingdom) and was known as the most important garden writer of the eighteenth century. *The Gardener's and Florist's Dictionary, Or a Complete System of Horticulture* (1724) was followed by a greatly improved edition entitled *The Gardener's Dictionary* (1831), containing the methods of cultivating and improving the kitchen, fruit and flower garden. this book was translated into Dutch, French, and German and became a standard reference for a century in both England and America. In the 7th edition (1759), he adopted the Linnaean system of classification. The edition enlarged by Thomas MARTYN (1735–1825), professor of botany at Cambridge University (United Kingdom), has been considered the largest gardening manual to have ever existed. MILLER is credited with introducing about 200 American plants. The 16th edition of one his books, *The Gardeners Kalendar* (1775), gives directions for gardeners month by month and contains an introduction to the science of botany.

MILOHNIC, J. (1920–1974): He was a Croatian breeder at the Institute of Breeding and Crop Production, Zagreb. His main concern was cereal and fodder legume breeding. The winter vetch variety "Ratarka" has been of great local importance.

MISCHER, F. (1844–1895): He was a German chemist. He first described a method for purification of nuclei from the cytoplasm. From the nucleus, he isolated an acid compound, the "nuclein;" later, it was called "nucleic acid."

MITSCHURIN, I. W. (1855–1935): He was a Russian botanist and plant breeder who was born in Dolgoje near Rjasan (Russia, now Mitschrowka). He bred more than 300 varieties of fruit trees, grape vine, and berries. His varieties were particularly suited for the cold and continental climate of Russia. Not recognized under the King's regime, he received much attention by Soviet government after 1917. His success based on the so-called "mentor method," believing that modifications are induced by environmental factors and not by genes. Later, ▶ T. D. LYSENKO became a propagandist of MITSCHURIN's ideas, misleading Soviet biological sciences for decades; however, modern molecular studies from 2009 show evidences for genetic recombination in cells of crafted hybrids, for example, in tobacco.

MOENS, H. P. B. (1931–2008): He born in Sukabumi, Indonesia, to Dutch citizens, who had moved to Sukabumi, a Dutch colony of Indonesia. Later, he moved to Canada and enrolled in the Faculty of Forestry at the University of Toronto, where he received his BSc in 1959. He continued his education, earning an MSc (1961) and a PhD (1963). Before finishing his PhD thesis, MOENS was asked to join the faculty at the newly established York University (Toronto) first at Glendon College and later moving to the newly built Farquharson building at York's Keele campus, where he spent the rest of his very successful career. His adventures with the "synaptonemal complex" (SC, SCHLEGEL 2009) began shortly after he started working at York. Following a course in electron microscopy, he used his newly acquired skills to study the structural characteristics of the SC as a basic process of genetic recombination. First observed in 1956 by M. J. MOSES (Duke University, Durham, North Carolina) in meiotic prophase spermatocytes of crayfish, the SC was also found in species of plants. P. MOENS has been an internationally recognized expert on the SC, being involved in all of the steps and characterizing many of the structural and functional components of the complex, mainly in mice but also in other species. He was member of the Genetics Society of Canada, the Genetics Society of America, and the Canadian Society for Cell Biology and a fellow of Royal Society of Canada. He also was North-American scientific editor of the journal *Chromosoma* for several years and was the editor of the NRC Research Press journal *Genome* for a quarter of a century.

MORGAN, T. H. (1866–1945): He was an American geneticist who worked on fruit fly (Drosophila melanogaster), demonstrating that genes are located on chromosomes. Along with ▶ W. BATESON, he was the co-founder of modern genetics, an experimentalist, and originally a mutationist, opposed to natural selection, Lamarckism, orthogenesis, and the chromosomal theory of inheritance; however, he converted to the chromosomal theory of inheritance and to Darwinism after 1910. He discovered the linkage and recombination of genes with his students and was the first to show that variation derives from numerous small mutations.

MULLER, H. J. (1890–1967): He was an American geneticist best remembered for his demonstration that mutations and hereditary changes can be caused by X-rays striking the genes and chromosomes of living cells. He was born in New York City as son of German immigrants. He attended Columbia University from 1907 to 1909. A laboratory assistantship in zoology in 1912 allowed him to spend part of his time doing research on *Drosophila* at Columbia University. He produced a series of papers, now classic, on the mechanism of crossing-over of genes, obtaining his PhD in 1916. His dissertation established the principle of the linear linkage of genes in heredity. The discovery of artificially induced mutations in genes had far-reaching consequences, particularly in plant breeding. He was awarded the Nobel Prize for Physiology and Medicine in 1946.

MÜNTZING, A. (1903–1984): He was born in Gothenburg (Sweden). He developed interest in the work of ▶ H. NILSSON-EHLE when he was a student. The group around NILSSON-EHLE influenced his early species concept, when he was already a teacher at the University of Lund (Sweden). He also studied the theory of Ö. WINGE of the role of polyploidy in plant evolution. In his PhD thesis in 1930, he reported about the polyploidy in the genus *Galeopsis*. He was able to re-synthesize a Linnean species *Galeopsis tetrahit* ($2n = 4x = 32$) by combining *G. speciosa* ($2n = 2x = 16$) and *G. pubescens* ($2n = 2x = 16$). At this time, MÜNTZING was serving as a sugar beet breeder in Hilleshög; however, in 1931, he founded a cytogenetic department in Svalöf, in order to study and utilize polyploidy as a factor in breeding. NILSSON-EHLE stated the reason for its establishment as follows: "In recent years geneticists have found that important changes in the structure of organisms may be induced by intentionally changing the actual number of chromosomes (and thereby the number of genes) in the nuclei. It seems to be especially important to try to induce an increase in the chromosome number, since such an increase is often connected with an increased size of the plant organs and consequently with the production of increased vegetative yield." After tentative experiments to produce polyploids by applying the colchicine method, discovered by A. F. BLAKESLEE in 1937, he proposed it as a general and very useful method in plant breeding. MÜNTZING was successful in producing autopolyploids as tetraploid rye as well as allopolyploids as triticale. In 1993, he published a paper on apomictic and sexual seed formation in *Poa* and was the first to demonstrate apomictic seed formation in *Poa pratensis*. Later, ▶ E. ÅKERBERG found both apomictic and sexual types in this species and that the apomictic plants were of the apogamic type, combined with pseudogamy. When in 1938 ▶ NILSSON-EHLE retired as professor of genetics at the University of Lund, MÜNTZING succeeded him. Due to his concept of genetics, his chair at the University of Lund (Sweden) covered a broader field of genetics than just crops. He incorporated human genetics and developed a chromosome research branch, which was headed by A. LEVAN. He was very happy to see the success of triticale breeding during the 1970s–1980s, which he already started during the thirties of the last century.

NABA, K. (1595–1648): He was a Japanese Prunus expert and the first person to ever write a book devoted to it. He documented and described many cultivars for the first time.

NÄGELI, C. ▶ CORRENS, C. E.

NAVASHIN, S. G. (1857–1930): He was born in Tsarevshchina (Russia) in the family of a physician. In 1874, after finishing gymnasium in Saratov, he entered Medico-Surgical Academy in St. Petersburg. In 1878, he entered the Natural Sciences Faculty of Moscow University, where he continued to specialize in chemistry; profound knowledge of NAVASHIN in this field was noticed by K. A. TIMIRIAZEV, who read lectures on plant physiology at the university. The final master examination, which NAVASHIN passed successfully at St. Petersburg University, was on botany. It allowed him to read courses on mycology and phytopathology. He gave the introduction to fungi systematics. At the same time, he began his first mycological investigations. When he investigated heterospory in *Sphagnum* moss, NAVASHIN proved that filiform structures in the so-called "female flower" of *Sphagnum* species are hyphae of the new fungus species *Helotium schimperi* (discomycetes). He demonstrated that microspores of these mosses are, in fact, spores of smut fungus *Tilletia sphagni*. He also described new fungi species, including *Gymnosporangium tremelloides* and *Puccinia wolgensis*. In 1889, he again moved to St. Petersburg University as an assistant of Prof. I. P. BORODIN, where his attention was drawn to genus *Sclerotinia* on plants of the genus *Vaccinium* and the slime mould *Plasmodiophora brassicae*—agent of club root of cabbage. In 1894, NAVASHIN headed the Chair of Morphology and Systematics of Plants of St. Vladimir University in Kiev. He began to study thoroughly *P. brassicae* and described vegetative stage of this intracellular parasite and its sporogenesis. NAVASHIN, for the first time, concentrated his attention on cytology of sporogenesis and, in particular, determined the phenomenon of simultaneous nuclei division in plasmodium of *P. brassicae*. Studying the relationships between parasite and host–plant, he demonstrated for the first time that protoplasm of cabbage cells, affected by amoebas of slime mould, remains viable. After the discovery of double fertilization in angiosperms, which did credit to NAVASHIN (1898), he dedicated himself to embryological and karyological investigation of higher plants.

NEETHLING, J. H. (1887–1960): He was born in Lydenburg, Mpumalanga (South Africa). During the Anglo–Boer war, he accompanied his father on commando. After the war, he returned to the Boys High School in Stellenbosch to complete his studies and matriculated in 1906. In 1907, he gained the B.A. degree at the Victoria College. He was one of the youngsters selected by General L. BOTHA to further their studies in agriculture overseas. In 1911, he was awarded the BSc degree by the University of Cornell (United States). In the same year, he completed the MSc degree under Profs. WEBBER and GILBERT. Prior to returning to South Africa, he received tuition in hybrid maize breeding under the guidance of Prof. MOORE in Wisconsin and then was instructed in genetics by ▶ Prof. R. GOLDSCHMIDT in München (Germany) and by ▶ Prof. H. de VRIES in Amsterdam (the Netherlands). The existence and well-being of the small grain industry in South Africa was largely built on NEETHLING's knowledge and expertise. For almost a half century, the wheat cultivars grown in the South Western Cape Province emanated from his breeding program. The wheat cultivars "Union 17" and "Union 52," which were released in 1915, were followed by "Gluretty," "Hoopvol," "Koalisie," and especially "Pelgrim," "Vorentoe," and "Sterling." In 1949, the Wheat Board built a greenhouse and adjoining laboratory at the Welgevallen Experiment Station (South Africa) and named it in his honor, as, in 1982, the Agriculture Building of the University of Stellenbosch (South Africa). During his tenure, 4 DSc and 29 MSc degrees were conferred to him. NEETHLING retired in 1949 and was succeeded by one of his students Prof. ▶ F. X. LAUBSCHER.

NEWMAN, L. H. ▶ GOULDEN, C. H.

NEWTON, I. ▶ HOOKE, R.

NIEDERHAUSER, J. S. (1917–2005): He was an American potato scientist and 1990 World Food Prize winner of the University of Arizona. He was a pioneer in international cooperation for the improvement of agricultural productivity and was known for developing potato varieties resistant to late blight disease. In 1946, NIEDERHAUSER joined the newly formed Rockefeller Foundation Mexican Agricultural Program. He spent 15 years working in Mexico on maize, wheat, and bean production. During this time, he began to study potato production in Mexico. His work over the next decades focused on the improvement of potato production in many developing countries. Owing to the success of this work, the International Potato Center, now supported by CGIAR, the Consultative Group on International Agricultural Research was established in Lima (Peru) in 1971. In 1978, he established the Regional Cooperative Potato Program in Mexico, Central America, and the Caribbean. One of NIEDERHAUSER's important scientific contributions was the development of potato varieties with resistance to late blight disease. He discovered that the source of the pathogen responsible for the Irish potato famine came from Mexico, together with many wild inedible potato species that possessed a durable field resistance to the late blight fungus. NIEDERHAUSER's work resulted in the establishment of the potato as the fourth major food crop worldwide. As another result, potato production in Mexico increased from 134,000 tons in 1948 to greater than 1 million tons by 1982.

NILAN, R. A. (1923–2015): Robert Arthur NILAN was born in New Westminster, British Columbia (Canada). From the University of British Columbia in Vancouver, he earned a BSc degree in general studies in 1944 and an MSc degree in plant science in 1946. He received his PhD in genetics in 1951 from the University of Wisconsin in Madison. He moved to Pullman (Washington State College, United States) with his family. Initially, his focus was maize genetics, but later, barley became his passion. He and his colleagues created a world-renowned program in barley breeding and genetics. He published over 100 scientific research articles and trained 60 students for their Masters and PhD degrees. NILAN helped create the Genetics Department and served as department Chair for 9 years. In 1979, he was appointed dean of the College of Science, a position that he held for 12 years. During his long and distinguished career, he won several awards and honors, including appointment to the Danish Academy of Science, the Nilan Distinguished Professorship in Barley Research and Education, the Washington State University Foundation Outstanding Service Award, the College of Sciences Legacy of Excellence Award, and, most recently, the establishment of the Robert A. Nilan Endowed Chair. One of his proudest career accomplishments was being a founding member of the International Barley Genetics Symposium.

NILSSON-EHLE, H. (1873–1949): He was a Swedish botanist trained at the University of Lund. He was employed at the Swedish Seed Association, Svalöf, in 1888, where he acted as a director from 1890 to 1924. when the Svalöf Institute started in 1886, the first director was a German from Kiel, T. BRUUN von NEERGAARD. He was an agricultural engineer by training and had constructed a number of instruments for measuring, analyzing, and sorting spikes and seeds of cereals. The German ▶ RÜMKER (1889) described him as a pioneer in the development of more systematic and exact methods in plant breeding by introducing "measure, number, and weight." At that time, the breeders in Europe tried to apply the Darwinian theory to plant breeding, that is, the character of a plant could be improved by continuous selection. It began with repeated mass selection in heterogeneous local land races of wheat, oats, barley, and peas. Often, the results were disappointing. At this time, experienced botanists such as H. TEDIN and P. BOLIN were employed at the Swedish Seed Association. As early as 1890, they started with individual selection, in order to enable a better description of the range of variation available in the local varieties. Almost at the same time, the rediscovery of the Mendelian laws in 1900 influenced the breeding work. During the 1890s, a great number of crosses in different crops were performed and analyzed at Svalöf. It was clear that both parent plants had an equal influence on the progeny. P. BOLIN points out from his experience with barley crosses "that regarding the heredity of characters in the second and in the immediately following generations a definite regularity seems to exists: the types, which appear, represent all possible combinations of the characters from the parents, and can therefore be calculated in advance with almost mathematical accuracy" (BOLIN 1897). NILSSON-EHLE emphasized as early as 1904 the possibilities of counteracting economic losses caused by plant diseases through the production of resistant varieties. He pointed out that the attitude to the appearance of plant diseases was characterized by the opinion that the cause could be found in insufficient access to plant nutrients. In the years 1908–1911, he published his papers about the inheritance of quantitative characters in wheat and oats. He proved that these characters are inherited in the same way as qualitative characters. He strongly emphasized the need for careful planning of the cross-breeding method in the breeding work; only lines and varieties with well-known reactions and features should be chosen to achieve the expected goal. In the year 1915, he was appointed as professor of botany at the University of Lund. In 1917, a personal professorship in genetics was conferred upon him, which he held in conjunction with the directorship of the Swedish Seed Association from 1925. The study of the classical subject for plant breeding research, Pisum sativum, was also continued in Sweden. NILSSON-EHLE mainly used wheat and oats for his genetic studies. One of the problems studied was the wheat "speltoid" phenomenon; however, the definite solution to the problem was presented by J. MACKEY (1954), who at that time was working at Svalöf. He concluded that "speltoid" can be produced by simple deficiency, by deficiency duplication, and probably also by gene mutations or chromosome substitution. His main contribution was the discovery of the phenomenon of polymery in plants. He applied the knowledge to practical combination and population breeding of cereal and forest plants.

NOBBE, F. (1830–1922): The agrochemist Friedrich NOBBE was born in Bremen (Germany). He was a school teacher before he studied natural science in Jena and Berlin in 1854–1859. In 1861, he became a professor at the Chemnitz University of Applied Sciences. At the same time, he took over the editorship of the *Landwirthschaftlichen Versuchs-Stationen* (Agricultural Research Stations) and became a professor at the Academy for Forestry and Agriculture in Tharandt in 1868. In 1869, with the support of the agricultural district association in Dresden, he set up a phytophysiological testing station, which was expanded by a horticultural department in 1888. In 1869, he founded the first center of seed control, which became exemplary for all crops. His main research subjects were plant nutrition (using test solutions), nitrogen fixation in legumes, and vaccination methods in plants. His agricultural research station was later followed by the "Association of German Agricultural Research and Testing Institutes" (VDLUFA). He wrote monographs such as *On the Organic Performance of Potassium in the Plant* (with SCHRÖDER and ERDMANN, Chemnitz, 1870); *Manual of Seed Science* (Berlin 1876), and *Against Trade of Forestral Grass Seeds for Meadows* (Berlin 1876) and edited the fourth edition of DOBNER's *Botany for Forestry Men* (Berlin 1882). He was awarded with "Ritterkreuz des Schwedischen Nordstern-Ordens" (1880), "Ritterkreuz I. Klasse des Sächsischen Albrechts-Ordens" (1882), "Auswärtiges Mitglied der Königlichen Schwedischen Landwirtschaftsgesellschaft" (1888), "Königlicher Titel Geheimer Hofrat" (1889), "Ehrenmitglied des Kaiserlichen Russischen Forstinstituts zu St. Petersburg" (1893), "Ehrenmitglied des Landwirtschaftlichen Kreisvereins Dresden" (1894), "Ehrenmitglied der Royal Agricultural Society of England" (1896), and "Ehrenmitglied des Verbandes der landwirtschaftlichen Versuchs-Stationen im Deutschen Reich" (1903).

NOVER, I. (1915–1985): Born in Kassel (Germany), Inge studied at the Biological and Agricultural Faculties of the Universities Wroclaw (Poland) and Halle/S. (Germany) from 1934 to 1938. She obtained her PhD degree at the University of Halle/S. (Germany) in 1941, working on mildew in wheat. From 1948, she was appointed at the Phytopathological Institute of the University of Halle/S., where she remained until her retirement in 1976. During the 28 years' work on resistance to rusts, smuts, and mildew, she heavily contributed to resistance breeding in wheat, barley, and rye in Germany. She was one of the first plant pathologists to establish tester stock collections, evaluate wild collections, and transfer results of basic research to breeding programs.

OEHLKERS, F. (1890–1971): He was a German botanist studying the mitosis of several fungi. Influenced by O. RENNER of the Botanical Institute of Munich, in 1921, he began genetic studies on the genus Oenothera. From 1934, he investigated the physiology of meiosis. During the years 1942–1943, his laboratory at the Institute of Botany of the University Freiburg (Germany) discovered several chemical mutagens.

OKAMOTO, M. ▶ SEARS, E. R.

OSBORNE, D. J. (1930–2006): She was born in India, where her father was a colonial administrator. She attended The Perse School in Cambridge. Her BSc in chemistry and MSc in botany were from King's College London. Her PhD on the topic of plant growth regulators was from the University of London at Wye College, Kent, where her supervisor was R. Louis WAIN. Her first postgraduate position was in the Department of Biology of the California Institute of Technology, United States, as a Fulbright Scholar, where she worked with botanist Fritz WENT, among others. Daphne OSBORNE had a long and illustrious career as a British seed biologist, having focused on DNA, RNA, and protein synthesis, an interest that emanated from her earlier work on plant growth regulators. Her research in the field of plant physiology spanned five decades and resulted in over 200 papers, 20 of which were published in the journal *Nature*; her contributions to knowledge on deterioration of seeds and their propensity for repair on imbibition are legendary. OSBORNE is best known for her work on the gas ethylene, in particular for demonstrating that ethylene is a natural plant hormone and that it is the major regulator of ageing and the shedding of leaves and fruits. She also originated the concept of the target cell as a model for understanding plant hormone action. She made her mark not only as an insightful researcher but also by her unstinting personal interaction with students and peers worldwide.

OSMOMYSL, Y. ▶ FIFE, D.

PAP, A. (1897–1992): He was born as a child of Jewish farmer–tenant parents with the handicap of a double dislocation of the hip, for which there was no cure at the same level 100 years ago as today. He was saved from deportation by the Hungarian compatriot ▶ George REDEI, who at that time was an official in the Ministry of Agriculture (later a professor at the University of Missouri, United States). In the late 1940s and early 1950s, Pab was imprisoned for a year; after his release from prison, he was employed at the Agricultural Research Institute in Martonvasar (Hungary). During 4 years, he succeeded in obtaining the maize hybrid combination "Mv 5" and years later "Mv 1," the first inbred hybrid in Europe. In this work, he also used his own hybrid lines ("Mpf 511D") selected earlier during his life as a private farmer. His hybrids have brought the breakthrough. In a few years, Hungary became the most significant hybrid maize seed exporter in Europe, and this position has been maintained for decades. Pap was an excellent scientist; however, he wrote only a few scientific papers. The only Hungarian paper referred to by ▶ VAVILOV in his famous work *The Origin, Variation, Immunity and Breeding* was a publication of A. PAP. After he left Hungary following the revolution in 1956, his followers continued the maize breeding work that he had initiated, both at the Agricultural Research Institute of the Hungarian Academy of Sciences in Martonvasar and at the Cereal Research Institute in Szeged.

PAREY, P. ▶ FRUWIRTH, C.

PATEL, C. T. (1917–1990): He was a cotton scientist who developed the first commercial cotton hybrid in Central India, known as "Hybrid-4" ("Sankar-4"), in 1970, which was later cultivated commercially in the states of Gujarat and Maharashtra (India). Chandrakant T. PATEL was born in Sarsa in the Kaira District of Gujarat and obtained his MSc degree in plant breeding and genetics from Bombay University in 1954. He worked at Surat Agricultural University, and after two decades of continuous research efforts, he successfully developed an intraspecific hybrid by crossing "Gujarat-67" with "American Nectariless" varieties, known as "Hybrid-4," which produced about 800–1000 kg of cotton balls (kapas) per hectare (record was 6918 kg kapas/ha), as compared with common varieties with 213–304 kg lint per hectare. The fiber properties were excellent, and its adaptability makes it still popular. This was the first successful hybrid of commercially cultivated cotton. PATEL devised many innovative methods in plant breeding, such as nursery-cum-pot irrigation and the so-called telephone system. The Sardar Patel University in Vallabh bestowed an honorary DSc degree on him in 1978.

PAULI, W. ▶ DELBRÜCK, M.

PAULY, E. (1905–1989): She was a German breeder working in Quedlinburg (Germany). She bred several varieties of Matthiola, Callistephus, Antirrhinum, Petunia, Viola, and so on.

PAWLETT, T. L. ▶ WATSON, I. A.

PEARSON, K. ▶ FISHER, R. A.

POEHLMAN, J. M. (1910–1995): John Milton POEHLMAN was born on a farm in Macon County, Missouri (United States). He graduated from Macon High School in 1928. Then, he attended the University of Missouri and received a MSc in agriculture in 1931 and a PhD in botany in 1936. He has been on the faculty at the University of Missouri ever since, serving as a professor in the Departments of Field Crops and Agronomy, an adviser to over 250 students, and professor emeritus since 1980. Between 1936 and 1980, he did extensive research on genetics and breeding of wheat, oats, barley, and rice, which resulted in the release of over 20 new cultivars. He also performed research through grants from the university and the US Agency for International Development on the mungbean, a high-protein pulse crop grown in underdeveloped countries for food. Additionally, he served as a consultant for the World Bank, International Agricultural Development Service, and the USDA, addressing world hunger. In 1963, he travelled to Orissa University of Agriculture and Technology in Bhubaneswar, India, where he served for 2 years as research advisor, and he lectured in Romania in 1971 and China in 1982. He was most proud as author of several books, including four editions of *Breeding Field Crops*, *Breeding Asian Field Crops* (1st ed., 1959), *History of Field Crops in the University of Missouri*, and *The Mungbean*, and some others. He has also authored more than 150 journal and technical articles.

PESOLA, V. A. (1892–1983): Born in Turku (Finland); he studied botany and plant biology at the University of Turku; from 1918 to 1924 he worked in the Plant Breeding Station of Jaevenpaeae and specialized later in Jokioinen from 1924 to 1928; he became the head of the Plant Breeding Station in Jokioinen in 1928; since 1930 he has been director of Plant Breeding Station of the Center for Agronomic Science in Tikkurila; not only for his own country but also for the Scandinavian countries, for Canada and the Russia his work had a pioneered character in opening hundreds of kilometers for agriculture towards the North; he produced more than 20 varieties for practice in cold northern regions.

PIENAAR, R. de V. (1928–): He was born in Hamburg (Germany) and grew up in Johannesburg (South Africa). He graduated at the University of the Witwatersrand. After 2 years of postgraduate study in England, Sweden, and Holland, he completed his cytological studies on Eragrostis at Wits and received the PhD degree in 1953. PIENAAR started his career as an assistant scientific officer at the Department of Agriculture in 1954 and proceeded through the ranks, until he was appointed as professor of cytogenetics in 1964. On the retirement of ▶ LAUBSCHER, he was appointed as Head of Department in 1969. PIENAAR was an exceptionally hard worker, meticulous scientist, and dedicated teacher. His first appointment was related to initiating carrot, onion, and chincherinchee (Ornithogalum) breeding programs. Later, his field of interest and expertise was in wide crosses involving the genus Triticum and related species, in order to create new allopolyploids from which gene transfers could be effected to T. durum and T. aestivum from rye, Thinopyrum, Haynaldia, Agropyron, or Hordeum. For this purpose, he also created monosomic, nullisomic, and telosomic series of aneuploids in the variety "Pavon 76" and transferred the wheat crossability genes Kr1 and Kr2 to this cultivar. He was the first to successfully intercross the indigenous wild growing Thinopyrum with bread and durum wheats. This research was well supported by postgraduate students, which led to 13 MSc and 6 PhD degrees. With industry funding and technical support by H. S. ROUX, he initiated triticale, durum wheat, and rye breeding programs. This led to the release of six durum wheat and five triticale cultivars. In 1986, a grant from the Anglo American Corporation made possible the founding of a laboratory for biotechnological research in the Department of Genetics, as part of the Institute of Biotechnology of the Faculty of Agriculture. With funds bequeathed from the legacy of Prof. ▶ J. H. NEETHLING, the post of chief researcher to head this institute was created. The first incumbent of this position was Prof. G. F. MARAIS. PIENAAR was well recognized internationally, traveled extensively, and was awarded various awards, among these being a special award by CIMMYT "in recognition of his lifelong dedication to wheat improvement through building bridges between wheat and its relatives." In 1986, he was appointed as third MONSANTO-SEARS visiting professor at the University of Missouri, Columbia, Missouri (United States). Before his retirement in 1991, he initiated a haploid breeding research project with wheat and barley, which he continued for SENSAKO Ltd., until its take over by Monsanto in 1998.

PISAREV, V. E. (1883–1972): He was a Russian agronomist, crop breeder, and geographer and the founder of Tulun Experiment Station (Russia). From 1921, he was a scientific expert of the Department of Applied Botany and from 1925 deputy director of VIR (Saint Petersburg, Russia).

PRIDHAM, J. T. (1879–1954): He was a plant breeder born in Stanmore, Sydney (Australia). Educated at Sydney Grammar School and Hawkesbury Agricultural College, where he was dux in 1900. He joined the Department of Agriculture as an experimentalist in 1901. For 3 years, he assisted W. FARRER in his wheat-improvement program in Lambrigg, near Canberra. While experimenting at Bathurst Experiment Farm in 1904–1906, PRIDHAM began an oat-breeding program. He continued the work for 40 years, concentrating on developing varieties with early maturity, less tillering, and better adapted to wheat belt conditions. In 1906–1908, he was assistant manager at Cowra Experiment Farm and then worked with the Victorian Department of Agriculture in 1908–1911. At Longerenong Agricultural College, Victoria (Australia), in a plot of "Algerian," the dominant oat variety in Australia, PRIDHAM selected an earlier-maturing plant, which became the variety "Sunrise" and the most popular oat in New South Wales. Other varieties developed in later years included "Belar" (his most successful variety), "Mulga," "Guyra," "Lampton," "Lachlan," and "Weston." He was probably the first to carry out cross-breeding in oats in Australia and laid the foundation of oat improvement. As plant breeder at Cowra Experiment Farm, New South Wales, in 1911–1944, PRIDHAM also continued the wheat-improvement program started by FARRE. His first notable variety was "Canberra," selected and released in 1914 at Wagga Experiment Farm from a cross made by R. J. HURST, which proved a satisfactory early-maturing variety for dry areas of the wheat belt and for late sowing in good-rainfall areas. He also developed the hard, translucent variety, "Hard Federation" (1914), which demonstrated that it was possible to combine moderately good baking quality and satisfactory yield in an Australian wheat. Noting the drought resistance of the old wheat variety, "Steinwedel," he crossed it with "Thew" to produce "Bobin" in 1925. Other notable varieties that made significant contributions to the wheat industry included "Dundee," "Gular," "Aussie," "Wandilla," "Union," "Bena," "Baringa," and "Clarendon." He also did important pioneering work on barley on lines imported from North Africa. PRIDHAM was in charge of all cereal improvement work carried out at the various research stations, until the formation of the plant-breeding branch in 1938 by ▶ W. L. WATERHOUSE.

PUNNETT, R. C. ▶ STURTEVANT, A. H.

PUSTOVOYT, V. S. (1886–1972): He was a Russian breeder born in Taranovka, Charkovskaya guberniya (now: Ukraine). He introduced the sunflower as a crop plant that later spread all over the world. He was the pioneer in breeding first sunflower cultivars. He bred 34 cultivars; among them, two became worldwide fame: "Salut" and "Peredovik."

QUICK, J. S. (1940–2015): James S. QUICK was born on a farm near Starkweather, North Dakota. His interest in plant breeding began during his undergraduate experience with wheat genetics at North Dakota State University and the USDA. He received his BSc degree from North Dakota State University and his MSc and PhD degrees in plant breeding and genetics from Purdue University (United States). On completion of his PhD, he served in India with the Rockefeller Foundation in sorghum research for 3 years, and then, he returned as an associate professor. He was successful in durum wheat breeding and cultivar development. In 1981, QUICK joined Colorado State University as a professor and leader of the Wheat Investigations project. During the next 23 years, he made significant contributions in hard red winter wheat breeding to improved methodology, high-temperature tolerance, Russian wheat aphid resistance, and herbicide tolerance. He and his associates released more than 30 new wheat cultivars and several improved germplasm lines. QUICK created the Colorado Wheat Research Foundation in collaboration with Colorado wheat industry leaders. He was proud to have served as major professor for 23 PhD and MSc candidates. He served the Crop Science Society of America as an associate editor of *Crop Science* and as editor of the *Annual Wheat Newsletter* from 1983 to 1994. In 1996, QUICK became

Head of Soil and Crop Sciences at Colorado State University, until his retirement in 2003. He received numerous awards, such as Fellow of American Society of Agronomy (1985), Fellow of Crop Science Society of America (1986), Shepardson Instruction Development Grant Award (1985), Honor Society of Agriculture (1989), USDA NCISE–IARC Award for Study at CIMMYT, Mexico (1995), and Agronomic Research Award, American Society of Agronomy (2002), among others.

RAATZ, W. (1864–1919): He was a German sugar beet breeder in Klein Wanzleben. He developed efficient selection methods for increasing yield and sugar content by classification of E (= high yielding and normal sugar content), N (=normal yielding and normal sugar type), Z (=high sugar type and normal yielding), and ZZ types (=very high sugar type and normal yielding).

RABINOVYCH, S. V. (1932–): She is a Russian plant breeder. In 1954, she received a degree at Kharkov Agricultural Institute and was appointed as an agronomist, plant breeder, and seed producer. She worked at the Myrgorod State Variety Test Station in 1954, the Forage Production Department of Research Institute for Livestock Farming of Ukraine in 1955–1957, and for more than 45 years at the Institute of Plant Production of the Ukrainian Academy of Agricultural Sciences. Her principal directions of scientific activity were the collection, study, conservation of plant genetic resources, and their introduction into breeding process, particularly of wheat, rye, and triticale.

RATCHINSKY, T. (1929–1980): He was a Bulgarian wheat breeder working for long at the Institute of Wheat and Sunflower, General Toshevo/Varna. Born in Vratsa, he started his career at the Agricultural Institute in Sofia, until he was appointed in 1957 at the Agricultural Institute, Knesha, and in 1963 at the Institute of Wheat and Sunflower, General Toshevo. He bred important winter wheat varieties for Bulgaria, such as "Rusalka," "Jubilje," "Ogosta," "Rubin," "Tcharodejka," "Dobrudja 1," "Ludogorka," "Vega," or "Slatija," which still serve as important gene pool in new programs. He was the first in Bulgaria who introduced the genes for short straw and earliness from the Italian wheat lines into native wheats.

REDEI, G. ▶ KISS, A.

RENNER, O. ▶ OEHLKERS, F.

RHOADES, M. M. (1903–1991): He was born in Graham (Missouri, United States) and spent his childhood in Downs, Kansas. He attended the University of Michigan, majoring in botany and mathematics. He was befriended by ▶ Prof. E. G. ANDERSON, who introduced him to plant genetics. After receiving his BSc and MSc degrees from Michigan, RHOADES studied his PhD at Cornell University under ▶ R. A. EMERSON, a maize geneticist. He was part of a brilliant group of maize cytogeneticists, which included ▶ B. McCLINTOCK, ▶ C. BURNHAM, and ▶ G. BEADLE. RHOADES interrupted his PhD work for 1 year to visit the Caltech as a teaching fellow. At this time, he worked on Drosophila under the guidance of ▶ A. H. STURTEVANT and T. DOBZHANSKY, with occasional support from ▶ T. H. MORGAN and C. B. BRIDGES. Following completion of his PhD in 1932, he stayed at Cornell as an experimentalist in plant breeding until 1935. In that year, he joined the USDA as a research geneticist and was stationed at Iowa State University until 1937. In 1937, the USDA transferred him to the Arlington Experimental Farm outside Washington, D.C. At the farm, his basic cytogenetic research flourished, with considerable support from both his supervisor M. T. JENKINS and bureau chief F. D. RICHEY. He returned to academics in 1940 as associate professor at Columbia University. He was promoted to professor in 1943 and remained at Columbia until 1948, when he was appointed professor at the University of Illinois. He spent 10 years of teaching and research at Illinois and next served as chairman and professor of the botany department at Indiana University from 1958 to 1968. In 1968, he resigned from the chairmanship and was given the rank of distinguished professor at Indiana University. During his career, RHOADES worked on a wide variety of topics in maize cytogenetics, including crossing over and basic cytogenetic principles, cytoplasmic male sterility, centromeric misdivision, the first transposon-type mutator system, a nuclear gene (*iojap*) that affects the chloroplast genome, meiotic mutations (including ameiotic 1), meiotic drive by abnormal maize chromosome 10, properties of heterochromatin, and the effect of B chromosomes on heterochromatin. He served as editor of *Maize Genetics Cooperation Newsletter* from 1932 to 1935 and again from 1956 to 1974. He was editor of the journal *Genetics* from 1940 to 1948. RHOADES and DEMPSEY demonstrated that the system producing the chromosomal breakage contains two components. It requires at least two B chromosomes plus a specific inbred genetic background to be effective. Under these circumstances, chromosomes with knobs undergo frequent chromosome breakage in the pollen. The breakage is visualized by the expression of a recessive phenotype in a homozygous recessive × homozygous dominant cross. The system came to be known as "high loss" due to the frequent elimination of dominant markers from knobbed chromosomes. A further value of this system was the discovery of new transposon systems (cf. ▶ B. McCLINTOCK). Chromosome breakage in maize seems to stimulate the activation of transposons. New transposons were reported and analyzed in 1989.

RICHEY, F. D. ▶ EMERSON, R. A. ▶ RHOADES, M. M.

RICK, C. M. (1915–2002): Charles M. RICK was born in Reading, Pennsylvania (United States). He took his BSc degree at Pennsylvania State University and earned his PhD at Harvard University in 1940. Afterward, he joined the faculty of the Vegetable Crops Department at Davis, where he remained for his career of more than 60 years. RICK used to be the world's foremost authority on tomato genetics. In addition to the thorough studies of tomato genes and chromosomes, he organized numerous plant-collecting expeditions to the Andes to sample the wide range of genetic variation found in the wild species but missing from the modern domestic tomato. RICK established and directed the C. M. RICK Tomato Genetics Resource Center at the Davis University of California (United States), which serves as a permanent bank of genetic material for tomato.

RIMPAU, W. (1842–1903): He was a German agronomist and breeder working in Schlanstedt. He elaborated important scientific and practical fundamentals of cereal breeding and described for the first time the self-sterility of rye. One of the most famous commercial grain cultivars at this time was the "Schlanstedter Roggen" (Schlanstedt rye), which was produced by him through 20–25 years of meticulous mass selection (MEINEL 2003). In 1883, he selected from a two-rowed barley a plant with branched spikes, which could be bred true by constant selection. In 1877, he discovered within a red-glumed and awnless landrace of wheat three new types of spikes: (a) spikes showing awns, (b) white-glumed plants, and (c) a compactum-type of spike linked with stiffy straw. He was also first to produce, in 1888, a fertile octoploid wheat–rye hybrid, which can be taken as the birth of the triticale research. He already knew that in 1876 the botanist A. S. WILSON had presented stalks of two wheat–rye hybrid plants produced by him in 1875 to the Edinburgh Botanical Society (United Kingdom). He also mentioned attempts on production of wheat–rye hybrids by BESTEHORN and CARMAN (1882).

RÖMER, T. (1883–1951): He was a German agronomist, with strong contributions to plant breeding at the University of Halle/S (Germany). He initiated the utilization of statistic tests in agricultural and breeding research, programs on resistance, and quality breeding in cereals. Under his guidance, more than 20 varieties of winter wheat, spring wheat, winter barley, spring barley, and oats pea were developed. He was one of the editors of the *Handbuch der Züchtung landwirtschaftlichen Kulturpflanzen* (P. Parey Verl., Berlin), with five volumes.

ROSEN, J. A. (1877–1949): He was born in Moscow and studied agronomic sciences in Russia and Germany. He was exiled to Siberia for his political involvement with the Russian Social-Democratic Party (Mensheviks). In 1903, he emigrated to the United States, where he completed his agronomic training. Joseph A. ROSEN gained international renown in the field of agriculture after he developed a new variety of winter rye, which was named ROSEN's rye after him and which was widely used by US farmers (cf. SCHLEGEL 2014). When he was a Russian student at Michigan Agricultural College in 1908, he was sent for the original rye seed back to his father in his homeland. Successive selections of desirable plant were made in this population by F. A. SPRAGG, and, in 1912, the first bushel of ROSEN's rye went from Michigan station to Carleton Herren of Albion. At that time, probably the most famous of ROSEN's rye growers were George and Louis HUTZLER of South Manitou Island in northern Lake Michigan. Here, in perfectly isolated island fields, these growers worked with ROSEN's rye of the finest, purest stock. Other growers could renew their supplies when cross-pollination contaminated their seed crop. During the beginning of the twentieth century, ROSEN's rye was grown worldwide. First in 1934, rye from Poland was again imported to the United States, in order to widen the spectrum of rye varieties and the gene pool of cropped rye. Later, he became an official of the American Jewish Joint Distribution Committee. In the 1920s and 1930s, he organized and coordinated relief activities for impoverished Jews in the former Soviet Union. Joseph ROSEN was a director of the American Jewish Joint Agricultural Corporation (Agro-Joint), which tried to develop Jewish settlements and assisted with organization of Jewish factories, cooperatives, schools, and healthcare facilities.

ROUX, H. S. ▶ PIENAAR, R. de V.

RUDORF, W. (1891–1965): He was born in Rotingdorf (Westphalia, Germany), where his father was a practicing farmer. He visited the elementary and private secondary schools in Werther (1898–1907) and, for the last 4 years, in Bielefeld, where he passed his qualifying examination (Abitur) in 1913. In 1920, he took up university education at the Agricultural College of Berlin and graduated with a diploma in agricultural sciences in 1923. After moving to the University of Halle/S. (Germany), he was recruited by ▶ T. RÖMER as candidate for a doctor's degree and he finished his thesis in 1926 on "Statistical Analyses of Variation in Varieties and Lines of Oats." From 1927 to 1929, he became an assistant of the directing manager of the large agricultural estates of WENTZEL in Teutschenthal and Salzmünde (Germany) and concluded this period with an inaugural dissertation (D.Sc.) at the University of Halle/S. in the field of agronomy and plant breeding entitled "Contribution to Breeding for Immunity Against *Puccinia glumarum tritici.*" In 1929, RUDORF was invited for professorship at the University of La Plata in Argentina. There, he founded and established the Instituto Fitotécnico in Santa Catalina and served as its director until 1933. After a three-year period as an ordinary professor and director of the Institute of Agronomy and Plant Breeding of the University of Leipzig, RUDORF was appointed director of the "Kaiser Wilhelm Institut für Züchtungsforschung" in Müncheberg (Germany) following the late ▶ E. BAUR. At the same time, he became professor of breeding research at the University of Berlin. In 1945, the "Kaiser Wilhelm Institut" was transferred to Western Germany and found an almost ten-year interim shelter at a state property in Voldagsen. Only in 1955 was the new homestead prepared for the (now) Max Planck Institute for Breeding Research in Köln–Vogelsang (Germany). RUDORF retired I 1961. His particular interest continued to be directed toward plant pathogen resistance and wide crosses with related species to achieve progress in wheat resistance. After the death of G. STELZNER in the last days of the World War II, he took over the charge of direction of the Potato Division of the Institute and engaged himself actively in experimental work, not only on *Phytophthora* and virus but also on beetle resistance. With his own experiments, RUDORF also participated in breeding for combined *Gloeosporium* and virus resistance of *Phaseolus* beans. He conducted extensive investigations on medicinal plants, such as *Datura, Digitalis,* and *Mentha*, and described experimental-induced mutants with potential breeding value, for example, *compactum* forms of *Festuca pratensis*, unifoliata mutants of *Medicago sativa,* and leafy, finely branched types of *Melilotus albus*. He also developed a lime-tolerant alfalfa variety, a highly vigorous and productive white clover, and the low-coumarin melilot variety "Acumar." He created synthetic rapeseed from interspecific crosses of turnip × cabbage and gained considerable progress with soybean in Germany by combining earliness, yield, and other important traits of performance. RUDORF was one of the first in Germany to recognize the crucial importance of heterosis in breeding for yield characters, and he effectively supported hybrid breeding in maize by elaborating efficient breeding designs. He devoted himself to many public duties and cooperation within the German Agricultural Society (DLG) and the Federal Association of Plant Breeders (BDP). He founded the most successful German study group on potato breeding and seed potato production and continued at his Cologne Institute the seminars for practical plant breeders initiated by his predecessor ▶ E. BAUR. He set out in 1958 for a several months' lecture and study trip through North and South America and lectured as honorary professor at the University of Göttingen (1946–1955) and the University of Köln (1956–1961), thereby supervising 13 PhD theses. Together with ▶ T. RÖMER, he edited the 5 (1st ed.) and 6 (2nd ed.) volumes of the *Handbook of Plant Breeding*, published by P. PAREY, Berlin (1st edition in 1941–1950; 2nd edition in 1958–1962) and also contributed many articles to this reputed monograph. He published more than 120 original papers in high-ranked international journals.

RÜMKER, K. H. von (1859–1940): He was a German agronomist who contributed a lot to scientific agronomy and to the development of new crop varieties. In 1889, RÜMKER applied the method of distinction of hereditary variation between and within certain systematic groups to breeding of races ("Sorten"). Their heredity could be modified by mass selection; however, new races originated from "spontaneous variations." The mutation theory of ▶ VRIES also contained this kind of hierarchy. New elementary species originated by what VRIES called "progressive mutation;" this corresponded to the creation of a new sort of "pangene."

RUSSELL, W. A. ▶ BOJANOWSKI, J. S.

RUTGER, J. N. (1940–): He received his MSc Agriculture degree from the University of California, Davis, in 1962 and his PhD in 1964. Since 1960, he served at the Department of Agronomy and Range Sciences (1970–1979, Research Geneticist, 1979–1980, Research Leader and Location Leader, and 1980–1988, USDA Agricultural Research Leader). He was honored for his research on induced mutations in rice genetics. He was very successful in developing early-maturity mutants, endosperm mutants, elongated uppermost internode mutants, genetic male steriles, low-phytic acid mutants, giant embryo mutants, and semi-dwarf basmati and jasmine germplasm and varieties at Agricultural Research Service (USDA) in Davis (United States). He developed methods of asexual seed production in rice (apomixis) and searched for transposable genetics elements to use as "tags" for cloning genes whose primary products were unknown. This latter research goal represented a necessary step to applying the then-emerging molecular techniques to rice improvement. In 1993, as the first director of the Dale Bumpers National Rice Research Center, Mississippi, RUTGER recruited staff and developed the center into a world-class facility, which included the Rice Genomics Facility and the Genetic Stocks Oryza Collection. In 1982, he was elected as fellow of the American Society of Agronomy. In 1986, he received the Rice Industry Research Award for the variety "Calrose 76," first semi-dwarf in 1976, followed by "M-201."

SAGERET, M. (1763–1851): He was a French naturalist and agronomist and a member of the Socie'te' Royale de Centrale d'Agriculture de Paris. By the botanist BROGNIART, a plant genus *Sageretia* was named after him. His hybridization experiments within the family of *Cucurbitaceae* were important, where he—for the first time in the history of plant hybridization— arranged the parental characters in an opposite scheme. In the segregation studies, the term "dominant" is clearly introduced.

SÁGI, F. (1927–2006): After school and study at the Eötvös Loránd University in Budapest (Hungary), he was first employed as a teacher in a secondary school on countryside. In 1956, he got the chance to start a scientific career at the Horticultural Research Institute in Fertőd (Hungary), where he later became head of the Laboratory of Plant Physiology (1968–1973). Initially, he worked on fruit chemotaxonomy, followed by investigations on the hormonal regulation of morphogenesis in berries. He heavily contributed to the understanding of indoleacetic acid metabolism in plants. Drom 1973, he worked for the Cereal Research Institute in Szeged, for more than 10 years as head of the Central Research Laboratory, to assist the wheat breeding program with novel analytic techniques; thus, he intensively studied the genetic and physiological mechanism of hybrid vigor and demonstrated experimentally the role of mitochondrial complementation in the background of heterotic effects in wheat. In addition, he contributed to research on a cell and tissue culture system for production of somaclonal variants and for regeneration of haploid plants in wheat as well as development of a protoplast-to-plant system, which were essential for establishing the molecular breeding capacity in the Szeged Institute during the 1980s. He was co-initiator of the periodical journal, *Cereal Research Communications* (Hungary).

SAKAMURA, T. ▶ KIHARA, H.

SAUNDERS, C. E. (1867–1937): He was born in London (United Kingdom). He selected, tested, and introduced the famous wheat variety "Marquis" to the Canadian West, the beginning for the large commercial production of high-quality bread wheat in Canada. When he grew up, he already assisted his father in his many varied interests, such as plant hybridization and entomology. In 1903, his father, recognizing his meticulous standards and perseverance, appointed him to the Experimental Farms Service as experimentalist. SAUNDERS immediately applied scientific methods to his new task and spent summers selecting individual heads of wheat from breeding material that had previously been selected in mass. From a cross of "Hard Red Calcutta" × "Red Fife," made in 1892 by his brother A. P. SAUNDERS, a new variety, "Markham," resulted; however, "Markham" did not produce uniform offspring, even though many plants had desirable characteristics. SAUNDERS again carefully selected individual heads from early plants that had stiff straws. He emphasized that seeds from each plant was grown separately, with no mixing of strains. Selection was rigorous, and only the top lines were kept. He determined which lines had strong gluten by chewing a sample of kernels, and he introduced the baking of small loaves to measure volume. The best strain was named "Marquis." In 1907, all surplus seeds were sent to Indian Head (Saskatoon) for further testing. According to SAUNDERS's annual reports, the response of "Marquis" to Saskatchewan conditions was extremely good, considering earliness and baking quality.

SAVITSKY, V. F. (1902–1965): He was a Russian-born plant breeder. He emigrated after the World War II to the United States. He successfully bred monogerm beets and is considered to be the father of monogermity of cultivated sugar beet. His daughter, Helen SAVITSKY, became a recognized beet cytogeneticist. She produced a first trisomic series of beet.

SCHELL, J. S. (1936–2003): He was one of the pioneers of plant genetic engineering technology and discoverer of the transfer mechanism of bacterial genes into plants. In 1976, he demonstrated the plasmid of *Agrobacterium tumefaciens* as tumor-inducing vehicle in plants, when he was working at the University of Gent (Belgium). After the so-called "Green Revolution," he proposed the term "Gene Revolution." The following studies led to the first transgenic plants. Therefore, ▶ J. G. MELCHERS proposed him as new director of the "Max-Planck-Institut für Züchtungsforschung," Köln-Vogelsang (Germany). He was a member of the New York Academy of Sciences (United States) and of the National Academy of Sciences (India), chairman of the Council of the European Molecular Biology Organization (EMBO) in Heidelberg (Germany) and of the Scientific Advisory Board of the Otto-Warburg Centers in Rehovot (Israel), honorary member of the Academy of Arts and Sciences in Cambridge (United Kingdom), and member of the Royal Swedish Academy, Stockholm.

SANFORD, J. C. (1950–): John C. SANFORD is an American plant geneticist who started to work as a research scientist at Cornell University in the 1980s, after receiving his BSc in horticulture (1976) from the University of Minnesota–St. Paul, his MSc in plant breeding and genetics (1978) from the University of Wisconsin–Madison, and his PhD in plant breeding and genetics, (1980) from the University of Wisconsin–Madison. He co-invented the "gene gun" approach to genetic engineering of plants. This technology has had a major impact on agriculture and breeding around the world. Almost all the early transgenic crops were transformed with the "gene gun," especially maize and soybeans. His book *Genetic Entropy and the Mystery of the Genome* (2005) completely re-evaluated the evolutionary genetic theory, showing the problems underlying classic neo-Darwinian theory. After his opinion, Darwinian theory fails on every level; it fails because mutations arise faster than selection can eliminate them. He published over 100 scientific papers, plus several dozen patents and two books. He was awarded with Adjunct Associate Professor of Botany (Duke University, Durham, North Carolina), Distinguished Inventor Award (1995), and Distinguished Inventor Award (1990) of Central New York Patent Law Association, among others.

SAX, K. ▶ KIHARA, H.

SCHINDLER, O. (1876–1936): He was a German gardener and from 1922 to 1936 the director of the State Experiment and Research Station for Horticulture in Pillnitz (Germany). His particular interest was focused on the development of better strawberries and apple trees. In 1925, he bred the strawberry variety "Mieze Schindler" that is still successfully grown in Germany, besides the apple root stock "Pi 80."

SCHNELL, F. W. W. (1913–2006): Wolfgang SCHNELL was a German professor of applied genetics and plant breeding at the University Hohenheim (Germany). He attended high school (Gymnasium) in Celle, Halle/S., and Leipzig. After having passed the final examinations (Abitur) in 1931, he served his apprenticeship in agriculture on a farm near Hamburg (1932–1934); thereafter, he studied agricultural sciences in Berlin, Munich, and Göttingen (1935–1939). In 1949, he received a doctoral degree in agricultural economics at the University of Göttingen. Then, he changed the subject and attended a 2-year training course in plant breeding at the Max Planck Institute in Voldagsen (later Cologne). In 1952, he got an appointment at a Max Planck Institute branch station in Scharnhorst, close to Hannover. There, he was responsible for the cross-pollinated cereals maize and rye. He focused on the general principles of plant breeding methodology and the genetic foundations of heterosis. In 1958, he spent 6 months at the North Carolina State University in Raleigh (hosted by C. C. COCKERHAM, R. E. COMSTOCK, and H. F. ROBINSON). The visits greatly stimulated his research, since he became acquainted with cutting0-edge research in statistics, quantitative genetics, and breeding methodology. Five years later, he earned his DSc (habilitation) at the University of Göttingen. In the same year (1963), he was appointed full professor and director of the newly established Institute of Plant Breeding at the University of Hohenheim (Stuttgart, Germany). There, he headed the Chair of Applied Genetics and Plant Breeding, until his retirement in 1981. He intensively promoted German hybrid breeding in maize and started in 1953 the first hybrid breeding in rye, which is now successfully established in European agriculture.

SCHRIBAUX, E. ▶ TSCHERMAK-SEYSENEGG, E. von.

SEARS, E. R. (1910–1991): He was born in Bethel, Oregon (United States). He obtained his BSc degree from the University of Oregon and graduated at Harvard University. He was appointed by USDA as a research geneticist and started his long association with the University of Missouri in 1936 until his retirement in 1980. Both by his theoretical and practical contributions, he became the "father of wheat cytogenetics." The discovery of a low level of female fertility in a wheat haploid and the recovery of aneuploid progeny led to the construction of a vast range of aneuploids that is unequalled in its versatility, practicality, and creativity in any other species known to man. He developed the concept that chromosomes from three species contributed their genomes to bread wheat. This concept of so-called "homoeology" is now fundamental to perception of all allopolyploid species. Since 1950, he changed the course of experimental manipulation of crop plants by producing interspecific chromosome additions, substitutions, and translocations. For the first time, he was able to transfer a gene for leaf rust resistance from an alien chromosome into common wheat. His joint work with Japanese M. OKAMOTO led to deeper understanding of how genes regulate chromosome pairing in polyploid wheat. The recognition of how homoeologous chromosomes are prohibited from pairing and thus immediately allowing a high degree of fertility in a polyploid plant became a cornerstone for modern chromosome manipulations. Despite the influence on molecular genetics and biotechnology, his basic contributions to wheat cytogenetics heavily influenced crop genetics in general.

SEGURA, J. ▶ BEADLE, G. W.

SENGBUSCH, R. von (1898–1985): He was born in Riga (Latvia). From 1918 to 1919, he studied agronomy at the University of Halle/S. (Germany). He received his PhD degree at the same university in 1924 under the supervision of ▶ Prof. T. RÖMER. After several services in Klein Wanzleben, Berlin, Müncheberg, Petkus, and Göttingen, he moved to Max Planck Institute of Breeding Research Hamburg–Volksdorf, where he became emeritus in 1968. His main merit was the selection of low-alkaloid lupines. His rapid seed testing method was the basis of modern screening procedures. Already in 1931, he started selection for nicotine-free tobacco and cannabinol-free hemp. He contributed to the sex-specific inheritance of hemp, spinach, and asparagus. His screening method was also successfully applied to the selection of low-alkyl resorcinol plants in rye. From 1945, he started strawberry breeding. In 1954, the variety "Senga Sengana" was released, showing high yields and good industrial adaptability. It became the all-round variety in the world for many years and is still grown in many countries. SENGBUSCH represented the prototype of a scientist—educated, well-experienced, and focused to the target.

SHARKOV, T. ▶ IVANOV, I. V.

SHIRREFF, P. (1791–1876): P Shirress, the third son of J. SHIRREFF, was born in Mungoswells (Scotland). He became a famous Scottish breeder in Haddington, where he carried out his cereal research. He was one of the best-known oats and wheat breeders of the middle decades of the nineteenth century in the United Kingdom. He published a book *Improvement of Cereals* and developed a series of varieties by selection of useful genotypes within landraces ("Mungoswells," 1819; "Shirreff's Bearded Red," "Shirreff's Bearded White," and "Pringle," 1857; and "Shirreff's Squarehead," 1882). He was one of the first who made use of hybridization between wheat varieties. His cross of "Talavera" × "Shirreff's Bearded White" resulted in the varieties "King Richard" (1850) and "King Red Chaff White." Also, the oat varieties "Early Fellow," "Fine Fellow," "Long Fellow," and "Early Angus" were released around 1865 and were widespread in the United Kingdom.

SHULL, G. H. (1874–1954): He was an American botanist and plant geneticist who discovered the heterosis effect in maize. He acquired his doctorate in 1904 and worked thereafter primarily at the Carnegie Institute in Cold Spring Harbor, New York, and at Princeton University (United States). He developed a method of breeding maize that made the seed capable of thriving under various soil and climatic conditions, as a result of which yields per acre were increased 25%–50%. The production of the so-called "hybrid corn" has been connected with his name forever. Soon after the re-discovery of the Mendelian laws, he started genetic studies together with Prof. DAVENPORT at Cold Spring Harbor (New York, United States) on beans, poppy, *Melandrium*, and *Digitalis* and studied mutations in *Oenothera*. He developed his first hybrids before 1910, but commercial production did not begin until 1922. He also founded the American journal *Genetics* in 1916.

SKOVMAND, B. (1945–2007): He was born in Copenhagen (Denmark) and was a highly respected and well-liked person internationally. He came to the University of Minnesota (United States) through a student agricultural training program in 1966. There, he earned his B.Sc., M.Sc., and PhD degrees in plant pathology. Sir Bent SKOVMAND, the Knight of Denmark, fought for the good of humanity up until his passing. With his great scientific knowledge and unique personality, he has left his mark especially in the scientific world within his two favorite areas, wheat breeding and plant genetic resources. He advanced in biology and specialized in plant diseases, obtaining his PhD in 1976. Throughout his university years, jobs became more sophisticated, and his last years were financed through a Rockefeller scholarship. His first assignment was as a plant breeder at CIMMYT in Mexico, working on the then novel triticale cereal. He became head of the program but later specialized in wheat and was in charge of the breeding program for wheat in Turkey, a UNDP project. Back at CIMMYT, he became head of wheat breeding and the gene bank, with its thousands of accessions. In 2003, he was appointed as director of the Nordic Gene Bank, situated in the southern part of Sweden, for agricultural and horticultural plants and their wild relatives. In 2005, he received the international Crop Science Award. In the same year, the Agricultural University of Copenhagen, Denmark, awarded him an honorary post as professor.

SOMORJAI, F. (1900–1981): He retired as a director of the Cereal Research Institute, Szeged (Hungary). Born in Nagykoros, he took a degree in agriculture in Keszthely and Budapest. In 1927, he moved to the Cereal Research Institute, Szeged. He took part in first Hungarian experiments on rice production. He bred the "Szegedi Yellow Dent Corn," "Szegedi Angustifoliate blue-grass," "Szegedi wheat-grass," and "Ujszegedi winter oat."

STELZNER, G. ▶ RUDORF, W.

SPRAGUE, G. F. ▶ BOJANOWSKI, J. S.

STAKMAN, E. C. ▶ WATERHOUSE, W. L.

STRAMPELLI, N. (1866–1942): He was a famous Italian breeder. His wheat varieties "Ardito," "Mentana," "Villa Glori," "Damiano," and "San Pastore" were distributed around the world prior to the "Green Revolution."

STRUBE, F. (1847–1897): He was a German agronomist and breeder working in Schlanstedt. He bred several wheat, oat, and rye varieties. He founded one of the most successful breeding companies, which is still active in Germany.

STRUBE, H. (1878–1919): He was a German breeder in Schlanstedt. After the death of his father, F. STRUBE, he became head of seed breeding "Fa. Friedrich Strube Saatzucht." In 1911, he established an experimental station in Guty (today Poland), concerned with seed reproduction for the eastern seed market, and examined his own seeds regarding winter hardiness. In 1913, he founded an examination station on the outskirts of the "Lüneburger Heide" (German heathland region) to examine the cultivation on light habitat.

STUBBE, H. (1902–1989): He was a German geneticist who contributed a lot to mutation research and its application in plant breeding. He became the first director of the "Kaiser Wilhelm Institut für Kulturpflanzenforschung" in Vienna (Austria). This institute was transferred after the World War II to Germany, near Gatersleben. Based on this institute, the further "Institute of Crop Plant Research" was established in Gatersleben (now: Institute of Plant Genetics and Crop Plant Research). One of the main ideas was to establish a world crop plant collection (later called: gene bank) and its utilization for breeding approaches.

STUBBS, A. (1913–2003): Annabelle STUBBS was inspired and encouraged by the successful fuchsia hybridizer Roy WALKER. Annabelle began a breeding program that resulted in the introduction of 73 new fuchsia cultivars between 1970 and 1997. Her "Pink Marshmallow" (1971) won popularity contests throughout the fuchsia world. In the early 1970s, she moved to Fort Bragg on the northern California coast, where she established Annabelle's Fuchsia

Gardens. She, along with Bill BARNES, Cliff EBELING, and Barbara SNYDER, was the founding member of the Mendocino Coast Branch of the American Fuchsia Society. Later, she established a home nursery in Oceanside, called Fire Mountain Fuchsias, and more fuchsias were born, including the cultivar "Fire Mountain" in 1980. While in Oceanside (California), Annabelle was the driving force for the creation of a beautiful fuchsia garden at the San Diego Wild Animal Park near Escondido. During the 1980s, she moved to Eureka. Here, the fuchsia "Eureka Red" (1991) was bred.

STURTEVANT, A. H. (1891–1970): He was the youngest of six children. His grandfather, J. STURTEVANT, was a Yale graduate, a Congregational minister, and one of the founders and later president of Illinois College in Jacksonville, Illinois (United States). His father taught mathematics for a while at that college but later took up farming, first in Illinois and later in southern Alabama, where the family moved when STURTEVANT was 7 years old. He went to a country school and later to a public high school in Mobile (Alabama). At the age of 17 years, he entered Columbia University. He developed interest in genetics as a result of tabulating the pedigrees of his father's horses. He continued this interest at Columbia and also collected data on his own pedigree. On his brother's suggestion, he read the textbook on Mendelism by R. C. PUNNETT (1907). He saw at once that Mendelism could explain some of the complex patterns of inheritance of coat colors in horses that he and others before him had observed. His 16-year older brother, Edgar, encouraged him also to write an account of his findings and take it to ▶ T. H. MORGAN, who at that time was professor of zoology at Columbia. MORGAN encouraged Sturtevant to publish the paper, and it was submitted to the *Biological Bulletin* in 1910. The other result of STURTEVANT's interest was to explain the relation between inversion sequences in different species, when he was given a desk in the famous fly room at Columbia University, where, only 3 months before, MORGAN had found the first white-eyed fly. During his Caltech period, he collaborated especially with ▶ S. EMERSON, ▶ T. DOBZHANSKY, and ▶ G. BEADLE. Later, STURTEVANT developed a keen interest in the history of science. His classic book *A History of Genetics* was published in 1965.

SUTTON, W. (1877–1916): He was an American geneticist. Together with T. BOVERI, he demonstrated during the years 1902–1904 that chromosomes segregate as genes.

SWAMINATHAN, M. S. ▶ ANDERSON, R. G.

SWINGLE, W. T. (1871–1952): He was born in 1871 in Canaan Township in Wayne County, Pennsylvania (United States). In 1873, the family moved to Manhattan (Kansas), where they farmed nearby. In 1885, SWINGLE began to study at the Kansas State Agricultural College in Topeka. At the age of 16 years, he wrote an essay on cereal-rust fungi. From 1888 to 1891, he was assistant botanist at the Kansas Agricultural Experiment Station. In 1890, he received his BSc degree from the Kansas State College of Agriculture in Manhattan. In 1891, he had already published 21 collaborative works with Professor William A. KELLERMAN and six of his own works. By the end of 1891, he moved to Washington, D.C., to work at the USDA in the new Vegetable Pathology Department headed by Beverly T. GALLOWAY, dealing with rust diseases of citrus fruits; therefore, his research work was mainly in Florida. By a cross of grapefruit "Duncan" × "Dancy," he produced his hybrid variety, the "Minneola" orange. Further tangelo varieties were "Sampson" (1897), "Thornton" (1899), and "Orlando" (1911). He was also involved in the development of the "Murcott" orange (mandarin–sweet orange hybrid) and kumquat varieties. From 1895 to 1896, he studied with Eduard STRASBURGER at the University of Bonn (Germany). In 1896, SWINGLE received his MSc degree from Kansas State Agricultural College and in 1922 his PhD In 1899, he conducted research on figs in Naples and published research on the culture of figs and dates. In 1900, he brought 405 plants of the date palms varieties "Deglet Noor" and "Rhars" from Algeria to the United States. They were planted in Tempe, Arizona. He also imported fig wasps (*Blastophaga* sp.) from Algeria as pollinator for fig plantations in California. In

January 1941, SWINGLE retired and became a consultant at the University of Miami for the botany of tropical plants and head of the "Swingle Plant Research Laboratory;" the genus *Swinglea* Merr. (Rutaceae) has been named after this great plant breeder and botanist.

TATUM, E. ▶ BEADLE, G. W.

TAVCAR, A. (1895–1979): Since 1922, he was Head of the Department of Genetics and Plant Breeding of Agricultural Faculty, University of Zagreb (Croatia). During his fruitful life, he was occupied with research on maize, wheat, barley, rye, and some other crops. He recognized the importance of plant resistance to certain wheat diseases (1927) and winter hardiness (1929–1930) for stable grain production. He studied the relation between morphological and some physiological characteristics of wheat plant (1930–1934). Between the two World Wars, by means of individual selection from the cultivar "Sirban Prolifik," he released the cultivar "Maksimirski Prolifik 39," and by means of crossing and pedigree selection in segregating progeny, he released the cultivars "Maksimirska Brkulja 530," "Maksimirska Brkulja 540," and "Maksimirska Brkulja 24;" those varieties were grown in the western part of Croatia and Bosnia between 1930 and 1959. In the 1950s, he started a pioneer work in Yugoslavia on mutation breeding of crop plants. Gamma rays-induced mutants, produced by ^{60}Co irradiation, were tested under field conditions nationwide.

TEDIN, T. ▶ NILSSON-EHLE, H.

TIMIRIAZEV, K. A. ▶ NAVASHIN, S. G.

TIMOFEEFF-RESSOVSKY, N. ▶ GREBENSCIKOV, I. S.

TROLL, H.-J. (1906–2004): He was professor of plant breeding at the University of Leipzig (Germany). After school in Flensburg (1922), he studied at "Landwirtschaftliche Hochschule," Berlin (1924–1927), where he received his PhD. degree at the "Kaiser Wilhelm Institut für Züchtungsforschung," Müncheberg, in 1930 (his supervisor was ▶ Prof. E. BAUR). His name is closely associated with lupine breeding in Germany. He was able to select a white-seeded strain (No. 8) in *Lupinus luteus*—the basis of several varieties, such as "Weiko." TROLL was the first university teacher who stimulated the interest in plant breeding, genetics, and research of the author of this book.

TSCHERMAK-SEYSENEGG, E. von (1871–1962): He was born in Vienna (Austria) as the third child in a well-known Viennese family of scientists. His father, G. TSCHERMAK (1836–1927), was professor of mineralogy; his grandfather, E. FENZL (1808–1879), was a professor of botany and director of the botanical garden in Vienna; Tschermak received his basic education from the padres of the clerical gymnasium at Kremsmünster. He began to study agricultural science in 1891 at the former "Hochschule für Bodenkultur" (now: University of Agricultural Sciences, Vienna) and simultaneously biology at the University of Vienna. TSCHERMAK interrupted his studies and worked for 1 year (1892–1893) as an agricultural volunteer in a manorial estate in Freiberg/Saxonia. Stimulated by friends, he dislocated his study place from Vienna to Halle/S. (Germany), where he continued to study agriculture from 1893 until 1895 (BSc in 1894, PhD in 1896). His friendship with ▶ K. von RÜMKER opened to him a scientific approach to plant breeding. To support his professional career, his father sponsored his training in a breeding station for vegetables and ornamentals in STENDAL (1896–1897) and in Quedlinburg (1897–1898). He was confronted for the first time with practical breeding activities, such as mass selection in legumes, cabbages, lettuce, and ornamentals. Later, he acquired experience in crossing techniques, a skill that significantly determined his whole scientific life as a plant breeder. Back to Vienna in 1898, he found a position as a university assistant. Simultaneously, he traveled to Gent (Belgium) and Paris (France), where he saw huge horticultural estates and legume-breeding stations. He received the permission to start crossing experiments with wallflowers (*Erysimum cheiri*) and garden pea (*Pisum vulgare*) at the botanical garden in Gent. In the library, he found DARWIN's book *The Cross and Self-Pollination in the Kingdom of Plants*, which inspired him to search about seldom pollination effects leading to xenia, neighboring pollination phenomenon (geitonogamy), and crossing effects between individuals of the same species (heteromorphic xenogamy). After the experimental time in Gent, the seeds of his experiments were sent home to Vienna, while he traveled to Dutch botanist ▶ VRIES in Amsterdam. VRIES showed him his mutants of *Oenothera lamarckiana* but did not tell him about his recent experiments with peas. After another study tour to France (1898), where he met ▶ H. L. de VILMORIN in Paris (and saw his huge facilities for vegetable and mushroom breeding) and Prof. E. SCHRIBAUX, who led the small Institute of Plant Breeding in the Agricultural College in Grignon, near Paris. When TSCHERMAK returned to Vienna, he continued the work with the progenies of his pea crosses. He observed the phenomenon of xenia in seed pods of the same F1 plants, whereby seed color and seed shape resemble the difference in parental characters at this early stage after hybridization. These observations and the results of some backcrossing procedures in which parental characters appeared in a 1:1 segregation scheme when the hybrids were crossed again with their parental types formed the basis of his DSc thesis (1900). In this publication, he demonstrated and discussed some of the results of his studies—similar data as had already been achieved half a century ago by ▶ G. MENDEL (1866). Soon after, TSCHERMAK became fully aware that these were the fundamental principles of inheritance. To prevent his leaving for a new professorship at Brno (Technical University) and Wroclaw (University), the Viennese authorities of the "Hochschule für Bodenkultur" promoted him to the position of an assistant professor (1903) and founded in 1906 a separate Chair of Plant Breeding for him. It was the first established Chair for Plant Breeding in Europe. At the experimental farm of the "Hochschule für Bodenkultur" in Groß-Enzersdorf, he founded the first plant breeding station in 1903. The Prince of Liechtenstein showed interest in this prospective field of agricultural developments and founded in 1913 a new institute for plant breeding, the "G. Mendel Institute," in Lednice. There, he started selection of early pea types from crosses "pois acacia" × "pois à cinq" and nearly all his work with ornamentals. His cooperation (since 1904) with the nestor of the Moravian plant breeders, E. von PROSKOWETZ, extended his activities, particularly in the field of spring barley breeding. Several further institutions were founded by TSCHERMAK, such as Kvasice (1904) and 18 experimental field and plant breeding stations in Moravia, Bohemia, Western Hungary, and Lower Austria (FEICHTINGER 1932, TSCHERMAK 1958). Rhe list of his varieties reflects his tremendous breeding work: rye—"Tschermak's Marchfelder Roggen" (1926)—"Prof. Tschermak Roggen" (1927); winter wheat—"Weisser begrannter Marchfelder" (1909, land race reselection from Marchfeld area, later registered in 1927)—"Brauner begrannter Marchfelder"—"Hochschulweizen" (1928, from the Hungarian landrace Dioszeger wheat)—"Zborowitzer Kolbenweizen" (reselection of Rimpau's Bastard)—"Russischer Rotweizen" (Red Zborowitz wheat × White Russian wheat)—"Moraviaweizen" (Edelepp × Marchfelder)—"Non plus ultra" (Svalöf's Grenadier × Banatian wheat)—"Non plus ultra I, II, and III" (selections of the above-named cross)—"Glasweizen"—"Schilfweizen" (French Bon fermier × Blé gros bleu)—"Excelsiorweizen" (Banater × Extra Squarehead Master); spring wheat—"Znaimer" (Znaimer × Tucson 1925); 2-rowed winter barley—"Kirsche" (Kirsche × 2-rowed Hanna × 4-rowed Hanna's Riesen 1927); spring barley—"Hanna" or "Kwassitzer Original Hanna Pedigree-Gerste" (reselection of a landrace in the Hanna district, released 1908)—"Hanna Kargyn" (Hanna × Turkish landrace from Kargyn, world champion in the London Exhibition 1927)—"Hanna Kaisarie" (Hanna × Turkish landrace from the district Kaisarie)—"Hanna × Chevalier"—"Hanna × Schwarzenberggerste"—"Hanna × Hannchen;" spring oat—"Tschermak's Frühhafer" (1931, Svalöf's Siegeshafer × Hungarian 60-day oat selection)—"Tschermak's Gelbhafer" (Lochow's Gelbhafer × Svalöf's Goldregen)—"Tschermak's Weisshafer" (Russian Ligowo × Savlöf's Goldregen); garden peas—"Tschermak's veredelte Victoria Erbse" (pois acacia × pois à cinq cosses)—"Saxa × Buxbaum × Saxa;" garden beans—"Tschermak's fadenlose frühe Buschbohne" (Wiener Busch × weissgründige Heinrich's Riesen)—"Tschermak's Feuerbohne" (*Phaseolus vulgaris × P. multiflorus*); faba bean—"Tschermak's weissblühende Ackerbohne;" oil pumpkin—"Tschermak's schalenloser Ölkürbis" as well as numerous vegetable varieties. He published more than 100 scientific papers and also numberless articles in agricultural newspapers and garden journals to keep the agronomists and the gardeners in the mood to buy seeds or root stockings of new varieties. TSCHERMAK also carried out experimentally phylogenetic research. In his different fertile intergeneric hybrids, he suspected that their existence was caused by unreduced gametes of F1 hybrids, for example, wheat × rye or *Aegilops ovata* ($2n = 4x = 28$) × *Triticum dicoccoides* ($2n = 4x = 28$). The latter hybrid was called "Aegilotricum" ($2n = 8x = 56$). It was the first induced and cytologically approved intergeneric goat wheat hybrid (TSCHERMAK and BLEIER 1926). It marked the early beginning of synthetic polyploids for agricultural utilization. The list of genera and species included demonstrates the excellent crossing techniques that he developed (e.g. *Triticum × Secale, Triticum × Agropyron, Aegilops × Triticum, Haynaldia × Triticum, Agropyron × Secale,* radish (*Raphanus sativus f. radicula*) × Wild radish (*Raphanus raphanistrum*), *Secale montanum × S. cereale* (perennial rye), *Hordeum trifurcatum × H. compositium, Avena sativa × A. fatua, A. sativa × A. brevis, Triticum vulgare × T. spelta, T. vulgare × T. durum, T. vulgare × T. turgidum, T. vulgare × T. compositum, T. vulgare × T. dicoccum, T. vulgare × T. polonicum, T. turgidum × T. villosum,* and other species, such as *Phaseolus vulgaris × P. multiflorus, Matthiola incana × M. tricuspicata, Verbascum olympicum × V. phoeniceum, Beta trigyna × B. lomatogon*; TSCHERMAK 1958).

TSUCHIYA, T. (1923–1992): He was a Japanese geneticist born in the Oita Prefecture (Japan). He graduated from the Gifu Agricultural College in 1943, with BSc degree at Kyoto Imperial University in 1947, majoring in genetics. He taught biology, cytogenetics, and human genetics in two institutions in Japan before accepting a position as cytogeneticist at the prestigious Kihara Institute for Biological Research near Yokohama. The experience of working with the internationally recognized geneticist ▶ Prof. H. KIHARA was the turning point of TSUCHIYA's career. KIHARA became TSUCHIYA's lifelong mentor and friend. While working at the Kihara Institute, he completed his DSc degree at the Kyoto University in 1960. In 1963, he moved to the University of Manitoba in Winnipeg (Canada) as a postdoctoral fellow, where he worked for 5 years in the Department of Plant Science. In 1968, TSUCHIYA joined the Department of Agronomy at Colorado State University. He filled the position vacated by the retirement of D. ROBERTSON. His research involved the genetics, cytology, cytogenetics, and cytotaxonomy of barley, sugar beet, melon species, triticale, tree species (e.g., giant sequoia), and *Alstroemeria,* the Peruvian lily. In the last years of his life, he elucidated the genetic and cytogenetic effects in seeds subjected to long-term storage.

VALLEGA, J. (1909–1978): He was born in Italy. His scientific career started in 1931 at the University of Buenos Aires (Argentina) as a plant pathologist. From 1934 to 1943, he was head of the Phytotechnical Institute of Santa Catalina of the University of La Plata (Argentina). He pioneered research on the physiological specialization of rusts on cereals and flax. In 1956, he founded "Robigo," an Argentinean newsletter on rust diseases.

VANDERPLANK, J. E. (1909–1997): James Edward VANDERPLANK was born in 1908 in Eshowe, Zululand. After the primary and secondary schooling, he took a BSc in botany and chemistry from the University of South Africa in 1927, an MSc in botany in 1928, and an MSc in chemistry in 1932. He was awarded the PhD in botany, with an emphasis on physiology, by the University of London in 1935, and a DSc in chemistry by the University of South

Africa in 1944. His research began in plant pathology in Pretoria in 1941, and he became Chief of the Division of Plant Protection of the South African Department of Agriculture in 1958 and of the Plant Protection Research Institute in 1962, a position he held until his retirement in 1973. He was recognized as possibly the greatest plant pathologist who ever lived and who transformed plant pathology into a scientific discipline (THRESH 1998). He developed the concepts and coined the terms "vertical resistance" and "horizontal resistance." He influenced plant pathology mainly through his books and other writings. The first book was *Plant Diseases: Epidemics and Control* (1963), followed by five others, including *Principles of Plant Infection* (1975) and the more controversial *Genetic and Molecular Basis of Plant Pathogenesis* (1978) and *Host–Pathogen Inter-actions in Plant Disease* (1982), extended and elaborated on some of VANDERPLANK's basic ideas, re-arguing the bases of resistance and of epidemic development. Some of his papers were innovative in using simple mathematical concepts to provide fresh insights into the epidemiology and control of diseases. The early papers and the 1963 book were written while Prof. VANDERPLANK was employed in the South African Department of Agriculture. He joined the department in 1941 after postgraduate degrees in botany (Natal University) and chemistry (Rhodes, London, and South African Universities). He was director of the Plant Protection Research Institute, Pretoria, at the time of his formal retirement in 1971. During this period, he was closely involved with the development of potato varieties and the South African seed potato industry. His varieties are still widely grown and account for about 60% of current production. In 1979, he received an honorary degree at the University of Giessen (Germany), including two further honorary degrees, Captain Scott medals of the South African Biological Society, the medal of the South African Association for the Advancement of Science, a Fellowship of the American Phytopathological Society, the Ruth Allen Award (1979), and the Stakman Award of the University of Minnesota (1985).

VAVILOV, N. I. (1887–1943): He was a Russian plant geneticist born in Moscow. He studied at the Moscow Agricultural Institute (now: Timirjasev Academy of Agriculture). In 1907, he graduated from the Agricultural Institute and continued to work at the Department of Agriculture headed by Prof. PRYANISHNIKOV. From 1911 to 1912 he was employed at the Institute of Applied Botany in St. Petersburg that was directed by R. REGEL (1867–1920). From 1913, he visited several institutions in England, France, and Germany. In 1917, he became a professor of genetics at the Agricultural Faculty of the University of Saratov. During the Civil War (1918–1920), Saratov became the scientific stronghold for the Department of Applied Botany (Bureau till 1917). In 1920, VAVILOV was elected Head of the Department, and he soon moved to St. Petersburg, together with his students and associates. In 1924, the department was transformed into the Institute of Applied Botany and New Crops (since 1930, All-Union Institute of Plant Production, VIR) and occupied the position of the central nationwide institution responsible for collecting the world plant diversity and studying it for the purposes of plant breeding. In 1927, for the first time, those ideas were published during the International Congress of Genetics in Berlin (Germany). VAVILOV was recognized as the foremost plant geographer of contemporary times. To explore the major agricultural centers in Russia and abroad, VAVILOV organized and took part in over 100 collecting missions. His major foreign expeditions included those to Iran (1916), United States, Central and South America (1921, 1930, and 1932), the Mediterranean, and Ethiopia (1926–1927). For his expedition to Afghanistan in 1924, he was awarded the N. M. PRZHEVALSKII Gold Medal of the Russian Geographic Society. From 1931 to 1940, VAVILOV was its president; these missions and the determined search for plants were based on the VAVILOV's concepts in the sphere of evolutionary genetics, that is, the "Law of Homologous Series in Variation" (1920) and the theory of the "Centers of Origin of Cultivated Plants" (1926). To the original five world centers delineated by him, four others were added later:

1. China, including Western and Central China (Japanese millet, buckwheat, soybean, rice, common millet, foxtail millet, velvet bean, adzuki bean, turnip, Chinese yam, Chinese radish, ginger, kiwi, Chinese olive, Chinese hickories, Chinese quince, Chinese hazelnut, Oriental persimmon, loquat, litchi, apricot, peach, Japanese plum, Chinese pear, jujube, abutilon hemp, ramie, hemp, tea, and camphor); total 136 species.

2. India, including Indo-Malaya, Indo-China, Burma, and Assam (pigeon pea, guar, millet, rice bean, urad, mung bean, cucumber, Indian mustard, black mustard, sesame, mango, tamarind, jute, sun hemp, kenaf), and Indochina (Job's tears, Jacobean, rice, winged bean, mat bean, taro, winged yam, starfruit, sour and sweet oranges, pomelo, grapefruit, durian, mangosteen, mango, banana, plantain, rambutan, litchi, longan, Manila hemp, sugar palm, and sugarcane); total 172 species.

3. Central Asia, including Northeastern India, Afghanistan, Turkmenistan, and Anatolia (common wheat, rye, oats, onion, garlic, carrot, spinach, walnut, apple, pistachio, apricot, pear, and jujube); total 42 species.

4. West Asia, including Transcaucasia region, Iran, Turkmenistan, and Asia Minor (emmer wheat, einkorn wheat, rye oats, barley, vetches, alfalfa, clovers, plums, pea, lentil, and fig); total 83 species.

5. Near East and Mediterranean coastal and adjacent regions, including regions surrounding the Mediterranean Sea (vegetables, rape, lupins, beets, clovers, pea, lentil, flax, olive, broad bean, serradella, chickpea, barley, grass pea, lentil, lupine, pea, rye, hard wheat, einkorn, bitter vetch, broad bean, carrot, parsnip, parsley, lettuce, rapeseed, black mustard, safflower, linseed, poppy, almond, hazelnut, melon,[1] quince, fig, walnut, apple, date palm, cherry, European plum, pomegranate, pear, grapes, celery, asparagus, and artichoke); total 84 species.

6. East Africa including Ethiopia, Eritrea, and Somalia (coffee, ricinus, sorghum millet, barley (*Hordeum vulgare convar. labile*), emmer wheat, and linseed); total 38 species

7. Southern Mexico–Central America and Antilles (maize, *Phaseolus* beans, sweet pepper, sweet potato, cotton, sisal, cacao, pumpkins, amaranth, sword bean, huauzontle, tepary bean, scarlet runner bean, Lima bean, common bean, maize, pepper, chili pepper, squash, tomato, vanilla, sapodilla, *Annona*, papaya, avocado, guava, sisal, and upland cotton); total 49 species.

8. South America with Bolivia, Ecuador, Peru, Chile, Bolivia, and parts of Brazil (maize, potato, tomato, cotton, peanuts, bananas, tobacco, rubber tree, quinoa, jack bean, achira, yam, sweet potato, manioc, ulluco, cashew, pineapple, Brazil nut, passion fruit, and tobacco); total 62 species.

9. North America (lupines, strawberry, sunflower, and American grape);

some crops originated outside those centers are, for example, *Avena sativa* and *A. strigosa* (Europe), *Cola acuminata* (Western Africa), and *Ribes grossularis* and *Rubus idaeus* (Europe). VAVILOV, the symbol of glory of the national science is, at the same time, the symbol of its tragedy. As early as the beginning of the 1930s, his scientific programs were being deprived of governmental support. In the stifling atmosphere of a totalitarian state, the institute headed by him turned into a resistance point to the pseudo-scientific concepts of ▶ T. D. LYSENKO. As a result of this controversy, VAVILOV was arrested in 1940, and his closest associates were also sacked and imprisoned. He died in the Saratov prison from dystrophia and was buried in a common prison grave.

VETTEL, F. (1894–1965): He was a successful German cereal breeder in Hadmersleben. Some well-known wheat varieties, such as "Heines Teverson," "Heine II," "Heine IV," and "Heine VII," were bred by him.

---

[1] Recent studies trace the origin of cucumbers and melons to Asia, particularly to the Himalaya region.

VILMORIN, P.-L.-F. de (1816–1860): He was a famous French plant breeder. As ▶ J. GOSS in England (1822), he observed during his experiments between 1856 and 1860 the 3:1 segregation for flower color in pea and *Lupinus hirsutus*. He was the first to introduce the principle of individual progeny testing, and in France, he improved sugar beet by continuous selection.

VOGEL, O. A. ▶ KRONSTAD, W. E.

VORONOV, Y. N. (1874–1931): He was a Russian geobotanist and taxonomist and an expert in subtropical plant diversity. From 1918 to 1921, he was the director of Tiflis Botanical Gardens (Georgia). In 1925, he became a research scientist of the Institute of Applied Botany of VIR (Saint Petersburg, Russia).

VRIES, H. de (1848–1935): He was a Dutch botanist who, in 1900, independent of, but simultaneously with, the biologists ▶ E. CORRENS and ▶ E. TSCHERMAK von SEYSENEGG rediscovered ▶ G. Mendel's historic paper on principles of heredity. He became particularly known by his mutation theory. New elementary species originated from what VRIES (1901) called "progressive mutation." This corresponded to the creation of a new sort of "pangene;" within a species, there could occur "retrogressive" and "degressive" mutations, which correspond to the modification of existing pangenes. Retrogression meant the disappearance of characters through inactivation of pangenes, whereas degression meant the re-appearance of a character. The hereditary differences behaved differently in hybridization. Progressive mutations led to what he called "unisexual" hybrids, whereas the hybrids of retrogressive as well as degressive mutations were "bisexual;" only the latter type of hybrid was subject of MENDEL's law. Unisexual hybrids were constant and nonsegregating. VRIES' speculative explanation for this behavior was that progressive mutations of the new type of pangenes would be unpaired in the hybrid. There is no "antagonist," that is, a modified pangene of the same kind with which it could pair up; therefore, progressive mutation was VRIES' main interest, not Mendelism. He also thought that individual selection is the only method needed in plant breeding. VRIES gave currency to modern use of the term "mutation." He provided major inspiration for research on spontaneous change of hereditary factors. His book *Mutationstheorie* from 1901 made a greater impact in some ways than the rediscovery of MENDEL's laws.

WALLACE, H. A. ▶ BORLAUG, N. E.

WARBURG, O. (1859–1938): He was a famous German taxonomist and plant geographer. He explored the tropical flora of South and East Asia. Later in his career, he became an expert on tropical agriculture and settlement in German colonies.

WARD, M. ▶ WATERHOUSE, W. L.

WATERHOUSE, W. L. (1887–1969): He was born in Maitland (Australia) in 1887. He attended Chatswood Public School and Sydney Boy's High School. He was awarded the first Farrer Research Scholarship for a study on "The effects of superphosphate on the wheat yield in New South Wales," published as a science bulletin in 1918. Later, he went to Imperial College in London (United Kingdom) and obtained his diploma. On the return voyage to Australia via the United States, he spent some time in the Department of Plant Pathology at the University of Minnesota, St. Paul; there, he came under the influence of E. C. STAKMAN. While returning to Australia in 1921, he was appointed lecturer and demonstrator in plant pathology, genetics, plant breeding and agricultural botany at the University of Sydney. In 1929, he received the DSc degree. When the early basic studies of the genus *Puccinia* began in 1921, there was a great speculation throughout the world as to the cause of pathogenic variability. M. WARD in England had received some support for his proposals that pathogenicity may be increased as a result of organisms growing on a "bridging" host. It was suspected that the alternate host of this fungus played some role, because it had been known for a long time that there was a connection between the barberry plant and the cereal stem rusts, barberry eradication programs in both Europe and

North America being successful in reducing rust damage in cereal crops. In Australia, it was known that species of barberry were introduced as early as 1859, but prior to 1921, it was widely accepted that the local wheat stem rust organism had lost the ability to infect them. This belief was dispelled by WATERHOUSE, who was able to show that the basis for the confusion lay in the variability within the organism itself. *Puccinia graminis* f. sp. *tritici*, the rust attacking wheat, was found to comprise six strains, which were characteristically Australian. These strains could be separated as dikaryons by their ability or inability to attack a group of 12 different wheats, but he showed that as monokaryons, there were also differences. One strain, which he called race 43, was avirulent on barberry plants; others such as 45 and 46 grew normally on these plants. This was the first evidence of differentiation at the monokaryotic level. In 1928, the real advance was to be made, because he took yet another rust race "34," which when used to infect barberry gave rise in its progeny to two new races, 11 and 56, outstandingly different from the parent, which of course was also recovered. Race 34 was apparently heterozygous for the genes for virulence on the wheats "Einkorn" and "Mindum," and hence, segregation had occurred. A new race, 21, was found by him on *Agropyron monticola* in 1948. He had earlier observed the glasshouse mutation both for spore color and virulence in *P. graminis tritici,* but in 1942, he had his first clear demonstration of the importance of mutation in the field. About 1940, there was a great satisfaction with the contribution that wheat breeders had made toward a solution of rust control. "Thatcher," a hard spring wheat, was thriving in North America and MACINDOE had released "Eureka" in New South Wales in 1938. It was commonly accepted that stem rust was no longer to be feared. WATERHOUSE had released his wheats "Fedweb" and "Hofed," yet while their resistance was still effective, he was always on guard for the fungus to attack any wheat that was currently resistant. The blow fell in 1942, when a rusty crop of "Eureka" was found near Narrabri. WATERHOUSE and WATSON showed that the fungus had changed and a new strain had arisen, which, except for its ability to attack "Eureka," was identical with that against which MACINDOE had obtained resistance. It was a stepwise mutation in the fungus for virulence on plants with $Sr6$, the gene for resistance in "Eureka." Most of the changes that have been observed in the virulence of *P. graminis* involved single gene mutations, in which a particular host gene for resistance became ineffective. It is now know that certain strains of rust, when mixed on a compatible host, will hybridize as dikaryons. WATERHOUSE recognized four formae speciales of *P. graminis: tritici, avenae, secalis,* and *lolii.* The ecological relationships between host and pathogen, which first showed up with the release of "Eureka" (gene $Sr6$) wheat in 1938, were studied by him for 15 years. As "Eureka" became increasingly popular, the mutant strain of 1942 specific for it increased in prevalence, until finally the variety was so susceptible that farmers rejected it in favor of "Gabo," a wheat with a different gene for resistance ($Sr11$). WATERHOUSE's varieties "Hofed" and "Fedweb" were never widely grown, as they had been developed without adequate yield testing. This had come from crosses in which he had attempted to transfer the rust resistance of the tetraploid wheats "Khapli Emmer" (*T. dicoccum*) and "Gaza" (*T. durum*) to hexaploid wheats. He had been attracted to the vigor of material in which "Gaza" had been backcrossed to a special accession of "Bobin," received from J. T. PRIDHAM. This accession is not unlike "Gular," a wheat of good quality and quite unlike the "Bobin" grown commercially. "Gabo," the wheat that will be remembered best and that evolved from WATERHOUSE's early studies, was registered in 1945. When the Rockefeller Foundation Program began to develop in Mexico, "Gabo" had shown its superiority in rust resistance, earliness, and yielding ability. Current Mexican lines can be attributed to the presence of this insensitivity gene, which first became important in "Gabo." It is seldom that a single wheat variety contributes so much in so many different environments. Both "Gabo" and its sib, "Timstein," were widely used in the pedigree of the Mexican wheats, and "Cajeme," "Mayo," "Nainari" and many others can be traced back to them.

WATSON, I. A. (1914–1986): He was educated at the Universities of Sydney, Australia (B.Sc., 1938) and Minnesota, United States (PhD, 1941). In 1938, he became assistant lecturer in agriculture at the University of Sydney. By T. L. PAWLETT Scholarship, he worked at the Department of Plant Pathology and Plant Genetics, the University of Minnesota, St Paul (United States) from 1939 to 1941. After his return to Australia, he was again appointed as assistant lecturer in agriculture, University of Sydney (1941–1944), lecturer in agriculture (1944–1946), senior lecturer in agriculture (1946–1955), head of Section of Genetics and Plant Breeding, Faculty of Agriculture (1952–1965), and associate professor in genetics and plant breeding (1955–1962). He shortly returned to the University of Minnesota, from 1955 to 1956; finally, he became professor of agricultural botany at the University of Sydney (1962–1965), head of Department of Agricultural Botany (1966–1976), dean of the Faculty of Agriculture (1966–1967), director of the Plant Breeding Institute, and head of the Department of Agricultural Botany (1974–1977) before he retired. He was one of the most known Australian wheat rust researchers. The most important aspect of his research has been the development of theories and explanations for the origin of genetic variability in the wheat rust pathogen by asexual means. He established the classification system for leaf and stem rusts that are prevalent in Australia and New Zealand. Based on his studies, he developed and implemented the theory of multiple gene resistance (horizontal resistance) as a means of achieving lasting resistance to wheat rusts. As junior partner of ▶ W. WATERHOUSE, he released since 1940 the varieties "Gabo" (1945), "Kendee" (1946), "Saga" (1951), and "Koda" (1955). After 1960, under his leadership, the varieties "Mendos" (1964), "Gamut" (1965), "Timgalen" (1967), "Gatcher" (1969), "Songlen" (1975), "Timson" (1975), and "Shortim" (1977) were registered. After his retirement in 1977, the varieties "Sunkota" (1981), "Suneka" (1982), and "Sunstar" (1983) were released. As a matter of interest, the origin of "Sunstar" goes back to his idea to create a purple-seeded feed wheat; however, the idea was not well received by the industry, and finally, white prime hard wheat variety was selected from that particular cross.

WATSON, J. D. (1928–): He is an American from Chicago. He entered the University of Chicago at age of 15 years to study zoology and graduated with a BSc In 1947, he went to Indiana University in Bloomington (United States) for graduate studies. There, he studied bacteriophages under ▶ S. E. LURIA (1912–1991) and received his PhD in zoology in 1950. He then spent a year in Copenhagen (Denmark) as a postdoctoral researcher, working on viruses. During this period, he went to a conference in Napoli (Italy), where he saw X-ray diffraction picture of DNA by ▶ M. WILKINS. After this pivotal incident, he soon arranged to move to the Cavendish Laboratory of Cambridge University (England), where he hoped to study DNA. Together with ▶ F. H. C. CRICK, he was able to show that the hereditary substance of the chromosomes is deoxyribonucleic acid (DNA). He and F. CRICK shared the 1962 Nobel Prize in Physiology or Medicine with M. H. F. WILKINS.

WEISMANN, A. (1834–1914): He was a German biologist who had developed notions of the continuity of the inherited material (so-called "germ-plasm") from generation to generation, thus suggesting that acquired characteristics are not inherited.

WELLS, D. G. (1917–2005): He was an American winter wheat breeder at South Dakota State University since 1960, until his retirement in 1982. He received his doctoral degree from the Agronomy Department at the University of Wisconsin in 1949 and worked as a winter wheat breeder at Mississippi State College in Starkville (United States). In 1958, he accepted a work through the State Department in Washington, D.C., to be a member of a 10-men team to improve farm crops for Nigeria. He released eight wheat varieties ("Hume," "Winoka," "Bronze," "Gent," "Rita," "Nell," "Rose," and "Dawn") and eight germplasm lines that carried resistance to wheat streak mosaic virus from intermediate wheatgrass.

WETTSTEIN, F. von (1895–1945): He was an Austrian botanist and geneticist working at the universities of Göttingen, München, and Berlin (Germany). He promoted scientific plant breeding and was one of the most important biologists in Germany.

WIEBE, G. A. (1899–1975): He joined the staff of the USDA as a full-time employee in 1922, following his graduation from the University of Idaho. His entire career was spent with the department in cereal breeding and genetics, and he worked almost exclusively with barley since 1935. From 1946 until his retirement in 1969, he was leader of barley investigations for the USDA Agricultural Research Service. He was the principal founder of the North American Barley Research Conference and was one of the original organizers of the International Barley Genetics Symposium and served as its first president. His studies included many phases of barley improvement and genetics, and he was co-author of several *Summaries of Linkage Studies in Barley*. He was widely known for his work with barley classification, isogenic lines, composite crosses, and breeding methods and traveled extensively in the United States and abroad, consulting with breeders and geneticist. After retirement, WIEBE continued his personal research in Arizona, Idaho, Montana, and Maryland until his death. In 1970, he personally collected over 1100 barley lines in Ethiopia. He was the recipient of the MSc and PhD degrees from the University of California, an honorary degree from the University of Idaho, and the USDA's Superior and Distinguished Service Awards.

WILFAHRT, H. ▶ HELLRIEGEL, H.

WILKINS, H. F. ▶ WATSON, J. D.

WILLIAMSON, J. ▶ BACKHOUSE, W. O.

WILSON, A. S. ▶ RIMPAU, W.

WILSON, J. A. (–2012): James (Jim) A. WILSON, born on a farm of Oklahoma, was a true pioneer in the wheat industry and the father and visionary of hybrid wheat. He received his BSc and MSc degrees from Oklahoma State University, followed by a PhD from Texas A&M University. His career began as wheat project leader at the Fort Hays Branch Experiment Station, Fort Hays, Kansas, in 1957. In Kansas, he initiated among the first studies on hybrid breeding and cytoplasmic sterility in wheat. Later, he joined DeKalb AgResearch in 1961, with the opportunity to restart private wheat breeding in the United States, with emphasis on breeding and marketing of hybrid wheat. He published several papers on hybrid breeding and restoration systems while with DeKalb. His project released the first commercially available hybrid wheat to growers in 1978. He was also responsible for the development and release of the first semi-dwarf wheat cultivars for production in Kansas. When DeKalb closed their hybrid wheat project in 1981, in the same year, Wilson founded his own Trio Research Inc. as a family-owned corporation in Wichita, Kansas. This corporation was initially formed to be a service company for a new biotech company, Agrigenetics Research Associates (ARA). His first Trio wheat cultivar, "T67," was released in 1993, followed by "T81" in 1995. His most recent release "T158" has been grown on significant acreage in the Central Plains and contributed as parents to many more. Federal seed authorities recently approved Trio Research for the marketing of the first synthetic wheat cultivars in the United States, developed using a mixture of male and female plants. Trio was acquired by Limagrain Cereal Seeds in 2010. Dr. Wilson then served as a consultant for Limagrain Cereal Seeds, continuing his research and breeding on hybrid wheat, until his death.

WINGE, Ö. (1886–1960): He was a Danish botanist. He demonstrated the utility of using characters of chromosomes and noted different chromosome numbers in plants. He formulated the theory that polyploid series in nature arise through species hybridization and the summation of the chromosome complements of the intercrossed species. This theory was later verified by MÜNTZING's synthesis of *Galeopsis tetrahit* ▶ A. MÜNTZING.

WOLSKI, T. (1924–2005): Tadeusz WOLSKI brought triticale to agricultural bloom. He was born in Warsaw (Poland). Tadeusz's mother was the daughter of Alexander JANASZ, who founded the A. Janasz Company in 1880, the largest and most important private Polish plant breeding company, specializing in the breeding of grain and sugar beet. The whole family formed an excellent community of specialists in plant breeding and achieved excellent results with new high-yielding cereal and sugar beet varieties. After completing the high school (Abitur), he began agricultural studies under the guidance of Prof. Jan ROSTAFIŃSKI. At the same time, he was involved in Janasz's breeding programs, in particular in the Dańków breeding station, which was the headquarters of the company. After the end of the war, he finished his studies. In 1948, he finally left the agricultural college in Warsaw with the titles of engineer of agriculture and master of agrotechnics. In 1948, he presented his diploma thesis "Monograph on DANKÓW Wheat Breeding," followed by the PhD thesis "Studies on Inbreeding in Rye" of the Academy of Agricultural Sciences in Poznań in 1967. In the same year, he was appointed lecturer at the Institute for Plant Breeding in Radzików. The title of full professor was awarded to him in 1989. In 1960, WOLSKI became head of the section cereal breeding in the enterprise, which later was named "Danko Hodowla Roślin." Danko company became the leading wheat breeder in Poland. In rye, he created an excellent variety that met all these targets, named "Dankowskie złote;" it became the most popular variety in Poland, with a market share of more than 90% for several decades. A similar, slightly shorter variety with the name "Danko" became a hit in foreign markets, in Europe, and in other parts of the world, even as far as Canada and South Korea. With "Amilo" rye, he created a real masterpiece, that is, a variety with excellent baking quality. In 1982, he decided to launch his own breeding program of triticale. It was his great merit to have developed breeding methods that made it possible to transform triticale as a botanical curiosity into a high-yielding and healthy cereal plant, which totally fulfils the original objective, to deliver grain of highest feeding value, such as "Salvo" (1984), "Jago" (1988), "Malno" (1990), "Moniko" (1991), "Bolero" (1991), "Dagro" (1991), "Grado" (1991), "Remiko" (2012), "Leontino" (2012), "Fredro" (2010), and "Mikado" (2012), among others.

WOOD, T. B. ▶ HUNTER, H.

WRICKE, G. (1928–2009): He was born in Niederwerbig, Brandenburg (Germany) as the son of a farmer. He finished his training as agriculturist after 1945 and began his studies in agriculture at the Humboldt University, Berlin. He finished his studies in 1951 at the Technical University of Berlin. ▶ Prof. Hans KAPPERT inspired him to focus his interests on plant breeding. He received his PhD degree in 1953 ("Inheritance of polyembrony in Flax, *Linum usitatissimum*"). He found his interest in quantitative genetics during the following years at the Institut für Vererbungs-und Züchtungsforschung (Berlin-Dahlem), where he studied gene–dosage effects in *Arabidopsis thaliana*. From 1956 to 1963, WRICKE was employed at the Department of the Plant Breeding for the company F. von Lochow-Petkus GmbH in Klausheide. During this period, he performed fundamental studies on selection parameters for fertility and kernel yield in tetraploid rye. The concept of ecovalence as a measure of phenotypic stability of crop cultivars was developed during this period. In 1964, WRICKE accepted an offer by ▶ Prof. Hermann KUCKUCK, Technical University of Hannover, to join the Institute of Applied Genetics as assistant professor. After the retirement of H. KUCKUCK in 1969, he received a call for the Chair of Applied Genetics in Hannover, which he headed until his retirement in 1996. Under his guidance, research was conducted on nematode resistance in *Beta* beets and on sex inheritance in asparagus. Additional research activities were focused on genetics of self-incompatibility in rye and on (biochemical molecular) marker-assisted selection. textbooks on *Populationsgenetik*, *Quantitative Genetics and Selection in Plant Breeding*, and *Genetic Markers in Plant Breeding, Advances in Plant Breeding* are important parts of his numerous publications.

Before his death, he founded the "Günter and Anna Wricke Foundation," awarding promising young scientists for outstanding scientific accomplishments in the field of applied genetics and breeding research in agricultural, horticultural, and forest plants.

YATES, F. (1902–1994): He was one of the pioneers of twentieth-century statistics as branch of applied mathematics, concerned with the collection and interpretation of quantitative data and the use of probability theory to estimate population parameters. In 1931, he was appointed as assistant statistician at Rothamsted Experimental Station (United Kingdom). In 1933, he became head of statistics when ▶ R. A. FISHER obtained a position at the University College London; at Rothamsted, he worked on the design of experiments, including contributions to the theory of analysis of variance, a statistical method for making simultaneous comparisons between two or more means, and a statistical method that yields values that can be tested to determine whether a significant relation exists between variables. After the World War II, he worked on sample survey design and analysis. He became an enthusiast of electronic computer, a machine for performing calculations automatically. In 1954, he obtained an "Elliott 401" for Rothamsted and contributed to the initial development of statistical computing. His main contributions to experimental statistics were the contingency tables (involving small numbers and the χ^2 test, 1934), the design and analysis of factorial experiments (1937), statistical tables for biological, agricultural, and medical research (with ▶ R. A. FISHER, 1938), systematic sampling (1948), selection without replacement from within strata (with probability proportional to size, together with P. M. GRUNDY, 1953), and sampling methods for censuses and surveys (1960). In 1966, he was awarded a Royal Medal from the Royal Society. He retired from Rothamsted in 1968 and became a senior research fellow at Imperial College.

ZHUKOVSKY, P. M. (1888–1975): He was a Russian botanist, wheat taxonomist, and pioneer of breeding research. From 1915 to 1925, he worked at Tiflis Botanical Gardens (Georgia). From 1925, he was a research scientist of the Institute of Applied Botany (Saint Petersburg, Russia). From 1951 to 1965, he was appointed as director of VIR (Saint Petersburg, Russia), taking over the heritage of ▶ N. I. VAVILOV.

Bibliography

Abd-Elsalam, K. A. and Alghuthaymi, M. A. 2015. Nanodiagnostic tools in plant breeding. *J. Nanotech. Mater. Sci.* 2: 1–2.

Abid, N., Khatoon, A., Maqbool, A., Irfan, M., Bashir, A., Shahid, M., Saeed, A., Brinch-Pedersen, H. and Malik, K. A. 2017. Transgenic expression of phytase in wheat endosperm increases bioavailability of iron and zinc in grains. *Transgenic Res.* 26: 109–122.

Aeberhard, M. 2005. *Geschichte der alten Traubensorten. Ein historisch-ampelographischer Rückblick.* Verl. Arcadia, Solothurn, Switzerland.

Ahloowalia, B. S., Maluszynski, M. and Nichterlein, K. 2004. Global impact of mutation-derived varieties. *Euphytica* 135: 187–204.

Åkerman, A. and MacKey, J. 1948. The breeding of self-fertilized plants by crossing. Some experiments during 60 years breeding at the Swedish Seed Association. In: Å. Åkerman and O. Tedin (Eds.), *Svalöf 1886–1946: History and Present Problems.* C. Bloms Boktryckerei, Lund, Sweden.

Allard, R. W. 1988. Genetic change associated with the evolution of adaptedness in cultivated plants and their wild progenitors. *J. Hered.* 79: 225–238.

Allard, R. W. 1999a. History of plant population genetics. *Ann. Rev. Genet.* 33: 1–27.

Allard, R. W. 1999b. *Principles of Plant Breeding.* John Wiley & Sons, New York, pp. 254.

Ammerman, A. J. and Cavalli-Sforza, L. L. 1984. *The Neolithic Transition and the Genetics of Populations in Europe.* Princeton University Press, Princeton, NJ.

Anderson, E. N. 1988. *The Food of China.* Yale University Press, New Haven, CT.

Anderson, W. T. 1996. *Evolution Isn't What Is Used to Be.* W. H. Freeman and Company, New York.

Anonymous. 1989. Wheat research and development in Pakistan. CIMMYT Report. ISBN 968·6127·31·3; http://libcatalog.cimmyt.org/download/cim/13994.pdf.

Anonymous. 1998. State of the world's plant genetic resources for food and agriculture. Report of UN Food and Agricultural Organization (FAO), Rome, Italy.

Anonymous. 2015. NIAB MAGIC population resources. http://www.niab.com/magic.

Anonymous. 2016a. GM crops list, GM approval database-ISAAA.org. www.isaaa.org (Retrieved January 30, 2016).

Anonymous. 2016b. New OSGATA policy on principles of organic plant breeding. http://www.osgata.org/2016/policy-on-principles-of-organic-plant-breeding/.

Anonymous. 2017. List of grape varieties. https://en.wikipedia.org/wiki/List_of_grape_varieties.

Arakawa, T., Chong, K. X. and Langridge, W. H. R. 1998. Efficacy of a food plant-based oral cholera toxin B subunit vaccine. *Nat. Biotechnol.* 16: 292–297.

Artsaenko, O., Kettig, B., Fiedler, U., Conrad, U. and Düring, K. 1998. Potato tubers as a biofactory for recombinant antibodies. *Mol. Breed.* 4: 313–319.

Asimov, I. 1967. *The Egyptians.* Houghton-Mifflin, Boston, MA.

Barton, J. H. and Berger, P. 2001. Patenting agriculture. Issues in science and technology. http://issues.org/17-4/barton.

Barton, K. A., Whiteley, H. R. and Yang, K. S. 1987. *Bacillus thuringiensis* delta-endotoxin expressed in transgenic *Nicotiana tabacum* provides resistance to lepidopteran insects. *Plant Physiol.* 85: 1103–1109.

Bartos, P. and Bares, I. 1971. Leaf and stem rust resistance of hexaploid wheat cultivars Salzmünder Bartweizen and Weique. *Euphytica* 20: 435–440.

Bateson, W. 1894. *Materials for the Study of Variation Treated with Special Regard to Discontinuity in the Origin of Species.* Mac Milland & Co., London, UK.

Bateson, W. 1895. The origin of cultivated *Cineraria. Nature* 51: 605–607.

Bateson, W. 1899. Hybridization and cross-breeding as a method of scientific investigation. *J. Roy. Hort. Soc.* 24: 59–66.

Bateson, W. 1907. The progress of genetics since the rediscovery of Mendel's papers. *Prog. Rei Bot.* 1: 368–418.

Bateson, W. 1909. *Mendel's Principles of Heredity.* Cambridge University Press, New York.

Baur, E. 1922. *Einführung in die experimentelle Vererbungslehre.* Verl. Gebr. Bornträger, Berlin, Germany.

Beadle, G. W. 1970. Alfred Henry Sturtevant (1891–1970). *Yearbook Am. Philos. Soc.* 1970: 166–171.

Beal, W. J. 1887. *Grasses of North America for Farmers and Students.* Thorp and Gedfrey, Lansing, MI.

Beaven, E. S. 1935. Discussion on Dr. Neyman's paper. *J. Royal Statist. Soc.* 2: 159–161.

Becker, H. C. 2007. In memoriam: Hansferdinand Linskens (1921–2007). *Theor. Appl. Genet.* 116: 1–2.

Bell, E. 1946. Development in sterile culture of stem tips and subjacent regions of *Tropaeolum majus L.* and of *Lupinus* albus. *L. Am. J. Botany* 33: 301–318.

Bernardo, R. 1998. A model for marker-assisted selection among single crosses with multiple genetic markers. *Theor. Appl. Genet.* 97: 473–478.

Berrall, J. S. 1966. *The Garden: An Illustrated History.* Viking Press, New York.

Bettoni, J. C., Costa, M. D., Gardin, J. P. P., Kretzschmar, A. A. and Pathirana, R. 2016. Cryotherapy: A new technique to obtain grapevine plants free of viruses. *Rev. Bras. Frutic.* 38. doi:10.1590/0100-29452016833.

Bevan, M. W., Flavell, R. B. and Chilton, M. D. 1983. A chimeric antibiotic resistance gene as a selectable marker for plant cell transformation. *Nature* 304: 184–187.

Bhalla, P. L. and Singh, M. B. 2004. Engineered allergens for immunotherapy. *Curr. Opin. Allergy Clin. Immunol.* 4: 569–573.

Biffin, R. H. 1905. Mendel's laws of inheritance and wheat breeding. *J. Agric. Sci.* 1: 4–48.

Bingham, E. T., Groose, R. W., Woodfield, D. R. and Kidwell, K. K. 1994. Complementary gene interactions in alfalfa are greater in autotetraploids than diploids. *Crop Sci.* 34: 823–829.

Birchler, J. A., Auger, D. L. and Riddle, N. C. 2003. In search of the molecular basis of heterosis. *Plant Cell* 15: 2236–2239.

Bitocchi, E., Rau, D., Rodriguez, M. and Murgia, M. L. 2016. Crop improvement of *Phaseolus* ssp. through interspecific and intraspecific hybridization. In: A. S. Mason (Ed.), *Polyploidy and Hybridization for Crop Improvement.* CRC Press, Boca Raton, FL.

Blakeslee, A. F. and Avery, A. G. 1937. Methods of inducing doubling of chromosomes in plants. *J. Hered.* 28: 393–411.

Blohmeyer, E. 1877. Vom Versuchsfeld des Landw. Inst. Leipzig. Frühlings Landw. Ztg. 402.

Bolin, P. 1897. Nagra iakttagelser ofver vissa karaktarers olika nedarfningsformaga vid hybridisiering hos korn. *Sveriges Utsadesforen. Tidskr.* 7: 137–147.

Bolley, H. L. 1901. Flax wilt and flaxsick soil. *N. Dak. Agr. Exp. Sta. Bull.* 50: 27–60.

Bordonos, M. G. 1940. The characteristics of single germ beet hybrids. Bull. USSR All Union Res. Inst. Sugar Beet Industry, Kiev (Ukraine): 238–240.

Borlaug, N. E. 1953. New approach to the breeding of wheat varieties resistant to *Puccinia graminis tritici. Phytopathology* 43: 467.

Borlaug, N. E. 1997. Feeding a world of 10 billion people: the miracle ahead. *Plant Tissue Cult. Biotechnol.* 3: 119–127.

Bos, I. and Caligari, P. 1995. *Selection Methods in Plant Breeding.* Chapman and Hall, New York.

Botstein, D., White, R., Skolnick, M. and Davis, R. W. 1980. Multiple forms of enzymes: Construction of genetic linkage map in human, using restriction fragment length polymorphism. *Am. J. Hum. Genet.* 32: 314–331.

Boveri, T. 1902. Über mehrpolige Mitosen als Mittel zur Analys des Zellkerns. *Verhandl. physikalisch-medizinischen Gesell., Würzburg* 35: 67–90.

Boyer, J. S. 1982. Plant productivity and environment. *Science* 218: 443–448.

Brandt, P. 2004. *Transgene Pflanzen.* Birkhäuser Verl., Basel, Switzerland.

Brar, D. S. and Khush, G. S. 1994. Cell and tissue culture for plant improvement. In: A. S. Basra (Ed.), *Mechanism of Plant Growth and Improved Productivity. Modern Approaches.* Marcel Dekker Inc., New York.

Breasted, J. H. 1951. *A History of Egypt. From the Earliest Times to the Persian Conquest.* 2nd ed., fully revised. C. Scribner, New York.

Brettell, R. I. S., Banks, P. M., Cauderon, Y., Chen, X., Cheng, Z. M., Larkin, P. J. and Waterhouse, P. M. 1988. A single wheatgrass chromosome reduces the concentration of barley yellow dwarf virus in wheat. *Ann. Appl. Biol.* 113: 599–603.

Bridges, C. B. 1916. Non-disjunction as a proof of the chromosome theory of heredity. *Genetics* 1: 587–596.

Briggs, F. N. 1938. The use of the backcross in crop improvement. *Am. Nat.* 72: 285–292.

Brim, C. A. 1966. A modified pedigree method of selection in soybeans. *Crop Sci.* 6: 20–23.

Brown, A. H. D., Munday, J. and Oram, R. N. 1989. Use of isozyme-marked segments from wild barley (*Hordeum spontaneum*) in barley breeding. *Plant Breed.* 100: 280–288.

Brown, D. C. W. and Thorpe, T. A. 1995. Crop improvement through tissue culture. *World J. Microbiol. Biotechnol.* 11: 409–415.

Bruce, A. B. 1910. The Mendelian theory of heredity and the augmentation of vigor. *Science* 32: 627–628.

Brush, S. B. 2004. *Farmers' Bounty: Locating Crop Diversity in the Contemporary World.* Yale University Press, New Haven, CT.

Bytebier, B., Deboeck, F., De Greve, H., van Montagu, M. and Hernalsteens, J. P. 1987. T-DNA organization in tumor cultures and transgenic plants of monocotyledon *Asparagus officinalis*. *Proc. Natl. Acad. Sci. U.S.A.* 84: 5345–5349.

Caldwell, B. E., Schillinger, J. A., Barton, J. H., Qualset, C. O., Duvick, D. N. and Barnes, R. F. 1989. Intellectual property rights associated with plants. *Am. Soc. Agron.* 52: 1–206.

Capelle, W. 1949. Theophrast über Pflanzenentartung. *Museum Helveticum* 6: 57–84.

Carleton, M. A. 1907. Development and proper status of agronomy. *Proc. Am. Soc. Agron.* 1: 17–24.

Carlson, P. S., Smith, H. H. and Dearing, R. D. 1972. Parasexual interspecific plant hybridization. *Proc. Natl. Acad. Sci. U.S.A.* 69: 2292–2294.

Carter, G. F. 1977. A hypothesis suggesting a single origin of agriculture. In: C. A. Reed (Ed.), *The Origins of Agriculture*. Mouton Publishers, The Hague, the Netherlands, pp. 89–133.

Castle, W. E. 1903. Laws of heredity of Galton and Mendel and some laws govering race improvement by selection. *Proc. Am. Acad. Arts Sci.* 39: 323–342.

Castle, W. E. 1946. Genes, which divide species or produce hybrid vigor. *Proc. Natl. Acad. Sci. U.S.A.* 32: 145–149.

Ceccarelli, S. 2009. Evolution, plant breeding and biodiversity. *J. Agric. Environ. Int. Dev.* 103: 131–145.

Ceccarelli, S. 2014. GM crops, organic agriculture and breeding for sustainability. *Sustainability* 6: 4273–4286. doi:10.3390/su6074273.

Celsus, A. C. 1935. *De Medicina*. Vol. 1. Loeb Classical Library 292. Harvard University Press, Cambridge, MA.

Chai, J.-F., Zhang, C.-M., Ma, X.-Y. and Wang, H.-B. 2016. Molecular identification of ω-secalin gene expression activity in a wheat 1B/1R translocation cultivar. *J. Integr. Agric.* 15: 2712–2718.

Chamberlain, B. H. 1932. *Kojiki: Records of Ancient Matters*. J. L. Thompson & Co., Kobe, Japan.

Chan, K. X., Mabbitt, P. D., Phua, S. Y., Mueller, J. W., Nisar, N., Gigolashvili, T., Stroeher, E., Grassl, J. et al. 2016. Sensing and signaling of oxidative stress in chloroplasts by inactivation of the SAL1 phosphoadenosine phosphatase. *Proc. Natl. Acad. Sci. U.S.A.* 113: 4567–4576.

Chaplin, S. M. and Steward, F. C. 1948. Effect of coconut milk on the growth of explants from carrot root. *Science* 108: 655–657.

Chase, S. S. 1963. Analytic breeding in *Solanum tuberosum* L. – A scheme utilizing parthenotes and other diploid stocks. *Can. J. Genet. Cytol.* 5: 359–363.

Chase, S. S. 1964. Analytic breeding of amphipolyploid plant varieties. *Crop Sci.* 4: 334–338.

Chee, M., Yang, R., Hubbell, E., Berno, A., Huang, X. C., Stern, D., Winkler, J., Lockhart, D. J., Morris, M. S. and Fodor, S. P. A. 1996. Accessing genetic information with high-density DNA arrays. *Science* 274: 610–614.

Chen, F. and Hayes, P. M. 1989. A comparison of *Hordeum bulbosum*- mediated haploid production efficiency in barley using *in vitro* floret and tiller culture. *Theor. Appl. Genet.* 77: 701–704.

Chen, F. H., Dong, G. H., Zhang, D. J., Liu, X. Y., Jia, X., An, C. B., Ma, M. M. et al. 2015. Agriculture facilitated permanent human occupation of the Tibetan Plateau after 3600 BP. *Science* 347: 248–250. doi:10.1126/science.1259172.

Chilton, M. D., Drummond, M. H., Merio, D. J., Sciaky, D., Montoya, A. L., Gordon, M. P. and Nester, E. W. 1977. Stable incorporation of plasmid DNA into higher plant cells: The molecular basis of crown gall tumorigenesis. *Cell* 11: 263–271.

Cialokowska, A. M., Arens, P. and Tuyl van, J. M. 2016. The role of polyploidization and interspecific hybridization in the breeding of ornamental crops. In: A. S. Mason (Ed.), *Polyploidy and Hybridization for Crop Improvement*. CRC Press, Boca Raton, FL.

Clarke, J. A. and Bayles, B. B. 1942. Classification of wheat varieties grown in the United States. Bulletin USDA 459.

Clarke, J. G. G. and Hutchinson, J. 1977. *The Early History of Agriculture*. British Academy, Oxford University Press, Oxford, UK.

Coe, E. H. 1959. A line of maize with high haploid frequency. *Am. Nat.* 91: 381–385.

Coffman, W. R. 1982. Gallery of cereal breeders – Neal F. Jensen. *Cer. Res. Comm.* 10: 247–257.

Collard, B. C. Y., Jahufer, M. Z. Z., Brouwer, J. B. and Pang, E. C. K. 2005. An introduction to markers, quantitative trait loci (QTL) mapping and marker-assisted selection for crop improvement: The basic concepts. *Euphytica* 142: 169–196.

Collins, G. N. and Kempton, J. H. 1911. Inheritance of waxy endosperm in hybrids of Chinese maize. *4th International Conference on Genetics*, (Paris), (eds.) P. Lévêque de Vilmorin, Masson et cie, Paris, France, pp. 347–356.

Comai, L., Young, K., Till, B. J., Reynolds, S. H., Greene, E. A., Codomo, C. A., Enns, L. C. et al. 2004. Efficient discovery of DNA polymorphisms in natural populations by EcoTILLING. *Plant J.* 37: 778–786.

Comstock, R. 1978. Quantitative genetics in maize breeding. In: D. Walden (Ed.), *Maize Breeding and Genetics*. Wiley, New York, pp. 191–206.

Comstock, R. 1996. *Quantitative Genetics with Special Reference to Plant and Animal Breeding*. Iowa State University Press, Ames, IA.

Coons, G. H. 1953. Breeding for resistance to diseases. U. S. Dep. Agric., Washington, D.C., Yearbook of Agriculture, (ed.) A. Stefferud.: 1–34.

Correns, C. 1900. Gregor Mendels Regel über das Verhalten der Nachkommenschaft der Bastarde. *Ber. Deutsch. Bot. Gesell.* 18: 158–168.

Correns, S. 1899. Untersuchungen über Xenien bei *Zea mays. Ber. Dtsch. Bot. Ges.* XVII: 410–417.

Costa, M. C. D., Farrant, J. M., Oliver, M. J., Ligterink, W., Buitink, J. and Hilhorst, H. M. W. 2016. Key genes involved in desiccation tolerance and dormancy across life forms. *Plant Sci.* 251: 162–168.

Cowling, W. A., Li, L., Siddique, K. H. M., Henryon, M., Berg, P., Banks, R. G. and Kinghorn, B. P. 2016. Evolving gene banks: Improving diverse populations of crop and exotic germplasm with optimal contribution selection. *J. Exp. Bot.* doi:10.1093/jxb/erw406.

Cox, D. R. 1951. Some systematic experimental designs. *Biometrika* 38: 312–323.

Cramp, L. H. E., Evershed, R. P., Lavento, M., Halinen, P., Mannermaa, K., Oinonen, M., Kettunen, J., Perola, M., Onkamo, P. and Heyd, V. 2014. Neolithic dairy farming at the extreme of agriculture in northern Europe. *Proc. Roy. Soc., Lond.* 281. doi:10.1098/rspb.2014.0819.

Crossway, A., Oakes, J. V., Irvine, J. M., Ward, B., Knauf, V. C. and Shewmaker, C. K. 1986. Integration of foreign DNA following microinjection of tobacco mesophyll protoplasts. *Mol. Gen. Genet.* 202: 179–195.

Cubero, J. I. 2002. *El libro de Agricultura de Al Awam*. Ed. Consejeria de Agricultura y Pesca, Junta de Andalucia, Sevilla, Spain.

Curtis, B. C., Rajaram, S. and Macpherson, H. G. 2002. *Bread Wheat—Improvement and Production*. FAO of United Nations, Rome, Italy.

Dalley, S. 2013. *The mystery of the Hanging Garden of Babylon: An elusive world wonder traced*. Oxford University Press, Oxford, UK.

Darby, W. J., Ghalioungui, P. and Grivetti, L. 1977. *Food: The Gift of Osiris*. 2 vols. Academic Press, London, UK.

Darwin, C. 1883. *The Variation of Animals and Plants under Domestication*. 2 vols. 2nd ed. D. Appleton & Co., New York (1st ed., London, John Murray, 1868).

Davenport, C. B. 1908. Degeneration, albinism and inbreeding. *Science* 28: 454–455.

de Vries, H. 1900a. Das Spaltungsgesetz der Bastarde. *Ber. Dtsch. Bot. Ges.* XVIII: 83–90.

de Vries, H. 1900b. Sur la loi de disjunction des hybrides. *Compts Rendus des Seances de l'Academie des Sciences* 130: 845–847.

de Vries, H. 1901. *Die Mutationstheorie*. Verl. Veit & Co., Leipzig, Germany.

de Vries, H. 1903. *Die Mutationstheorie*, Vol. 2. Veit & Co., Leipzig, Germany, p. 356.

Decaisne, J. 1855. Le jardin fruitier du muséum, ou monographie des arbres fruitiers cultivés dans cet établissement. Examen critique de la doctrine de van Mons. *Journal de la Société Impériale et Centrale d'Horticulture* 1: 218–240.

Delaunay, L. N. 1931. Resultate eines dreijährigen Röntgenversuchs mit Weizen. *Züchter* 3: 129–137.

Desta, Z. A. and Ortiz, R. 2014. Genomic selection: Genome-wide prediction in plant improvement. *Trends Plant Sci.* 19: 592–601.

Dittrich, M. 1959. *Getreideumwandlung und Artproblem*. G. Fischer Verl., Jena, Germany.

Dolezel, J., Greilhuber, J., Lucretti, S., Meister, A., Lysak, M. A., Nardi, L. and Obermayer, R. 1998. Plant genome size estimation by flow cytometry: Inter-laboratory comparison. *Ann. Bot.* 82: 17–26.

Dramer, I. N. 1981. *History Begins at Sumer*. University of Pennsylvania Press, Philadelphia, PA.

Driscoll, C. J. and Anderson, L. M. 1967. Cytogenetic studies of Transec—A wheat-rye translocation line. *Can. J. Genet. Cytol.* 9: 375–380.

Duchnesne, M. 1766. *Histoire naturelle des Fraisiers*. Didot le jeune et C. J. Panckoucke, Paris, France.

Dudley, P. 1724. An observation on Indian corn. *Philosophical Transactions, Royal Soc., London* 6: 204–205.

Dudley, J. W. 1984. A method for identifying populations containing favorable alleles not present in elite germplasm. *Crop Sci.* 26: 1053–1054.

Duhamel de Monceau, H.-L. 1778. *La physique des arbres*, Vol. II. Par la Meme, Paris, France.

Duke, J. A. 1983. *Medicinal Plants of the Bible*. Trado-Medic Books, Owerri, Nigeria.

Durant, W. 1954. *The Story of Civilization. Part I. Our Oriental Heritage*. Simon and Schuster, New York.

Duvick, D. N. 1997. What is yield? In: G. O. Edmeades, M. Banziger, H. R. Mickelson and C. B. Pena-Valdivia (Eds.), *Proceeding of the Symposium on Developing Drought- and Low N-Tolerant Maize*. CIMMYT, El Batan, Mexico, pp. 332–335.

Duvick, D. N. and Cassman, K. 1999. Post-green revolution trends in yield potential of temperate maize in the north-central United States. *Crop Sci.* 39: 1622–1630.

Duvick, D. N., Smith, J. S. C. and Cooper, M. 2004. Long-term selection in a commercial hybrid maize breeding program. *Plant Breed. Rev.* 24: 109–151.

D'Aost, M.-A., Lerouge, P., Busse, U., Bilodeau, P., Trépanier, S., Gomord, V., Faye, L. and Vézina, L.-P. 2004. Efficient and reliable production of pharmaceuticals in alfalfa. In: R. Fischer and S. Schillberg (Eds.), *Molecular Farming: Plant-made Pharmaceuticals and Technical Proteins*, John Wiley & Sons, Weinheim, Germany.

East, E. M. 1908. Inbreeding in corn. *Rep. Conn. Agric. Exp. Stat.* 1907: 419–428.

East, E. M. 1936. Heterosis. *Genetics* 21: 375–397.

East, E. M. and Jones, D. F. 1920. Genetic studies on the protein content of maize. *Genetics* 5: 543–610.

East, E. M. and Mangelsdorf, A. J. 1925. A new interpretation of the hereditary behavior of self-sterile plants. *Proc. Nat. Acad. Sci. U.S.A.* 11: 166–171.

Edwards, A. 1891. *Pharaohs, Fellahs and Explorers.* Cambrige Library Collection. Harper & Brothers, New York.

Edwards, I. A. 2001. Hybrid wheat. In: A. P. Bonjean and W. J. Angus (Eds.), *The World Wheat Book. A History of Wheat Breeding.* Levoisier, Paris, France, pp. 1019–1943.

Ehdaie, B., Whitkus, R. W. and Waines, J. G. 2003. Root biomass, water-use efficiency and performance of wheat-rye translocations of chromosome 1 and 2 in spring bread wheat "Pavon." *Crop Sci.* 43: 710–717.

Eisenstein, M. 2013. Plant breeding: Discovery in a dry spell. *Nature* 501: 7–9.

Elliot, F. C. 1958. *Plant Breeding and Cytogenetics.* McGraw-Hill, New York.

Ellur, R. K., Khanna, A., Yadav, A., Pathania, S., Rajashekara, H., Singh, V. K., Gopala Krishnan, S. et al. 2016. Improvement of Basmati rice varieties for resistance to blast, and bacterial blight diseases using marker-assisted backcross breeding. *Plant Sci.* 242: 330–341.

Endo, T. R. and Gill, B. S. 1996. The deletion stocks of common wheat. *J. Hered.* 87: 295–307.

Epstein, E. and Rains, D. W. 1987. Advances in salt tolerance. *Plant Soil* 99: 17–29.

Estaghvirou, S. B. O., Ogutu, J. O., Schulz-Streeck, T., Knaak, C., Ouzunova, M., Gordillo, A. and Piepho, H.-P. 2013. Evaluation of approaches for estimating the accuracy of genomic prediction in plant breeding. *BMC Genomics* 14: 860. doi:10.1186/1471-2164-14-860.

Evans, D. A. and Sharp, W. R. 1983. Single gene mutations in tomato plants regenerated from tissue culture. *Science* 221: 949–951.

Falconer, D. S. and Mackay, T. F. C. 1996. *Introduction to Quantitative Genetics*, 4th ed. Longman and Company, Harlow, UK.

Faris, J. D., Haen, K. M. and Gill, B. S. 2000. Saturation mapping of a gene-rich recombination hot spot region in wheat. *Genetics* 154: 823–835.

Farooq, S. and Azam, F. 2002. Molecular markers in plant breeding: Concepts and characterization. *Pak. J. Biol. Sci.* 5: 1135–1140.

Faust, M. and Timon, B. 1995. Origin and dissemination of peach. *Hort. Rev.* 17: 331–379.

Feichtinger, E. K. 1932. Die Entwicklung und die praktische Tätigkeit der Lehrkanzel für Pflanzenzüchtung an der Hochschule für Bodenkultur in Wien und der Pflanzenzuchtstation in Groß-Enzersdorf. *Zeitschrift für Pflanzenzüchtung* 17: 1–7.

Ferguson, A. R. and Wu, J.-H. 2016. Interploid and interspecific hybridization for kiwifruit improvement. In: A. S. Mason (Ed.), *Polyploidy and Hybridization for Crop Improvement.* CRC Press, Boca Raton, FL.

Fischer, R. and Schillberg, S. 2004. *Molecular Farming.* Wiley-VCH Verlag, Weinheim, Germany.

Fisher, R. A. 1918. The correlation between relatives on the sup position of Mendelian inheritance. *Trans. R. Soc. Edinb.* 52: 399–433.

Fisher, R. A. 1925. Theory of statistical estimation. *Proc Cambridge Philos. Soc.* 22: 200–225.

Fletcher, H. R. 1969. *Story of the Royal Horticulture Society (1804–1869).* Oxford University Press, Oxford, UK.

Flor, H. H. 1971. Current status of the gene-for-gene concept. *Ann. Rev. Phytopathol.* 9: 275–296.

Flores, H. E. and Filner, P. 1985. Metabolic relationships of putrescine, ABA, and alkaloids in cell and root cultures of *Solanaceae.* In: K.-H. Neumann, W. Barz and E. Reinhard (Eds.), *Primary and Secondary Metabolism of Plant Cell Cultures.* Springer Verlag, Berlin, Germany.

Focke, W. O. 1881. *Die Pflanzen-Mischlinge, ein Beitrag zur Biologie der Gewächse.* Verl. Gebr. Bornträger, Berlin, Germany.

Forster, B. P. and Shu, Q. Y. 2012. C01: Plant mutagenesis in crop improvement: Basic terms and applications. In: Q. Y. Shu, B. P. Forster and H. Nakagawa (Eds.), *Plant Mutation Breeding and Biotechnology.* doi:10.1079/9781780640853.0009.

Fortin, M. 1999. Syria: Land of Civilizations. Musée de la civilisation de Québec.

Frachetti, M. D., Spengler, R. N., Fritz, G. J. and Maryashev, A. N. 2010. Earliest direct evidence for broom-corn millet and wheat in the central Eurasian steppe region. *Antiquity* 84: 993–1010.

Fraley, R. T., Rogers, S. B., Horsch, R. B., Sanders, P. R., Flick, J. S., Adams, S., Bittner, M. L. et al. 1983. Expression of bacterial genes in plant cells. *Proc. Nat. Acad. Sci. U.S.A.* 80: 4803–4807.

Framond, A. J., Bevan, M. W., Barton, K. A., Flavell, R. and Chilton, M. D. 1983. Mini-Ti plasmid and a chimeric gene construct: New approaches to plant gene vector construction. *Advances in Gene Technology: Molecular Genetics of Plants and Animals. Miami Winter Symposium* 20: 159–170.

Frandsen, H. N. 1940. Some breeding experiments with timothy. *Imp. Agr. Bur. Joint Pub.* 3: 80–92.

Frankel, O. H. and Brown, A. H. D. 1984. Current plant genetic resources – A critical appraisal. In: V. L. Chopra, B. C. Joshi, R. P. Sharma and H. C. Bansal (Eds.), *Genetics: New Frontiers. Proceedings of the XV International Congress of Genetics*, Vol. 4. Oxford & IBH Publishing, New Delhi, India, pp. 3–13.

Fromm, M., Taylor, L. P. and Walbot, V. 1985. Expression of genes transferred into monocot Cand dicot plant cells by electroporation. *Nature* 319: 791–793.

Fruwirth, C. 1905. *Allgemeine Züchtungslehre der landwirtschaftlichen Kulturpflanzen.* Verlagsbuchhandlung P. Parey, Berlin, Germany.

Fujimura, T., Sakurai, M., Akagi, H., Negishi, T. and Hirose, A. 1985. Regeneration of rice plants from proto-plasts. *Plant Tissue Culture Lett.* 2: 74–75.

Fuller, D. Q. 2006. Agricultural origins and frontiers in South Asia: A working synthesis. *J. World Prehist.* 20: 1–86.

Fussell, B. 1992. *The Story of Corn.* Alfred Knopf, New York.

Gager, C. S. 1908. Effects of Rays of Radium on Plants. *Mem. New York Botan. Gard.* 4: 1–278.

Galil, J. 1968. An ancient technique for ripening sycomore fruit in East-Mediterranean countries. *Econ. Bot.* 22: 178–191.

Gall, J. and Pardue, M. 1969. Formation and detection of RNA-DNA hybrid molecules in cytological prepara-tions. *Proc. Natl. Acad. Sci. U.S.A.* 63: 378–383.

Galton, F. 1889. *Natural Inheritance.* Mac Millan & Co., London, UK.

Garbini, G. 1974. *Alte Kulturen des vorderen Orients.* Verl. Bertelsmann, München, Germany.

Gaut, B. S. and Doebley, J. F. 1997. DNA sequence evidence for the segmental allotetraploid origin of maize. *Proc. Natl. Acad. Sci. U.S.A.* 94: 6809–6814.

Gautheret, R. J. 1955. Sur la variabilitédes propriétés physiologiques des cultures de tissues végétaux. *Rev. Gen. Bot.* 62: 5–112.

Gavarett, J. 1840. *Principes generaux de statistique medicale ou development des regles qui doivent presider a son emploi.* Bechet Jeune & Labe, Paris, France.

Gayon, J. 2016. From Mendel to epigenetics: History of genetics. *Compt. Rend. Biol.* 339: 225–230.

Gengenbach, B. G. and Green, C. E. 1975. Selection of T-cytoplasm maize callus cultures resistant to *Helminthosporium maydis* race T pathotoxin. *Crop Sci.* 15: 645–649.

Gill, B. S., Lu, F., Schlegel, R. and Endo, T. R. 1988. Toward a cytogenetic and molecular description of wheat chromosomes. *Proceedings of the 7th International Wheat Genetics Symposium*, Cambridge, (eds.) T. E. Miller & R. M. D. Koebner, Institute of Plant Sci. Res., Cambridge, UK, pp. 477–481.

Glover, J. D. and Reganold, J. P. 2010. Perennial grains: Food security for the future. *Iss. Sci. Technol.* 26: 41–47.

Goldschmidt, R. 1911. *Einführung in die Vererbungswissenschaft.* W. Engelmann, Leipzig, Germany.

Goodspeed, T. H. and Clausen, R. E. 1928. Interspecific hybridization in *Nicotiana.* VIII. The *sylvestris-tomentosa-tabacum* hybrid triangle and its bearing on the origin of *Tabacum. Univ. Calif. Publ. Bot.* 11: 245–256.

Goodspeed, T. H. and Olson, A. R. 1928. The production of variation in *Nicotiana* species by x-ray treatment of sex cells. *Proc. Nat. Acad. Sci. U.S.A.* 14: 66–69.

Goor, A. and Nurock, M. 1968. *The Fruits of the Holy Land.* Israel University Press, Jerusalem, Israel.

Goss, J. 1822. On the variation in the colour of peas occasioned by cross-impregnation. *Trans. Hort. Soc. Lond.* V: 234–237.

Goulden, C. H. 1941. Problems in plant selection. *Proceedings of the 7th International Genetics Congress*, Edinburgh, 23–30 August 1939, Cambridge University Press: Cambridge, UK, pp. 132–133.

Grace, E. S. 1997. *Biotechnology Unzipped: Promises and Realities.* Joseph Henry Press, Washington, DC.

Grafius, J. E. 1965. A geometry of plant breeding. *Mich. Agric. Exp. Sta. Res. Bull.* 7: 59.

Grew, N. 1682. *Anatomy of Plants.* W. Rawlins, London, UK.

Guha, S. and Maheshwari, S. C. 1964. In vitro production of embryos from anthers of *Datura. Nature* 204: 497.

Gupta, P. K., Balyan, H. S., Varshney, R. K. and Gill, K. S. 2013. Development and use of molecular markers for crop improvement. *Plant Breed.* 132: 431–432.

Gustafsson, Å. 1947. Mutations in agricultural plants. *Hereditas* 33: 1–100.

Guzman, I., Hamby, S., Romero, J., Bosland, P. W. and O'Connell, M. A. 2010. Variability of carotenoid biosynthesis in orange colored *Capsicum* spp. *Plant Sci.* 179: 49–59.

Haberlandt, G. 1902. Kulturversuche mit isolierten Planzenzellen. *S. B. Weisen, Naturw., Wien* 111: 69–92.

Hacking, I. 1965. *Logic of Statistical Inference.* Cambridge University Press, Cambridge, UK.

Hallauer, R. A. 2007. History, contribution, and future of quantitative genetics in plant breeding: Lessons from maize. *Crop Sci.* 47: 4–19.

Hallo, W. W. 2000. *The Context of Scripture: Monumental Inscriptions from the Biblical World*, Vol. 3. Brill, Leiden, the Netherlands.

Hammer, K. 1984. Das Domestikationssyndrom. *Kulturpflanze* 32: 11–34.

Hanna, W. W., Burson, B. L. and Schwartz, B. M. 2016. Polyploidy and interspecific hybridization in *Cynodon, Paspalum, Pennisetum,* and *Zoysia.* In: A. S. Mason (Ed.), *Polyploidy and Hybridization for Crop Improvement.* CRC Press, Boca Raton, FL.

Hänsel, H. 1962. Die Bedeutung Tschermaks für Züchtungsforschung und praktische Pflanzenzüchtung. *Verhandlungen der Zoologisch-Botanischen Gesellschaft in Wien* 102: 13–17.

Hardy, G. H. 1908. Mendelian proportions in a mixed population. *Science* 28: 49–50.

Hareuveni, N. 1984. *Tree and Shrub in Our Biblical Heritage.* Neot Kedumin, Kiryat Ono, Israel.

Harlan, H. V. and Martini, M. L. 1938. The effect of natural selection in a mixture of barley varieties. *J. Agric. Res.* 57: 189–199.

Harlan, H. V. and Pope, M. N. 1922. The germination of barley seeds harvested at different stages of growth. *J. Hered.* 13: 72–75.

Harlan, J. R. 1992. *Crops and Man*, 2nd ed. ASA, Madison, WI.

Harland, S. C. 1936. Haploids in polyembryonic seeds of Sea Island cotton. *J. Hered.* 27: 229–231.

Harrington, J. B. 1937. The mass pedigree method in the hybridization for improvement of cereals. *J. Am. Soc. Agron.* 29: 379–384.

Harris, J. A. 1915. Studies on soil heterogeneity. *Am. Nat.* 49: 430–454.

Haverkort, A. J., Boonekamp, P. M., Hutten, R., Jacobsen, E., Lotz, L. A. P., Kessel, G. J. T., Vossen, J. H. and Visser, R. G. F. 2016. Durable late blight resistance in potato through dynamic varieties obtained by cisgenesis: Scientific and societal advances in the DuRPh. *Potato Res.* 59: 35–66.

Hayes, H. K. and Garber, R. J. 1919. Synthetic production of high protein corn in relation to breeding. *Agron. J.* 11: 309–318.

Hayes, H. K. and Johnson, I. J. 1939. The breeding of improved selfed lines of corn. *J. Am. Soc. Agron.* 31: 710–724.

Hayes, H. K., Parker, I. H. and Kurtzweil, C. 1920. Genetics of rust resistance in crosses of varieties of *Triticum vulgare* with varieties of *T. durum* and *T. dicoccum. Bull. Inst. Agric. Res.* (Washington, DC) 19: 523–542.

Hayes, H. K., Rinke, E. H. and Tsiang Y. S. 1946. Experimental study of convergent improvement and backcrossing in corn. Minnesota Agricultural Experiment Station Technical Bulletin 172.

Hazarika, M. H. and Rees, H. 1967. Genotypical control of chromosome behaviour in rye. X. chromosome pairing and fertility in autotetraploids. *Heredity* 22: 317–322.

Heffner, E. L., Sorrells, M. E. and Jannink, J.-L. 2009. Genomic selection for crop improvement. *Crop Sci.* 49: 1–12.

Helbaek, H. 1959. Domestication of food plants in the old world. *Science* 130: 365–372.

Helentjaris, T., Weber, D. and Wright, S. 1988. Identification of the genomic locations of duplicate nucleotide sequences in maize by analysis of restriction fragment length polymorphisms. *Genetics* 118: 353–363.

Heller, K. J. 2003. *Genetically Engineered Food.* Verl. Wiley-Vch GmbH & Co. KGaA, Weinheim, Germany.

Herbert, W. 1847. On hybridization amongst vegetables. *J. Hort. Soc.* 11: 267–273.

Herdt, R. W. and Capule, C. 1983. *Adoption, Spread, and Production Impact of Modern Rice Varieties in Asia.* International Rice Research Institute (IRRI), Los Banos, Philippines.

Herrera-Estrella, L., Depicker, A., van Montagu, M. and Schell, J. 1983. Expression of chimaeric genes transferred into plant cells using a Ti-plasmid-derived vector. *Nature* 303: 209–213.

Hesper, B. and Hogeweg, P. 1970. Bioinformatica: een werkconcept. *Kameleon* 1: 28–29.

Hiatt, A., Cafferkey, R. and Bowdish, K. 1989. Production of antibodies in transgenic plants. *Nature.* 342: 76–78.

Hoffmann, W. 1951. Ergebnisse der Mutationszüchtung. In: *Vorträge über Pflanzenzüchtung.* Bonn, Germany, Schriftenreihe der Gesellschaft für Pflanzenzüchtung e.V., Bonn, Deutschland: pp. 36–53.

Hogben, L. 1957. *Statistical Theory.* G. Allen & Unwin, London, UK.

Hopkins, C. G. 1899. Improvement in the chemical composition of the corn kernel. *Agric. Exp. Stn. Bull.* 55: 205–240.

Hort, A. 1916. *Theophrastus: Enquiry into Plants.* W. Heinemann & G. P. Putnam's Sons., London, UK, New York.

Hougas, R. W. and Peloquin, S. J. 1958. The potential of potato haploids in breeding and genetic research. *Am. Potato J.* 35: 701–707.

Hu, X. W. 1981. Archaian Liang Shu is sorghum. *China Agric. Hist.* 100: 77–84.

Huang, C. and Liang, J. 1980. Plant breeding achievements in ancient China. *Agron. Hist. Res.* 1: 1–10.

Hughes, M. B. and Babcock, E. B. 1950. Self-incompatibility in Crepis foetida L. subsp. rhoeadifolia Bieb. Schinze & Keller. *Genetica* 35: 570–588.

Hutchinson, J., Clark, J. G. G., Jope, E. M. and Riley, R. 1977. *The Early History of Agriculture.* Oxford University Press, Oxford.

Huxley, A. 1978. *An Illustrated History of Gardening.* The Lyons Press, New York.

Hyams, E. 1971. *A History of Gardens and Gardening.* Praeger, New York.

Hymowitz, T. 1970. On the domestication of the soybean. *Econ. Bot.* 23: 408–421.

Iltis, H. 1932. *Life of Mendel.* George Allen & Unwin, London, UK.

Inazuka, G. 1971. Norin 10. A Japanese semi-dwarf wheat variety. *Wheat Inform. Serv.* 32: 25.

Iriarte, J., Power, M. J., Rostain, S., Mayle, F. E., Jones, H., Watling, J., Whitney, B. S. and McKey, D. B. 2012. Fire-free land use in pre-1492 Amazonian savannas. PNAS. www.pnas.org/lookup/suppl/doi:10.1073/pnas.1201461109/-/DCSupplemental.

Itallie van Emden, W. 1940. *Interview with Beijerinck. Martinus Willem Beijerinck, His Life and His Work.* App. J. Gravenshage, the Netherlands.

Jain, S. M., Shahin, E. A. and Sun, S. 1988. Interspecific protoplast fusion for the transfer of atrazine resistance from *Solanum nigrum* to tomato (*Lycopersicon esculentum* L.). *Plant Cell Tissue Organ Cult.* 12: 189–192.

James, C. 2003. *Global Status of Commercialized Transgenic Crops.* ISAAA Briefs, Ithaca, NY.

Janick, J. 2001. Asian crops in North America. *Hort. Technol.* 11: 510–513.

Janick, J. 2002. Ancient Egyptian agriculture and the origins of horticulture, pp. 23–39. In: S. Sansavini and J. Janick (Eds.), Proceedings of the International Symposium on Mediterranean Horticulture Issues and Prospects. *Acta. Hort.* 582: 55–59.

Janick, J. 2004. Long-term selection in maize. *Plant Breed. Rev.* 24: 1–384.

Jannson, S. 2016. Future garden plants are here! https://www.blogg.umu.se/forskarbloggen/2016/09/future-garden-plants-are-here-a-diary-from-the-first-crispr-edited-plants-in-the-world/.

Jansen, R. C. and Nap, J. P. 2001. Genetical genomics: The added value from segregation. *Trends Genet.* 17: 388–391.

Jenkins, M. T. 1940. The segregation of genes affecting yield of grain in maize. *J. Am. Soc. Agron.* 32: 55–63.

Jenkins, M. T. and Brunson, A. M. 1932. Methods of testing inbred lines of maize in crossbred combinations. *J. Am. Soc. Agron.* 24: 523–530.

Jensen, N. F. 1952. Intra-varietal diversification in oats breeding. *Agron. J.* 44: 30–34.

Jensen, N. F. 1965. Multiline superiority in cereals. *Crop Sci.* 5: 566–568.

Jessen, K. F. W. 1864. *Botanik der Gegenwart und Vorzeit in culturhistorischer Entwicklung.* Republished in 1948 by The Chronica Botanica, Waltham, MA.

Jiang, G.-L. 2015. Molecular marker-assisted breeding: A plant breeder's review. In: J. M. Al-Khayri, S. M. Jain and D. V. Johnson (Eds.), *Advances in Plant Breeding Strategies: Breeding, Biotechnology and Molecular Tools.* doi:10.1007/978-3-319-22521-0_15.

Jiang, W. Z., Henry, I. M., Lynagh, P. G., Comai, L., Cahoon, E. B. and Weeks, D. P. 2017. Significant enhancement of fatty acid composition in seeds of the allohexaploid, *Camelina sativa*, using CRISPR/Cas9 gene editing. *Plant Biotechnol. J.* doi:10.1111/pbi.12663.

Johannsen, W. 1903. *Über die Erblichkeit in Populationen und in reinen Linien.* G. Fischer Verl., Jena, Germany.

Johannsen, W. 1926. *Elemente der exakten Erblichkeitslehre.* G. Fischer Verl., Jena, Germany.

Johnson, N. L. 1948. Alternative systems in the analysis of variance. *Biometrika* 35: 80–87.

Johnson, S. W. 1891. *How Crops Grow.* Orange Judd & Co., New York.

Jones, D. F. 1917. Dominance of linked factors as a means of accounting for heterosis. *Genetics* 2: 466–479.

Jones, D. F. 1920. Selection of self-fertilized lines as the basis of corn improvement. *J. Am. Soc. Agron.* 12: 77–100.

Jones, D. F. 1945. Heterosis resulting from degenerative changes. *Genetics* 30: 527–542.

Jones, L. R. and Gilman, J. C. 1915. The control of cabbage yellows through disease resistance. Wisconsin Agricultural Experiment Station, Research Bulletin 38.

Jung, J. H. and Altpeter, F. 2016. TALEN mediated targeted mutagenesis of the caffeic acid O-methyltransferase in highly polyploid sugarcane improves cell wall composition for production of bioethanol. *Plant. Mol. Biol.* 92: 131–142. http://link.springer.com/article/10.1007%2Fs11103-016-0499-y.

Kale, S. M., Jaganathan, D., Ruperao, P., Chen, C., Punna, R., Kudapa, H., Thudi, M. et al. 2015. Prioritization of candidate genes in *"QTL-hotspot"* region for drought tolerance in chickpea (*Cicer arietinum* L.). *Nature.* doi:10.1038/srep15296.

Kang, X.-Y. 2016. Polyploid induction techniques and breeding strategies in poplar. In: A. S. Mason (Ed.), *Polyploidy and Hybridization for Crop Improvement.* CRC Press, Boca Raton, FL.

Karp, A. 1991. On the current understanding of somaclonal variation. *Oxf. Surv. Plant Mol. Cell Biol.* 7: 1–58.

Karp, A. 1995. Somaclonal variation as a tool for crop improvement. *Euphytica* 85: 295–302.

Karp, A., Kresovich, S., Bhat, K. V., Ayand, W. G. and Hodgkin, T. 1997. Molecular tools in plant genetics resources conservation: A guide to the technologies. The International Plant Genetic Resources Institute (IPGRI), Rome (Italy). *Tech. Bull.* 2: 1–46.

Kasha, K. J. and Kao, K. N. 1970. High frequency haploid production in barley (*Hordeum vulgare* L.). *Nature* 225: 874–876.

Katayama, Y. 1935. Karyological comparison of haploid plants from octoploid Aegilotriticum and diploid wheat. *Jap. J. Bot.* 7: 349–380.

Kattermann, G. 1937. Zur Cytologie halmbehaarter Stämme aus Weizenroggenbastardierung. *Züchter* 9: 196–199.

Kaur, J., Fellers, J., Adholeya, A., Velivelli, S. L. S., El-Mounadi, K., Nersesian, N., Clemente, T. and Shah, D. 2017. Expression of apoplast-targeted plant defensing MTDef4.2 confers resistance to leaf rust pathogen Puccinia triticina but does not affect mycorrhizal symbiosis in transgenic wheat. *Transgenetic Res.* 26: 37–49.

Keeble, F. C. and Pellew, C. 1910. The mode of inheritance of stature and of time of flowering in peas (*Pisum sativum*). *J. Genet.* 1: 47–56.

Kemp, H. J. 1935. Mechanical aids to crop experiments. *Scientific Agric.* 15: 488–506.

Kempton, J. H. 1916. Lobed leaves in maize. *J. Hered.* 7: 508–510.

Kempton, R. and Fox, P. 1997. *Statistical Methods for Plant Variety Evaluation.* Chapman and Hall, New York.

Khlestkina, E. K., Huang, X. Q., Quenum, F. J.-B., Chebotar, S., Röder, M. S. and Börrner, A. 2004. Genetic diversity in cultivated plants – Loss or stability? *Theor. Appl. Genet.* 108: 1466–1472.

Kihara, Y. 1951. Triploid watermelons. *Proc. Am. Soc. Hort. Sci.* 58: 217–230.

Kimber, G. 1983. Gallery of cereal workers – Ernest Robert Sears. *Cer. Res. Comm.* 11: 175–178.

King, I. P., Reader, S. M. and Miller, T. E. 1988. Exploitation of the "cuckoo" chromosome (4S) of *Aegilops sharonensis* for eliminating segregation for height in semi-dwarf *Rht2* bread wheat cultivars. *Proceedings of the 7th Internation Wheat Genetics Symposium*, Cambridge, UK, pp. 373–341.

Kingsbury. N. 2009. *Hybrid: The History and Science of Plant Breeding.* University of Chicago Press, Chicago, IL. doi:10.1007/s40656-014-0010-5.

Kislev, M. E., Hartmann, A. and Bar-Yosef, O. 2006. Early domesticated fig in the Jordan Valley. *Science* 312: 1372–1374.

Klein, T. M., Wolf, E. D., Wu, R. and Sanford, J. C. 1987. High velocity micro-projectiles for delivering nucleic acids into living cells. *Nature* 327: 70–73.

Klümper, W. and Qaim, M. 2014. A meta-analysis of the impacts of genetically modified crops. *PLoS ONE* 9: e111629. doi:10.1371/journal.pone.0111629.

Knott, D. R. 1971. The transfer of genes for disease resistance from alien species to wheat by induced translocations. *Mutation Breeding and Disease Resistance.* Proceedings of a Panel organized by the Joint FAO/IAEA Division of Atomic Energy in Food & Agriculture and the FAO Plant Production and Protection Division, October 12–16, 1970, Vienna, Austria, 67–77.

Koivu, K. 2004. Novel sprouting technology for recombinant protein production. In: R. Fischer and S. Schillberg (Eds.), *Molecular Farming: Plant-made Pharmaceuticals and Technical Proteins.* John Wiley & Sons, Weinheim, Germany.

Komor, A. C., Kim, Y. B., Packer, M. S., Zuris, J. A. and Liu, D. R. 2016. Programmable editing of a target base in genomic DNA without double-stranded DNA cleavage. *Nature* 533: 420–424. doi:10.1038/nature17946.

Konzak, C. F., Randolph, L. F. and Jensen, N. F. 1951. Embryo culture of barley species hybrids: Cytological studies of Hordeum sativum × Hordeum bulbosum. *J. Hered.* 42: 125–134.

Kosambi, D. D. 1944. The estimation of map distance from recombination values. *Ann. Eugen.* 12: 172–175.

Kostoff, D. 1929. An androgenic *Nicotiana* haploid. *Z. Zellforsch.* 9: 640–642.

Krämer, C. 2006. *Rebsorten in Württemberg. Herkunft, Einführung, Verbreitung und die Qualität der Weine vom Spätmittelalter bis ins 19.* Jahrhundert. Jan Thorbecke Verl., Ostfildern, Germany.

Krens, F. A., Molendijk, L., Wullems, G. J. and Schilperoort, R. A. 1982. In vitro transformation of plant protoplasts with Ti-plasmid DNA. *Nature* 296: 72–74.

Kromdijk, J., Głowacka, K., Leonelli, L., Gabilly, S. T., Iwai, M., Niyogi, K. K. and Long, S. 2016. Improving photosynthesis and crop productivity by accelerating recovery from photoprotection. *Science* 354: 857–861. doi:10.1126/science.aai8878.

Kuckuck, H. 1959. Neuere Arbeiten zur Entstehung des hexaploiden Kulturweizens. *Zeitschrift für Pflanzenzüchtung* 41: 205–226.

Kynast, R. G., Riera-Lizarazu, O., Vales, M. I., Okagaki, R. J., Maquieira, S. B., Chen, G., Ananiev, E. V. et al. 2001. A complete set of maize individual chromosome additions to the oat genome. *Plant Physiol.* 125: 1216–1227.

Kyozuka, J., Hayashi, Y. and Shimamoto, K. 1987. High frequency plant regeneration from rice protoplasts by novel nurse culture methods. *Mol. Gen. Genet.* 206: 408–413.

Ladizinsky, G. 2000. A synthetic hexaploid (2n = 42) oat from the cross of *Avena strigosa* (2n = 14) and domesticated *A. magna* (2n = 28). *Euphytica* 116: 231–235.

Larkin, P. J. and Scowcroft, W. R. 1981. Somaclonal variation: A novel source of variability from cell cultures for plant improvement. *Theor. Appl. Genet.* 60: 197–214.

Larue, C. D. 1936. The growth of plant embryos in culture. *Bull. Torrey Bot. Club* 63: 365–382.

Laurie, D. A. and Bennett, M. D. 1988. The production of haploid plants from wheat × maize crosses. *Theor. Appl. Genet.* 76: 393–397.

Laxton, T. 1866. Observations on the variations effected by crossing in the color and character of the seed of peas. *Rep. Intern. Hort. Exhib. Bot. Congr.*, Truscott (Ed.), May 22–31, Son & Simmons, London, UK, 156–158.

Laxton, T. 1872. Notes on some changes and variations in the offspring of cross-fertilized peas. *J. Roy. Hort. Soc. Lond.* 3: 10–14.

Layard, A. H. 1853. *Discoveries among the Ruins of Nineveh and Babylon.* G. P. Putnam & Co., New York.

Lee, J.-M. and Oda, M. 2003. Grafting of herbaceous vegetable and ornamental crops. *Hort. Rev.* 28: 61–124.

Lee, M. 1998. Genome projects and gene pools: New germplasm for plant breeding? *Proc. Nat. Acad. Sci. U.S.A.* 95: 2001–2004.

Lehmann, E. 1916. Aus der Frühzeit der pflanzlichen Bastardierungskunde. *Arch. Gesch. Naturw. Techn.* 7: 78–81.

Lein, A. 1973. Introgression of a rye chromosome to wheat strains by Georg Riebesel – Salzmünde after 1926. Proc. EUCARPIA Symp. on Triticale, Leningrad: 158–167.

Lemieux, B., Aharoni, A. and Schena, M. 1998. Overview of DNA chip technology. *Mol. Breed.* 4: 277–289.

Leonard, J. N. 1973. *First Farmers.* Time Life Books, New York.

Levetin, E. and McMahon, K. 1996. *Plants and Society.* Wm. C. Brown, Dubuque, IA.

Lewontin, R. C. 1977. Evolution as a theory and ideology. *Enciclopedia Enaudi* 3: 9–64.

Li, J. and Yuan, L. 2000. Hybrid rice: Genetics, breeding, and seed production. *Plant Breed. Rev.* 17: 15–120.

Liljedahl, J. B., Hancock, N. I. and Butler, J. L. 1951. A self-propelled plot combine. *Agron. J.* 43: 516–517.

Lindström, E. W. 1929. A haploid mutant in tomato. *J. Hered.* 30: 23–30.

Liu, B. L. 1997. *Statistical Genomics: Linkage, Mapping, and QTL Analysis.* CRC Press, Boca Raton, FL.

Liu, D., Hao, M., Li, A., Zhang, L., Zheng, Y. and Mao, L. 2016. Allopolyploidy and interspecific hybridization for wheat improvement. In: A. S. Mason (Ed.), *Polyploidy and Hybridization for Crop Improvement.* CRC Press, Boca Raton, FL.

Logan, J. 1736. Some experiments concerning the impregnation of the seeds of plants. In: T. Shachtman (Ed.). 2014. *Gentlemen Scientists and Revolutionaries: The Founding Fathers in the Age of Enlightenment.* St. Martin's Press, New York.

Lonnquist, J. H. and McGill, D. P. 1956. Performance of corn synthetics in advanced generations of synthesis and after two cycles of recurrent selection. *Agron. J.* 48: 249–253.

Lörz, H. and Scowcroft, W. R. 1983. Somaclonal variation in protoplast-derived plants of Su/su heterozygotes *Nicotiana tabacum. Theor. Appl. Genet.* 66: 67–75.

Lörz, H. and Wenzel, G. 2005. *Molecular Marker Systems in Plant Breeding and Crop Improvment.* Springer Verlag, Berlin, Germany.

Löve, D. 1953. Cytotaxonomical remarks on the Gentianaceae. *Hereditas* 39: 225–235.

Lovelock, Y. 1972. *The Vegetable Book: An Unnatural History.* Allen & Unwinn, London, UK.

Luby, C. H. and Goldman, I. L. 2016. Freeing crop genetics through the open source seed initiative. *PLoS Biol.* 14: e1002441. doi:10.1371/journal.pbio.1002441.

Lumpkin, T. A. 2015. How a gene from Japan revolutionized the world of wheat: CIMMYT's quest for combining genes to mitigate threats to global food security. *Proceedings of the 12th International Wheat Genetics Symposium,* Y. Ogihara et al. (Eds.), September 8–14, 2013, Kihara Institute for Biological Research, Yokohama (Japan), 13–20.

Lynch, M. and Conery, J. S. 2000. The evolutionary fate and consequences of duplicate genes. *Science* 290: 1151–1155.

Maan, N. N. and Gordon, J. 1988. Compendium of alloplasmic lines and amphidiploids in the *Triticeae. Proceedings of the 7th International Wheat Genetics Symposium,* T. E. Miller and R. M. D. Koebner (Eds.), Institute of Plant Sciences. Cambridge, UK, 1325–1369.

MacKey, J. 1954. Neutron and X-ray experiments in wheat and a revision of the speltoid problem. *Hereditas* 40: 65–180.

Marchant, J. 1719. Observations sur la Nature des Plantes. *Hist. Acad. Roy. Sci.,* 57–58.

Martini, R. 1871. *Der mehrblütige Roggen.* Verl. A. W. Kafemann, Danzig, Germany.

Martini, S. 1961. *Giorgio Gallesio, Pomologist and Precursor of Gregor Mendel.* Fruit Varieties and Horticultural Digest, East Lancing, MI.

Marton, L., Wullems, G. J. and Molendijk, L. 1979. In vitro transformation of cultured cells from *Nicotiana tabacum* by *Agrobacterium tumefaciens. Nature* 277: 129–131.

Mascher, M., Schuenemann, V. J., Davidovich, U., Marom, N., Himmelbach, A., Hübner, S., Korol, A. et al. 2016. Genomic analysis of 6,000-year-old cultivated grain illuminates the domestication history of barley. *Nat. Genet.* 48: 1089–1093. doi:10.1038/ng.3611.

Matsuoka, Y., Vigouroux, Y., Goodman, M. M., Sanchez, J., Buckler, E. S. and Doebley, J. F. 2002. A single domestication for maize shown by multilocus microsatellite genotyping. *Proc. Natl. Acad. Sci. U.S.A.* 99: 6080–6084.

Matsuura, H. 1933. *A Bibliographical Monograph on Plant Genetics* (*Genic Analysis*)*, 1900–1929,* 2nd ed. Hokkaido Imperial University, Sapporo, Japan.

Matzke, M. A., Priming, M., Trnovsky, J. and Matzke, A. J. M. 1989. Reversible methylation and inactivation of marker genes in sequentially transformed tobacco plants. *EMBO J.* 8: 643–649.

Maul, E., Eibach, R., Zyprian, E. and Töpfer, R. 2015. The prolific grape variety (*Vitis vinifera* L.) 'Heunisch Weiss' (='Gouais blanc'): bud mutants, "colored" homonyms and further offspring. *Vitis* 5: 79–86.

McFadden, E. S. 1930. A successful transfer of emmer characters to vulgare wheat. *Agron. J.* 22: 1020–1034.

Medina-Filho, H. P. 1980. Linkage of *Aps-1, Mi* and other markers on chromosome 6. *Rep. Tomato Genet. Coop.* 30: 26–28.

Meinel, A. 2003. An early scientific approach to heredity by the plant breeder Wilhelm Rimpau (1842–1903). *Plant Breed.* 122: 195–198.

Meinel, A. 2008. *Aufbruch in die wissenschaftliche Pflanzenzüchtung. Der Beitrag von Wilhelm Rimpau (1842–1903).* Gesell. f. Pflanzenzüchtung, Göttingen, Germany, pp. 231.

Melchers, G. 1960. Haploide Blütenpflanzen als Material der Mutationszüchtung. *Züchter* 30: 129–134.

Mendel, G. 1866. Versuche über Pflanzen-Hybriden. *Verhandlungen des Naturforschenden Vereines,* Brünn, Verl. d. Vereins, Brünn, Czech Republic, 4: 3–47.

Metzger, J. 1841. *Die Getreidearten und Wiesengräser in botanischer und ökonomischer Hinsicht.* Akad. Verl. buchh. F. Winter, Heidelberg, Germany.

Michael, T. P. and VanBuren, R. 2015. Progress, challenges and the future of crop genomes. *Curr. Opin. Plant Biol.* 24: 71–81.

Miedaner, T., Herter, C., Goßlau, H., Wilde, P. and Hackauf, B. 2017. Correlated effects of exotic pollen-fertility restorer genes on agronomic and quality traits of hybrid rye. *Plant Breed.* 136: 224–229.

Mikkelson, T. R., Anderson, B. and Jørgensen, R. B. 1996. The risk of transgene spread. *Nature* 380: 31.

Miller, P. 1768. *The Gardeners Dictionary,* John (Ed.), Francis Rivington, London, UK.

Miller, T. E., Hutchinson, J. and Chapman, V. 1982. Investigation of a preferentially transmitted *Aegilops sharonensis* chromosome in wheat. *Theor. Appl. Genet.* 61: 27–33.

Mitter, N., Worrall, E. A., Robinson, K. E., Li, P., Jain, R. G., Taochy, C., Fletcher, S. J., Carroll, B. J., Lu, G. Q. and Xu, Z. P. 2017. Clay nanosheets for topical delivery of RNAi for sustained protection against plant viruses. *Nat. Plants* 3: 16207. doi:10.1038/nplants.2016.207.

Moldenke, H. N. and Moldenke, A. C. 1952. *Plants of the Bible.* Chronica Botanica Co., Waltham, MA.

Monaghan, F. and Corcos, A. 1986. Tschermak: A non-discover of Mendelism. I. An historical note. *J. Hered.* 77: 468–469.

Monaghan, F. and Corcos, A. 1987. Tschermak: A non-discover of Mendelism. II. A critique. *J. Hered.* 78: 208–210.

Moon, M. H. 1958. The botanical explorations. 1. China. *Bailey* 6: 1–9.

Moore, J. H. 1956. Cotton breeding in the Old South. *Agric. Hist.* 30: 95–104.

Moore-Colyer, R. J. 1995. Oats and oat production in history and pre-history. In: R. W. Welch (Ed.), *The Oat Crop: Production and utilization.* World Crop Series, Chapman & Hall, London, UK.

Moose, S. P. and Mumm, R. H. 2008. Molecular plant breeding as the foundation for 21st century crop improvement. *Plant Physiol.* 147: 969–977.

Morren, C. 1858. La Belgique horticole. J. Jardins, Serres, Vergers & Culture Maratcher 58: 1–12.

Morel, G. and Martin, C. 1952. Guérison de dahlias atteints d'une maladie à virus. *C.R. Acad. Sci. Paris* 235: 1324–1325.

Morel, G. and Wetmore, R. H. 1951. Tissue culture of monocotyledons. *Am. J. Bot.* 38: 138–140.

Morgan, T. H., Sturtevant, A. H., Muller, H. J. and Bridges, C. B. 1915. *The Mechanism of Mendelian Heredity.* H. Holt, New York.

Morran, L. T., Parmenter, M. D. and Phillips, P. C. 2009. Mutation load and rapid adaptation favour outcrossing over self-fertilization. *Nature* 462: 350–352. doi:10.1038/nature08496.

Mosella, L. C., Signoret, P. A. and Jonard, R. 1980. Sur la mise au point de techniques de microgreffage d'apex en vue de l'élimination de deux types de particules virales chez le pêcher (*Prunus persica* Batsch). *C.R. Acad. Sci. Paris* 290: 287–290.

Muller, H. J. 1927. Artificial transmutation of the gene. *Science* 66: 84–87.

Munk, L. 1972. Improvement of nutritional value in cereals. *Hereditas* 72: 11–128

Müntzing, A. 1937. Polyploidy from twin seedlings. *Cytologia Fujii Jubilaei* 1937: 211–227.

Murai, N., Sutton, D. W., Murray, M. G., Slightom, J. L., Merlo, D. J., Reichert, N. A., Sengupta-Gopalan, C. et al. 1983. Phaseolin gene from bean is expressed after transfer to sunflower via tumor-inducing plasmid vectors. *Science* 222: 476–482.

Murashige, T. and Skoog, F. 1962. A revised medium for rapid growth and bioassays with tobacco tissue cultures. *Physiol. Plants* 15: 473–497.

Nagata, T. and Bajaj, Y. P. S. 2001. *Somatic Hybridization in Crop Improvement.* Springer Verlag, Berlin, Germany.

Nakajima, K. 1991. Biotechnology for crop improvement and production in Japan. *Regional Expert Consultation on the Role of Biotechnology in Crop Production*, FAO Regional Office for Asia and the Pacific, Bangkok, Thailand.

Nakaya, A. and Isobe, S. N. 2012. Will genomic selection be a practical method for plant breeding? *Ann. Bot.* doi:10.1093/aob/mcs109.

Napoli, C., Lemieux, C. and Jorgensen, R. 1990. Introduction of a chimeric chalcone synthase gene into petunia results in reversible co-suppression of homologous genes in trans. *Plant Cell* 2: 279–289.

Naudin, C. 1863. Noevelles recherches sur l'hybridité dans les végétaux. *Ann. Sc. Natur. Botanique* 19: 180–203.

Navarro, L., Roistacher, C. N. and Murashigue, T. 1975. Improvement of shoot tip grafting in vitro for virus-free citrus. *J. Am. Soc. Hortic. Sci.* 100: 471–479.

Navarro, L. and Juarez, J. 1977. Tissue culture techniques used in Spain to recover virus-free citrus plants. *Acta Hort.* 78: 425–453.

Naville, E. 1898. *The Temple of Deir el Bahari: End of Northern Half and Southern Half of the Middle Platform.* Egypt Exploration Fund, London, UK.

Nawaz, Z. and Shu, Q. 2014. Molecular nature of chemically and physically induced mutants in plants: A review. *Plant Genet. Resour.* 12: 74–78.

Nebel, B. R. 1937. Mechanism of polyploids through colchicine. *Nature* 140: 1101.

Needham, J. and Bray, F. 1984. Science and civilisation in China. In: *Biology and Biological Technology. Part 2. Agriculture.* Cambridge University Press, Press Syndicate of University of Cambridge, Cambridge, UK.

Nene, Y. L. 2005. Rice research in south Asia through the ages. *Asian Agri-Hist.* 9: 85–106.

Neyman, J. and Pearson, E. S. 1933. On the problem of the most efficient tests of statistical hypotheses. *Philos. Transact. Royal Soc., Series A*, London 231: 289–337.

Nilsson-Ehle, H. 1908. Einige Ergebnisse von Kreuzungen bei Hafer und Weizen. Bot. *Notisen* 5: 257–294.

Nitsch, J. P. 1951. Growth and development in vitro of excised ovaries. *Am. J. Bot.* 38: 556–576.

O'Mara, J. G. 1947. The substitution of a specific *Secale cereale* chromosome for a specific *Triticum aestivum* chromosome. *Genetics* 32: 99–100.

Oehlkers, F. 1943. Die Auslösung von Chromosomen-Mutationen in der Meiosis durch Einwirkung von Chemikalien. *Z. ind. Abst. Verer.-lehre* 81: 313–341.

Oikawa, T., Maeda, H., Oguchi, T., Yamaguchi, T., Tanabe, N., Ebana, K., Yano, M., Ebitani, T. and Izawa, T. 2015. The birth of a black rice gene and its local spread by introgression. *Plant Cell* 27: 2401–2414.

Olby, R. C. 1966. *Origins of Mendelism*. Schocken Books, New York.

Olmo, H. P. 1976. Grapes. In: N. W. Simmonds (Ed.), *Evolution of Crop Plants*. Longman, Harlow, UK.

Olsson, G. 1986. *Svalöf 1886–1986—Research and Results in Plant Breeding*. LTS Vörlag, Stockholm, Sweden, p. 290.

Orel, V. 1996. *Gregor Mendel: The First Geneticist*. Oxford University Press, Oxford, UK.

Ortiz, R. 2016. Musa interspecific hybridization and polyploidy of breeding banana and plantain (Musaceae): In: A. S. Mason (Ed.), *Polyploidy and Hybridization for Crop Improvement* (Chapter 5), pp. 96–108. CRC Press, Boca Raton, FL.

Ortiz, R. 2016. *Musa* interspecific hybridization and polyploidy of breeding banana and plantain (Musaceae): In: A. S. Mason (Ed.), *Polyploidy and Hybridization for Crop Improvement*. CRC Press, Boca Raton, FL.

Orton, W. A. 1900. The wilt disease of cotton. USDA Division of Vegetable Physiology and Pathology Bulletin 27.

Osborn, T. C., Pires, J. C., Birchler, J. A., Auger, D. L., Chen, Z. J., Lee, H. S., Comai, L. et al. 2003. Understanding mechanisms of novel gene expression in polyploids. *Trends Genet.* 19: 141–147.

Pardue, M. L. and Gall, J. G. 1969. Molecular hybridization of radioactive DNA to the DNA of cytological preparations. *Proc. Natl. Acad. Sci. U.S.A.* 64: 600–604.

Parthasarathy, N. and Rajan, S. S. 1953. Studies on fertility of autotetraploids of *Brassica campestris* var. *toria*. *Euphytica* 2: 25–36.

Paschou, P., Drineas, P., Yannaki, E., Razou, A., Kanaki, K., Tsetsos, F., Padmanabhuni, S. S. et al. 2014. Maritime route of colonization of Europe. *Proc. Nat. Acad. Sci. U.S.A.* 111: 9211–9216. doi:10.1073/pnas.1320811111.

Paterson, A. H. 1996. *Genome Mapping in Plants*. Academic Press, New York.

Paterson, A. H. 1997. *Molecular Dissection of Complex Traits*. CRC Press, Boca Raton, FL.

Paterson, A. H., Lander, E. S., Hewitt, J. D., Peterson, S., Lincoln, S. E. and Tanksley, S. D. 1988. Resolution of quantitative traits into Mendelian factors by using a complete linkage map of restriction fragment length polymorphisms. *Nature* 335: 721–726.

Patterson, H. D. 1952. The construction of balanced designs for experiments involving sequences of treatments. *Biometrika* 39: 32–48.

Pearson, E. S. 1931. Analysis of variance in cases of non-normal variation. *Biometrika* 23: 114–133.

Pearson, E. S. 1937. Some aspects of the problem of randomization. *Biometrika* 29: 53–64.

Pearson, K. 1900. Mathematical contribution theory of evolution: On the law of reversion. *Proc. Roy. Soc. Lond.* 66: 140–164.

Perkins, J. H. 1997. *Geopolitics and the Green Revolution. Wheat, Genes and the Cold War*. Oxford Universit Press, New York.

Perry, L., Dickau, R., Zarrillo, S., Holst, I., Pearsall, D. M., Piperno, D. R., Berman, M. J. et al. 2007. Starch fossils and the domestication and dispersal of chili peppers (*Capsicum* spp. L.) in the Americas. *Science* 315: 986–988.

Person, C. O. 1959. Gene-for-gene relationships in parasitic systems. *Can. J. Bot.* 37: 1101–1130.

Petry, E. 1923. Zur Kenntnis der Bedingungen der Biologischen Wirkungen der Röntgenstrahlen. *Biochem. Z.* 128: 326–353.

Piperno, D. R. and Flannery, K. V. 2001. The earliest archaeological maize (*Zea mays* L.) from highland Mexico: New accelerator mass spectrometry dates and their implications. *Proc. Natl. Acad. Sci. U.S.A.* 98: 2101–2103.

Plackett, R. L. and Burman, J. P. 1946. The design of optimum multifactorial experiments. *Biometrika* 33: 305–325.

Potokina, E., Caspers, M., Prasad, M., Kota, R., Zhang, H., Sreenivasulu, N., Wang, M. and Graner, A. 2004. Functional association between malting quality trait components and cDNA array based expression patterns in barley (*Hordeum vulgare* L.). *Mol. Breed.* 14: 153–170.

Powell-Abel, P., Nelson, R., De, B., Hoffmann, N., Rogers, G., Fraley, T. and Beachy, R. N. 1986. Delay of disease development in transgenic plants that express the tobacco mosaic virus coat protein gene. *Science* 232: 738–743.

Punnett, R. C. 1907. *Mendelism*. Macmillan, Cambridge, UK.

Quételet, L. A. 1871. *Anthrometrie*. Bruxelles [etc.] C. Muquardt, Paris, France.

Rajaram, S., Maan, C. E., Ortiz-Ferrara, A. and Mujeeb-Kazi, A. 1983. Adaptation, stability and high yield potential of certain 1B/1R CIMMYT wheats. *Proc. 6th Int. Wheat Genet. Symp.*, Kyoto, ed. S. Sakamoto, *Kyoto* University Press, Kyoto (Japan): 613–621.

Randolph, L. F. 1932. Some effects of high temperature on polyploidy and other variations in maize. *Proc. Nat. Acad. Sci. U.S.A.* 18: 222–229.

Rao, S. R. and Ravishankar, G. A. 2000. Vanilla flavour: Production by conventional and biotechnological routes. *J. Food Sci.* 80: 289–304.

Redenbaugh, K. 1993. *Synseeds. Applications of Synthetic Seeds to Crop Improvement*. CRC Press, Boca Raton, FL.

Redenbaugh, K., Berner, T., Emlay, D., Frankos, B., Hiatt, W., Houck, C., Kramer, M. et al. 1993. Regulatory issues for commercialization of tomatoes with an antisense polygalacturonase gene. *In Vitro Cell Dev. Biol.* 29: 17–26.

Reed, C. A. 1977. *The Origins of Agriculture*. Mouton Publishers, The Hague, the Netherlands.

Reed, H. 1942. *A Short History of the Plant Sciences*. Ronald Press, Waltham, MA.

Reitemeier, A. 1904. *Geschichte der Züchtung landwirtschaftlicher Kulturpflanzen*. Bucdruckerei Fleischmann, Breslau, Poland.

Reitz, L. P. and Salmon, S. C. 1968. Origin, history, and use of Norin 10 wheat. *Crop Sci.* 8: 686–689.

Rhoades, M. M. 1931. Cytoplasmic inheritance of male sterility in *Zea mays*. *Science* 73: 340–341.

Rhoades, M. M. 1955. *The Cytogenetic of Maize*. G. G. Sprague (Ed.), Academic Press, New York.

Richey, F. D. 1920. Corn breeding. *Adv. Genet.* 3: 159–192.

Richey, F. D. 1946. Corn breeding: Gamete selection, the *Oenothera* method and related miscellany. *J. Am. Soc. Agron.* 39: 403–412.

Rick, C. M. and Butler, L. 1956. Cytogenetics of the tomato. *Adv. Genet.* 8: 267–382.

Rick, J. and Fobes, J. F. 1974. Association of an allozyme with nematode resistance. *Rep. Tomato Genet. Coop.* 24: 25.

Ridley, M. 1984. The horticultural abbot of Brunn. *New Sci.* 5: 24–27.

Riehl, S., Zeidi, M. and Conard, N. J. 2013. Emergence of agriculture in the foothills of the Zagros Mountains of Iran. *Science* 341: 65–67. doi:10.1126/science.1236743.

Rifkin, J. 1998. *The Biotech Century*. Victor Gollancz, London, UK.

Riley, R. 1960. The diploidisation of polyploid wheat. *Heredity* 15: 407–429.

Riley, R. and Chapman, V. 1958. The production and phenotypes of wheat-rye chromosome addition lines. *Heredity* 12: 301–315.

Riley, R., Chapman, V. and Johnson, R. 1968. Introduction of yellow rust resistance of *Aegilops comosa* into wheat by genetically induced homoeologous recombination. *Nature* 217: 383–384.

Rimpau, W. 1877. *Züchtung neuer Getreidearten*. Landw. Jahrb., Berlin, Germany.

Rimpau, W. 1883. Züchtung auf dem Gebiete der landwirtschaftlichen Kulturpflanzen. In: H. Thiel and E. von Wolff (Eds.), *Mentzel und von Lengerke's verbesserter landwirthschaftlicher Hülfs- und Schreib-Kalender*, pp. 33–92. Verl. P. Parey, Berlin, Germany.

Rimpau, W. 1891. *Kreuzungsprodukte landwirtschaftlicher Kulturpflanzen*. Verl. P. Parey, Berlin, Germany.

Rimpau, W. 1899. Monstrositäten am Roggen. *Dtsch. Landw. Pr.* XXVI: 878–901.

Röbbelen, G. 2000 + 2002. *Biographisches Lexikon zur Geschichte der Pflanzenzüchtung 1 + 2*. Gesell. f. Pflanzenz., Göttingen, Germany, No. 50 + 55.

Roberts, H. F. 1929. *Plant Hybridization before Mendel*. Princeton University Press, Princeton, NJ.

Robinson, J., Harding, J. and Vouillamoz, J. 2013. *Wine Grapes: A Complete Guide to 1,368 Vine Varieties, Including Their Origins and Flavours*. Penguin, London, UK.

Rodgers, A. D., III. 1965. *A Story of American Plant Sciences*. Hafner Press, New York.

Rodriguez-Ariza, M. O. and Moya, E. M. 2005. On the origin and domestication of *Olea europaea* L. (olive) in Andalusia, Spain, based on the biogeographical distribution of its finds. *Veget. Hist. Archaeobot.* 14: 551–561.

Rohde, E. S. 1927. *Garden—Craft in the Bible*. Herbert Jenkins, London, UK.

Rommens, C. M., Haring, M. A., Swords, K., Davies, H. V. and Belknap, W. R. 2007. The intragenic approach as a new extension to traditional plant breeding. *Trends Plant Sci.* 12: 397–403.

Ruckenbauer, P. 2000. E. von Tschermak-Seysenegg and the Austrian contribution to plant breeding. *Vortr. Pflanzenzüchtung* 48: 31–46.

Russell, J. E. 1966. *A History of Agricultural Science in Great Britain, 1620–1954.* Allen and Unwin, London, UK.

Rutger, J. N., Peterson, M. L., Hu, C. H. and Lehman, W. F. 1976. Induction of useful short stature and early maturing mutants in two japonica rice (*Oryza sativa* L.) cultivars. *Crop Sci.* 16: 631–635.

Sageret, M. 1826. Considération sur la Production des Hybrides, des Variantes et des Variétés en général, et sur celles de la famille des Cucurbitacées rn particular. *Ann. Sc. Natur. Paris* 8: 294–314.

Sailer, C., Schmid, B. and Grossniklaus, U. 2016. Apomixis allows the transgenerational fixation of phenotypes in hybrid plants. *Curr. Biol.* 26: 331–337. doi:10.1016/j.cub.2015.12.045.

Sallon, S., Solowey, E., Cohen, Y., Korchinsky, R., Egli, M., Woodhatch, I., Simchoni, O. and Kislev, M. 2008. Germination, genetics, and growth of an ancient date seed. *Science* 320: 1464–1472.

Sanborn, J. W. 1890. *Indian Corn. Rept. Maine Dept. Agric.* 33: 54–121.

Sapehin, A. A. 1935. X-ray mutants in soft wheat. *Bot. Zhur.* 20: 3–9.

Sauer, C. O. 1952. *Agricultural Origins and Dispersals.* American Geographical Society, New York.

Sax, K. 1922. Sterility in wheat hybrids. II. Chromosome behavior in partially sterile hybrids. *Genetics* 7: 513–552.

Sax, K. 1923. The association of size differences with seed-coat pattern and pigmentation in *Phaseolus vulgaris*. *Genetics* 8: 522–560.

Schaller, C. W. and Wiebe, G. A. 1952. Sources of resistance to net blotch of barley. *Agron. J.* 44: 334–336.

Schell, J., van Montagu, M., Holsters, M., Zambryski, P., Joos, H., Inze, V., Herrera-Estrella, L. et al. 1983. Ti plasmids as experimental gene vectors for plants. *Advances in Gene Technology: Molecular Genetics of Plants and Animals. Miami Winter Symposium* 20: 191–209.

Scheller, J., Gührs, K. H., Grosse, F. and Conrad, U. 2001. Production of spider silk proteins in tobacco and potato. *Nat. Biotechnol.* 190: 573–577.

Scheller, J., Henggeler, D., Viviani, A. and Conrad, U. 2004. Purification of spider silk-elastin from transgenic plants and application for human chondrocyte proliferation. *Transgenic Res.* 13: 51–57.

Schena, M., Shalon, D., Davis, R. W. and Brown, O. P. 1995. Quantitative monitoring of gene expression patterns with a complementary DNA micro array. *Science* 270: 467–470.

Schlegel, R. 1976. The relationship between meiosis and fertility in autotetraploid rye, *Secale cereale* L. *Tag. Ber. AdL, Berlin (Germany)* 143: 31–36.

Schlegel, R. 1996. Triticale – Today and tomorrow. In: H. Guedes-Pinto, N. Darvey and V. Carnide (Eds.), *Triticale Today and Tomorrow, Developments in Plant Breeding*, Vol. 5, Kluwer Academic Publishers, Amsterdam, the Netherlands, pp. 21–32.

Schlegel, R. 2005. *Rye (Secale cereale L.) – A Younger Crop Plant with Bright Future.* CRC Press, Boca Raton, FL.

Schlegel, R. 2016. Hybrid breeding boosted molecular genetics in rye. *Russ. J. Genet. Appl. Res.* 6: 569–583. doi:10.1134/S2079059716050105.

Schlegel, R. and Korzun, V. 1997. About the origin of 1RS.1BL wheat-rye chromosome translocations from Germany. *Plant Breed.* 116: 537–540.

Schlegel, R. and Meinel, A. 1994. A quantitative trait locus (QTL) on chromosome arm 1RS of rye and its effect on yield performance of hexaploid wheat. *Cer. Res. Comm.* 22: 7–13.

Schlegel, R., Melz, G. and Korzun, V. 1997. Genes, marker and linkage data of rye (*Secale cereale* L.). 5th updated inventory. *Euphytica* 101: 23–67.

Schlegel, R. and Mettin, D. 1975. Studies of valence crosses in rye (*Secale cereale* L.). IV. The relationship between meiosis and fertility in tetraploid hybrids. *Biol. Zbl.* 94: 295–302.

Schlegel, R., Kynast, R., Schwarzacher, T., Roemheld, V. and Walter, A. 1993. Mapping of genes for copper efficiency in rye and the relationship between copper and iron efficiency. *Plant Soil* 154: 61–65.

Schlegel, R., Melz, G. and Korzun, V. 1997. Genes, marker and linkage data of Rye (*Secale cereale* L.). 5th updated inventory. *Euphytica* 101: 23–67.

Schlegel, R., Vahl, U. and Müller, G. 1994. A compiled list of wheats carrying homoeologous group 1 wheat-rye translocations and substitutions. *Ann. Wheat Newslett. U.S.A.* 40: 105–117.

Schlegel, R. H. J. 2013. *Rye – Genetics, Breeding and Cultivation.* CRC Press, Boca Raton, FL.

Schlegel, R. H. J. 2014. *Dictionary of Plant Breeding*, 2nd ed. CRC Press, Boca Raton, FL, pp. 584.

Schleiden, M. J. 1838. Beiträge zur Phytogenesis. *Arch. Anat. Physiol. Wiss. Med.* 5: 137–176.

Schmalz, H. 1969. *Planzenzüchtung, Entwicklung—Stand—Aufgaben.* Deut. Landwirtschaftsverl. Berlin, Germany.

Schouten, H. J., Krens, F. A. and Jacobsen, E. 2006. Do cisgenic plants warrant less stringent oversight? *Nat. Biotech.* 24: 753.

Schranz, M. E. and Osborn, T. C. 2000. Novel flowering time variation in the resynthesized polyploid *Brassica napus*. *J. Hered.* 91: 242–246.

Schwanitz, F. 1967. *Die Evolution der Kulturpflanzen*. Bayr. Landw. Verl., München, Germany.

Schwann, T. 1839. *Mikroskopische Untersuchungen über die Übereinstimmung in der Struktur und im Wachstum der Thiere und Pflanzen*. Sanders, Berlin, Germany.

Sears, E. R. 1959. Aneuploids in common wheat. *Proceedings of the 1st International Wheat Genetics Symposium*, University of Manitoba, Winnipeg, Canada, pp. 221–228.

Sears, E. R. 1961. Identification of the wheat chromosome carrying leaf rust resistance from *Aegilops umbellulata*. *Wheat Inf. Serv.* 12: 12.

Sears, E. R. 1965. Nulli-tetrasomic combinations in hexaploid wheat. *Suppl. J. Hered.* 20: 29–45.

Sears, E. R. and Okamoto, M. 1958. Intergenomic chromosome relationships in hexaploid wheat. *Proc. 10th Int. Congr. Genet.* 2: 258–259.

Sebesta, E. E. and Wood, E. A. 1978. Transfer of greenbug resistance from rye to wheat with X-rays. *Agron. Abstr.* 70: 61–62.

Seibert, M. 1976. Shoot initiation from carnation shoot apices frozen to $-196°C$. *Science* 191: 1178–1179.

Semal, J. and Lepoivre, P. 1992. Biotechnologie et Agriculture: impact et perspectives. *Cahiers Agriculteurs* 13: 153–162.

Seton, A. 1824. On the variation in the color of peas from cross-impregnation. *Trans. Hort. Soc. Lond.* V: 236.

Shahin, E. H. and Spivey, R. 1986. A single dominant gene for *Fusarium* wilt resistance in protoplast-derived tomato plants. *Theor. Appl. Genet.* 73: 164–169.

Shamel, A. D. 1905. *The Effect of Inbreeding in Plants*. Yearbook USDA, Washington, DC, pp. 377–392.

Scheller, J., Gührs, K. H., Grosse, F. and Conrad, U. 2001. Production of spider silk proteins in tobacco and potato. *Nat. Biotechnol.* 190: 573–577.

Shepard, J. F., Bidney, D. and Shahin, E. 1980. Potato protoplasts in crop improvement. *Science* 208: 17–24.

Shepherd, K. W. and Islam, A. K. M. R. 1988. Fourth compendium of wheat-alien chromosome lines. *7th International Wheat Genetics Symposium*, T. E. Miller and R. M. D. Koebner (Eds.), Institute of Plant Sciences, Cambridge, UK, pp. 1373–1395.

Shinozaki, K. and Yamaguchi-Shinozaki, K. 2007. Gene networks involved in drought stress response and tolerance. *J. Exp. Bot.* 58: 221–227.

Shirreff, P. 1873. *Improvement of the Cereals and An Essay on the Wheat-Fly. Print for Private Circulation b.* W. Blackwood & Sons, Edinburgh, UK.

Shull, G. H. 1908. The composition of a field of maize. *Rept. Am. Breeders' Assoc.* 4: 296–301.

Shull, G. H. 1910. Hybridization methods in corn breeding. *Am. Breeders' Mag.* 1: 98–107.

Shull, G. H. 1912. The influence of inbreeding on vigor of *Hydatina senta*. *Biol. Bull.* 24: 1–13.

Shull, G. H. 1922. Über die Heterozygotie mit Rücksicht auf den praktischen Zuchterfolg. *Beitr. zur Pflanzenzüchtung.* 5: 134–158.

Shull, G. M. 1948. What is "heterosis?" *Genetics* 33: 439–446.

Shull, G. M. 1952. *Beginning of the Heterosis Concept*. Iowa State College Press, Ames, IA, pp. 14–48.

Shurtleff, W., Huang, H. T. and Aoyagi, A. 2014. *History of Soybeans and Soyfoods in China and Taiwan, and in Chinese Cookbooks, Restaurants, and Chinese Work with Soyfoods Outside China (1024 BCE to 2014): Extensively Annotated Bibliography and Sourcebook, Including Manchuria, Hong Kong and Tibet.* SOYINFO Center, Lafayette, CA, pp. 3015.

Sijmons, P. C., Dekker, B. M., Schrammeijer, B., Verwoerd, T. C., van den Elzen, P. J. and Hoekema, A. 1990. Production of correctly processed human serum albumin in transgenic plants. *Biotechnology* 8: 217–221.

Sinclair, T. R., Purcell, L. C. and Sneller, C. H. 2004. Crop transformation and the challenge to increase yield potential. *Trends Plant Sci.* 9: 70–75.

Singer, E., Holmyard, E. J. and Hall, A. R. 1954. *A History of Technology. Vol. 1. From Early Times to the Fall of Ancient Empires*. Oxford University Press, London, UK.

Singh, R. K. 2016. Breeding for salt tolerance in rice. Paper given as Power Point Presentation, at *Plant Breeding, Genetics, and Biotechnology Division (PBGB) at International Rice Research Institute (IRRI)*, Laguna, Philippines, pp. 1–58. Retrieved from http://webcache.googleusercontent.com/search?q=cache:AOQ41eZeu3gJ:www.knowledgebank.irri.org/ricebreedingcourse/Powerpoints/salt,_rksingh.ppt+&cd=2&hl=de&ct=clnk&gl=de.

Singh, S. P., Keller, B., Gruissem, W. and Bhullar, N. K. 2017. Rice *NICOTIANAMINE SYNTHASE 2* expression improves dietary iron and zinc levels in wheat. *Theor. Appl. Genet.* 130: 283–292.

Sjodin, C. and Glimelius, K. 1989. Transfer of resistance against *Phoma lingam* to *Brassica napus* by asymmetric somatic hybridization combined with toxin selection. *Theor. Appl. Genet.* 78: 513–520.

Smith, B. D. 1995. *The Emergence of Agriculture*. Scientific American Library, W. H. Freeman & Co., New York.

Smith, C. J. S., Watson, C. F., Ray, J., Bird, C. R., Morris, P. C., Schuch, W. and Grierson, D. 1988. Antisense RNA inhibition of polygalacturonase gene expression in transgenic tomatoes. *Nature* 334: 724–726.

Smith, H. H. 1943. Effects of genomic balance, polyploidy, and single extra chromosomes on size in *Nicotiana*. *Genetics* 28: 227–236.

Smith, M. E. 1996. *The Aztecs*. Blackwell, Oxford, UK.

Snape, J. W., Simpson, E., Parker, B. B., Friedt, W. and Foroughi-Wehr, B. 1986. Criteria for the selection and use of double haploid systems in cereal breeding programmes. In: W. Horn, C. J. Jensen, W. Odenbach and O. Schieder (Eds.), *Genetic Manipulation in Plant Breeding. Proceedings of the International Symposium on Eucarpia*. W. de Gruyter Verl., Berlin, Germany, pp. 217–229.

Solbrig, O. T. and Solbrig, D. J. 1994. *Farming and Crops in Human Affairs*. Island Press, Washington, DC.

Song, R. and Messing, J. 2003. Gene expression of a gene family in maize based on non-collinear haplotypes. *Proc. Natl. Acad. Sci. U.S.A.* 100: 9055–9060.

Sosna, M. 1966. G. Mendel memorial symposium 1865–1965. *Proceedings of a Symposium*, Brno, Czech Republic, August 4–7, 1965. Academia, Publishing House of the Czechoslovak Academy of Sciences, Prague, Czech Republic.

Soyk, S., Müller, N. A., Park, S. J., Schmalenbach, I., Jiang, K., Hayama, R., Zhang, L., Van Eck, J., Jiménez-Gómez, J. M. and Lippman, Z. B. 2016. Variation in the flowering gene *SELF PRUNING 5G* promotes day-neutrality and early yield in tomato. *Nat. Genet.* doi:10.1038/ng.3733.

Spengler, R., Frachetti, M., Doumani, P., Lynne Rouse, L., Cerasetti, B., Bullion, E. and Maryashev, A. 2014. Early agriculture and crop transmission among Bronze Age mobile pastoralists of Central Eurasia. *Proc. R. Soc. Lond. B* 281. doi:10.1098/rspb.2013.3382.

Spillman, W. J. 1901. Quantitative studies on the transmission of parental characters of hybrid offspring. *Proceedings of the 15th Annual Convention of Association of American Agricultural Colleges and Experiment Station*, Office of Experiment Station Bulletin 115, Washington, DC.

Sprague, G. F. and Jenkins, M. T. 1943. A comparison of synthetic varieties, multiple crosses and double crosses in corn. *J. Am. Soc. Agron.* 35: 137–147.

Sprague, G. F. and Tatum, L. A. 1942. General vs. specific combining ability in single crosses of corn. *J. Am. Soc. Agron.* 34: 923–932.

Sprengel, C. K. 1793. *Das entdeckte Geheimnis der Natur im Bau und in der Befruchtung der Blumen*. Ostwalds Klassiker der exakten Wissenschaften, Leipzig, Germany, p. 48.

Staab, J. 1997. Weinwirtschaft im frühen Mittelalter, insbesondere im Frankenreich und unter den Ottonen. In: C. Schrenk (Ed.), *Weinwirtschaft im Mittelalter - Quellen und Forschungen zur Geschichte der Stadt Heilbronn*, 9: 29–76. Heilbronn, Germany.

Stadler, L. J. 1928. Genetic effects of x-rays in barley. *Proc. Nat. Acad. Sci. U.S.A.* 68: 186–187.

Stadler, L. J. 1929. Chromosome number-and the mutation rate in *Avena* and *Triticum*. *Proc. Nat. Acad. Sci. U.S.A.* 15: 876–881.

Staropoli, N. 2016. FDA approves GMO potato that resists blight that caused Irish potato famine. Genetic Literacy Project. https://www.geneticliteracyproject.org/2016/01/14/fda-approves-gmo-potato-resists-blight-caused-irish-potato-famine.

Stebbins, G. L. 1971. *Chromosomal Evolution of Higher Plants*. Edward Arnold, London, UK.

Stegemann, S. and Bock, R. 2009. Exchange of genetic material between cells in plant tissue grafts. *Science* 324: 649–651.

Stein, E. 1922. Über den Einfluss von Radiumstrahlung auf Anthirrhinum. *Z. Ind. Abst. Verb.-lehre* 29: 1–15.

Stern, C. 1931. Zytologisch-genetische Untersuchungen als Beweise für die Morgansche Theorie des Faktorenaustauschs. *Biol. Zbl.* 51: 547–587.

Stern, C. and Sherwood, E. R. 1966. *The Origin of Genetics. A Mendel Source Book*. W. H. Freeman, San Francisco, CA.

Storey, W. K. 1995. Figs. In: N. W. Simmonds (Ed.), *Evolution of Crop Plants*. Longman, Harlow, UK.

Strasburger, E. 1910. Chromosomenzahl. *Flora* 100: 398–444.

Straub, S. C. K., Pfeil, B. E. and Doyle, J. J. 2006. Testing the polyploid past of soybean using a low-copy nuclear gene—Is *Glycine* (Fabaceae: Papilionoideae) an auto- or allopolyploid? *Mol. Phylogenet. Evol.* 39: 580–584.

Stubbe, H. 1965. *Kurze Geschichte der Genetik bis zur Wiederentdeckung der Vererbungsregeln Gregor Mendels*. G. Fischer Verl., Jena, Germany.

Stuber, C. W., Polacco, M. and Senior, M. L. 1999. Synergy of empirical breeding, marker-assisted selection, and genomics to increase crop yield potential. *Crop Sci.* 39: 1571–1583.

Student. 1938. Comparison between balanced and random arrangements of field plots. *Biometrika* 29: 363–379.

Sturtevant, A. H. 1913. The linear arrangement of six sex-linked factors in *Drosophila*, as shown by their mode of association. *J. Exp. Zool.* 14: 43–59.

Sturtevant, A. H. 1917. Crossing over without chiasmatype? *Genetics* 2: 301–304.

Sturtevant, A. H. 1965. *A History of Genetics*. Harper & Row, New York.

Sukekiyo, Y., Ogura, H., Kimura, Y. and Itoh, R. 1989. Development of a new rice variety Hatsuyume by protoplast breeding. *Proceedings of the 6th International Congress of SABRAO*, S. Iyama and G. Takeda (Eds.), pp. 497–500. SABRAO, Tsukuba, Japan.

Sun, Y., Zhang, X., Wu, C., He, Y., Ma, Y., Hou, H., Guo, X., Du, W., Zhao, Y. and Xia, L. 2016. Engineering herbicide-resistant rice plants through CRISPR/Cas9-mediated homologous recombination of acetolactate synthase. *Mol. Plant* 9: 628–631.

Suneson, C. A. 1956. An evolutionary plant breeding method. *Agron. J.* 48: 188–191. doi:10.2134/agronj1956.00021962004800040012x.

Suneson, C. A. and Wiebe, G. A. 1942. Survival of barley and wheat varieties in mixtures. *J. Am. Soc. Agron.* 34: 1052–1056.

Sutton, W. S. 1902. On the morphology of the chromosome group in *Brachystola magna*. *Biol. Bull. Mar. Biol. Lab.* 4: 1–16.

Sutton, W. S. 1903. The chromosomes in heredity. *Biol. Bull. Mar. Biol. Lab. Woods Hole* 4: 231–251.

Swingle, W. T. 1913. *The Date Palm and Its Utilization in the Southwestern States*. Bureau Plant Industry, USDA, Washington, DC.

Sybenga, J. 1964. The use of chromosomal aberrations in the autopolyploidization of autopolyploids. Proc. Symp. IAEA/FAO, Rome, Italy, pp. 741–749.

Tacker, C. O., Mason, H. S., Losonsky, G., Clements, J. D., Levine, M. M. and Arntzen, C. J. 1998. Immunogenecity in humans of a recombinant bacterial antigen delivered in a transgenic potato. *Nat. Med.* 4: 607–609.

Tammes, T. 1924. Das genotypische Verhältnis zwischen dem wilden *Linum angustifolium* and dem Kulturlein, *L. usitatissimum*. *Genetica* 5: 61–76.

Tang, S., Ding, L. and Bonjean, A. P. A. 2010. Rice production and genetic improvement in China. In: He, Z. and Bonjean, A. P. A. (Eds.), *Cereals in China*. Mexico: CIMMYT.

Tanksley, S. D. and Nelson, J. C. 1996. Advanced backcross QTL analysis: A method for the simultaneous discovery and transfer of valuable QTLs from unadapted germplasm into elite breeding lines. *Theor. Appl. Genet.* 92: 191–203.

Tanksley, S. D. and Orton, T. J. 1983. *Isozymes in Plant Genetics and Breeding*. Elsevier, New York.

Tanksley, S. D., Medino-Filho, H. and Rick, C. M. 1982. Use of naturally occurring enzyme variation to detect and map genes controlling quantitative traits in an interspecific backcross of tomato. *Heredity* 49: 11–25.

Tanksley, S. D., Rick, C. M. and Vallejos, C. E. 1984. Tight linkage between a nuclear male-sterile locus and an enzyme marker in tomato. *Theor. Appl. Genet.* 68: 109–113.

Thacker, C. 1979. *The History of Gardens*. University of California Press, Berkeley, CA.

Thresh, J. M. 1998. In memory of James Edward Vanderplank 1909–1997. *Plant Pathol.* 47: 114–115.

Tieman, D. T., Zhu, G., Resende, M. F. R., Lin, T., Nguyen, C., Bies, D., Rambla, J. L. et al. 2017. A chemical genetic roadmap to improved tomato flavor. *Science* 27: 391–394.

Till, B. J. 2014. Mining genetic resources via ecotilling. In: R. Tuberosa, A. Graner and E. Frison (Eds.), *Genomics of Plant Genetic Resources*. Springer, Dordrecht, the Netherlands, pp. 349–365.

Tiwari, J. K., Poonam, Srivastava, A. K., Singh, B. P. and Bag, T. K. 2015. Protoplast fusion in potato improvement. *Keanean J. Sci.* 2: 77–82.

Troyer, A. F. and Stoehr, H. 2003. Willet M. Hays, great benefactor to plant breeding and the founder of our association. *J. Hered.* 94: 435–441.

Tsai, H., Howell, T., Nitcher, R., Missirian, V., Watson, B., Ngo, K. J., Lieberman, M. et al. 2011. Discovery of rare mutations in populations: TILLING by sequencing. *Plant Physiol.* 156: 1257–1268.

Tsuchiya, T. and Gupta, P. K. 1991. *Chromosome Engineering in Plants*. A + B. Elsevier, Amsterdam, the Netherlands.

Tysdal, H. M., Kiesselbach, T. A. and Westover, H. L. 1942. Alfalfa breeding. University of Nebraska Agricultural Experiment Station Research Bulletin 124.

Udall, J. A. and Wendel, J. F. 2006. Polyploidy and crop improvement. *Crop Sci.* 46: 3–14.

van der Planck, J. E. 1963. *Plant Disease Epidemics and Control*. Academic Press, New York.

van Treuren, R. and van Hintum, T. J. L. 2014. Next-generation genebanking: Plant genetic resources management and utilization in the sequencing era. *Plant Genet. Resour.* 12: 298–307. doi:10.1017/S1479262114000082.

Vanholme, B., Cesarino, I., Goeminne, G., Kim, H., Marroni, F., Van Acker, R., Vanholme, R. et al. 2013. Breeding with rare defective alleles (BRDA): A natural *Populus nigra* HCT mutant with modified lignin as a case study. *New Phytol.* 198: 765–776.

Varshney, R. K., Graner, A. and Sorrells, M. E. 2005. Genomics-assisted breeding for crop improvement. *Trends Plant Sci.* 10: 621–630.

Vasil, V. and Vasil, I. K. 1965. Regeneration of tobacco and petunia plants from protoplasts and culture of corn protoplasts. *In Vitro* 10: 83–96.

Vattem, D. A. and Maitin, V. 2015. *Functional Foods, Nutraceuticals and Natural Products: Concepts and Applications*. DEStech Publications, Lancaster, PA.

Vavilov, N. I. 1926. Studies on the origin of cultivated plants. *Inst. Bot. Appl. Amelior. Plants, St. Petersburg (Russia)*, 16: 1–248.

Vavilov, N. I. 1928. Geographische Zentren unserer Kulturpflanzen (Geographic centers of our crop plants). *Zeitschr. Ind. Abst. Vererb.-Lehre, Suppl.* 1: 342–369.

Venn, J. 1888. Cambridge anthropometry. *J. Anthropol. Inst.* 18: 140–154.

Vogel, O. A. 1933. A three-row nursery planter for space and drill planting. *J. Am. Soc. Agron.* 25: 426–428.

von Gärtner, C. F. 1849. *Versuche und Beobachtungen über die Bastarderzeugung im Pflanzenreich*. K. F. Hering & Co., Stuttgart, Germany.

von Hagen, V. W. 1957. *Ancient Sun Kingdoms of the Americas*. The World Publishing Company, Cleveland, OH.

von Rümker, K. 1889. *Anleitung zur Getreidezüchtung auf wissenschaftlicher und praktischer Grundlage*. P. Parey Verl., Berlin, Germany, pp. 97–101.

von Sengbusch, R. 1930. Bitterstoffarme Lupinen. I. *Züchter* 2: 1–7.

von Tschermak, E. 1898. Gemüsesamenzucht in Deutschland. *Wiener Ldw. Zeitg.* 42: 343–344.

von Tschermak, E. 1900. Über künstliche Kreuzung bei *Pisum sativum. Zeitschrift für das landwirtschaftliche Versuchswesen in Österreich* 3: 465–555.

von Tschermak, E. 1901. Über Züchtung neuer Getreiderassen mittels künstlicher Kreuzung. I. Kritisch-historische Betrachtungen. *Zeitschrift für das landwirtschaftliche Versuchswesen in Österreich* 4: 1029–1060.

von Tschermak, E. 1906. Über Züchtung neuer Getreiderassen mittels künstlicher Kreuzung. II. Kreuzungsstudien am Roggen. *Zeitschrift für das landwirtschaftliche Versuchswesen in Österreich* 9: 699–743.

von Tschermak, E. 1932. Über einige Blütenanomalien bei Primeln und ihre Vererbungsweise. *Biologica Generalis* 8: 337–350.

von Tschermak, E. 1951a. The rediscovery of Gregor Mendel's work. *J. Hered.* 42: 163–171.

von Tschermak, E. 1951b. Wien, der Ausgangsort des praktischen Mendelismus. *Zeitschrift für Pflanzenzüchtung* 29: 262–275.

von Tschermak, E. 1958. *Leben und Wirken eines österreichischen Pflanzenzüchters*. Paul Parey Verlag, Berlin, Germany.

von Tschermak, E. and Bleier, H. 1926. Über fruchtbare *Aegilops*-Weizenbastarde. *Ber. Deut. Bot. Gesell.* 44: 110–132.

von Tschermak, E. and von Rümker, K. 1910. *Reisebericht über landwirtschaftliche Studien in Nord-Amerika mit besonderer Berücksichtigung der Pflanzenzüchtung*. Paul Parey Verlag, Berlin, Germany.

Walker, W. 1979. *All the Plants of the Bible*. Doubleday & Company, Garden City, NY.

Waltz, E. 2016. CRISPR-edited crops free to enter market, skip regulation. *Nat. Biotechnol.* 34: 582. doi:10.1038/nbt0616-582.

Wan, B., Zha, Z., Li, J., Xia, M., Du, X., Lin, Y. and Yin, D. 2014. Development of elite rice restorer lines in the genetic background of R022 possessing tolerance to brown planthopper, stem borer, leaf folder and herbicide through marker-assisted breeding. *Euphytica* 195: 129–142.

Wang, F., Wang, C., Liu, P., Lei, C., Hao, W., Gao, Y., Liu, Y.-G. and Zhao, K. 2016. Enhanced rice blast resistance by CRISPR/Cas9-targeted mutagenesis of the ERF transcription factor gene OsERF922. *PLoS ONE* 11: e0154027.

Wang, Y., Cheng, X., Shan, Q., Zhang, Y., Liu. J., Gao, C. and Qiu, J. L. 2014. Simultaneous editing of three homoeoalleles in hexaploid bread wheat confers heritable resistance to powdery mildew. *Nat. Biotech.* 32: 947–951.

Watson, J. D. and Crick, F. H. C. 1953. Molecular structure of nucleic acids. *Nature* 171: 737–738.

Webber, H. J. 1908. Plant breeding for farmers. *Cornell Univ. Bull.* 251: 289–332.

Weinberg, W. R. 1909. Über Vererbungsgesetze beim Menschen. *Z. Ind. Abst. Vererb.-lehre* 1: 377–392.

Welch, B. L. 1937. On the *z*-test in randomized blocks and Latin squares. *Biometrika* 29: 21–52.

Wellensiek, S. J. 1947. Rational methods for breeding cross-fertilizers. *Mededel. Landbouwhogeschool Wageningen* 48: 227–262.

Wendel, J. F., Stuber, C. W., Goodman, M. M. and Beckett, J. B. 1989. Duplicated plastid and triplicated cytosolic isozymes of triosephosphate isomerase in maize (*Zea mays* L.). *J. Hered.* 80: 218–228.

White, P. R. 1934. Potentially unlimited growth of excised tomato root tips in a liquid medium. *Plant Physiol.* 9: 585–600.

Wiegemann, A. F. 1828. *Über die Bastarderzeugung im Pflanzenreiche.* Verl. F. Vieweg, Braunschweig, Germany.

Wilks, W. 1900. Report of the first international conference on hybridisation and cross-breeding. *J. Royal Horticult. Soc.* 24: 1–346.

Willcox, G. 1998. Archaeobotanical evidence for the beginnings of agriculture in southwest Asia. In: A. Damania, J. Valkoun, G. Willcox and C. Qualset (Eds.), *The Origins of Agriculture and Crop Domestication.* ICARDA, Aleppo, Syria, pp. 25–38.

Wilson, K. 2011. Aussie transplant proves fruitful: The long road to GMO innovation. *Orchard & Vine*, pp. 18–19.

Wilson, W. 1900. *The Cell in Development and Heredity.* MacMillan, New York.

Winge, O. 1917. The chromosomes. Their number and general importance. *Comp. Rend. Trav. Lab. Carlsberg* 13: 131–275.

Winkler, H. 1916. Über die experimentelle Erzeugung von Pflanzen mit abweichenden Chromosomenzahlen. *Z. Bot.* 8: 417–424.

Wollny, E. 1885. *Saat und Pflege der landwirtschaftlichen Kulturpflanzen.* Verl. P. Parey, Berlin, Germany.

Wright, R. 1938. *The Story of Gardening.* Garden City Publ., New York.

Wright, S. 1921. Correlation and causation. Part 1: Method of path coefficients. *J. Agric. Res.* 20: 557–585.

Wright, S. 1931. Evolution in Mendelian populations. *Genetics* 16: 97–159.

Wunderlich, G. 1951. Die Bedeutung Tschermaks für den österreichischen Getreidebau. *Zeitschrift für Pflanzenzüchtung* 30: 478–483.

Xu, H., Swoboda, I., Bhalla, P. L., Sijbers, A., Chao, C., Ong, E., Hoeijmakers, J. H. J. and Singh, M. B. 1998. Human nucleotide excision repair gene ERCC1 homologue in plants and its preferential expression in male germ line cells. *Plant J.* 13: 823–829.

Xu, H., Swoboda, I., Bhalla, P. L. and Singh, M. B. 1999. Male gametic cell specific gene expression in flowering plants. *Proc. Nat. Acad. Sci. U.S.A.* 96: 2554–2558.

Yanagi, T. and Noguchi, Y. 2016. Strawberry (Plants in the genus *Fragaria*). In: A. S. Mason (Ed.), *Polyploidy and Hybridization for Crop Improvement.* CRC Press, Boca Raton, FL.

Yates, F. 1939. The comparative advantages of systematic and randomized arrangements in the design of agricultural and biological experiments. *Biometrika* 30: 440–466.

Yates, F. 1951. The influence of Statistical Methods for Research Workers on the development of the science of statistics. *J. Am. Stat. Assoc.* 46: 19–34.

Ye, X., Al-Babili, S., Kloeti, A., Zhang, J., Lucca, P., Beyer, P. and Potrykus, I. 2000. Engineering provitamin A (ß-carotene) biosynthetic pathway into (carotenoid-free) rice endosperm. *Science* 287: 303–305.

Zade, A. 1918. *Der Hafer – eine Monographie auf wissenschaftlicher und praktischer Grundlage.* G. Fischer Verl., Jena, Germany.

Zamecnik, P. C. and Stephenson, M. L. 1978. Inhibition of Rous sarcoma virus replication and cell transformation by a specific oligodeoxynucleotide. *Proc. Nat. Acad. Sci. U.S.A.* 75: 280–284.

Zarins, J. 1992. The early settlement of Southern Mesopotamia: A review of recent historical, geological, and archaeological research. *J. Am. Orient. Soc.* 112: 55–77.

Zeller, F. J. 1973. 1B/1R wheat-rye substitutions and translocations. *Proceedings of the 4th International Wheat Genetics Symposium*, E. R. Sears and L. M. S. Sears (Eds.), pp. 209–221. University of Missouri, Columbia, MO.

Zeller, F. J., Günzel, G., Fischbeck, G., Gertenkorn, P. and Weipert, D. 1982. Veränderungen der Backeigenschaften der Weizen-Roggen-Translokationssorten. *Getreide Mehl Brot* 36: 141–143.

Zeller, F. J. and Hsam, K. 1983. Broadening the genetic variability of cultivated wheat by utilizing rye chromatin. *Proceedings of the 6th International Wheat Genetics Symposium*, S. Sakamoto (Ed.), pp. 161–173. Kyoto University Press, Kyoto, Japan.

Zetsche, B., Gootenberg, J., Abudayyeh, O., Slaymaker, I., Makarova, K., Essletzbichler, P., Volz, S. et al. 2015. Cpf1 is a single RNA-guided endonuclease of a class 2 CRISPR-Cas system. *Cell.* doi:10.1016/j.cell.2015.09.038.

Zhang, L. 2016. Hybrid wheat time has come. http://soilcrop.tamu.edu/hybrid-wheat-time-has-come/.

Zhang, P., Friebe, B., Lukaszewski, A. J. and Gill, B. S. 2001. The centromere structure in Robertsonian wheat-rye translocation chromosomes indicates that centric breakage-fusion can occur at different positions within the primary constriction. *Chromosoma* 110: 335–344.

Zhang, X., Urry, D. W. and Daniell, H. 1996. Expression of an environmentally friendly synthetic protein-based polymer gene in transgenic tobacco plants. *Plant Cell Rep.* 16: 174–179.

Zhao, B., Lin, X., Jesse Poland, J., Harold Trick, H., Leach, J. and Scot, H. 2005. A maize resistance gene functions against bacterial streak disease in rice. *Proc. Natl. Acad. Sci. U.S.A.* 102: 15383–15388.

Zimmermann, U. 1982. Electric field-mediated fusion and related electrical phenomena. *Biochim. Biophys. Acta* 694: 227–277.

Zirxel, C. 1935. *The Beginnings of Plant Hybridization.* University of Pennsylvania Press, Philadelphia, PA.

Index

Note: Page numbers followed by f and t refer to figures and tables respectively.

Milton Keynes UK
Ingram Content Group UK Ltd.
UKHW051537141024
449569UK00028B/1511